An Introduction to Genetic Engineering

Third Edition

In this third edition of his popular undergraduate-level textbook, Desmond Nicholl recognises that a sound grasp of basic principles is vital in any introduction to genetic engineering. Therefore, as well as being thoroughly updated, the book also retains its focus on the fundamental principles used in gene manipulation. The text is divided into three sections: Part I provides an introduction to the relevant basic molecular biology; Part II, the methods used to manipulate genes; and Part III, applications of the technology. There is a new chapter devoted to the emerging importance of bioinformatics as a distinct discipline. Other additional features include text boxes, which highlight important aspects of topics discussed, and chapter summaries, which include aims and learning outcomes. These, along with key word listings, concept maps, and a glossary, will enable students to tailor their studies to suit their own learning styles and ultimately gain a firm grasp on this subject that students traditionally find difficult.

Desmond S. T. Nicholl is a Senior Lecturer in Biological Sciences at the University of the West of Scotland, Paisley, UK.

WITHDRAWN

QH
442
.N53
2008

An Introduction to Genetic Engineering

Third Edition

Desmond S. T. Nicholl

University of the West of Scotland, Paisley, UK

KALAMAZOO VALLEY
COMMUNITY COLLEGE
LIBRARY

CAMBRIDGE UNIVERSITY PRESS
Cambridge, New York, Melbourne, Madrid, Cape Town, Singapore, São Paulo, Delhi

Cambridge University Press
32 Avenue of the Americas, New York, NY 10013-2473, USA

www.cambridge.org
Information on this title: www.cambridge.org/9780521615211

© Cambridge University Press 1994, 2002
© Desmond S. T. Nicholl 2008

This publication is in copyright. Subject to statutory exception
and to the provisions of relevant collective licensing agreements,
no reproduction of any part may take place without
the written permission of Cambridge University Press.

First published 1994
Second edition 2002
Third edition 2008

Printed in Hong Kong by Golden Cup

A catalog record for this publication is available from the British Library.

Library of Congress Cataloging in Publication Data

Nicholl, Desmond S. T.
An introduction to genetic engineering / Desmond S. T. Nicholl. – 3rd ed.
  p., cm.
Includes bibliographical references and index.
ISBN 978-0-521-85006-3 (hardback) – ISBN 978-0-521-61521-1 (pbk.)
1. Genetic engineering. I. Title.
[DNLM: 1. Genetic Engineering. QU 450 N629i 2008]
QH442.N53 2008
660.65 – dc22   2007045103

ISBN   978-0-521-85006-3 hardback
ISBN   978-0-521-61521-1 paperback

Cambridge University Press has no responsibility for
the persistence or accuracy of URLs for external or
third-party Internet Web sites referred to in this publication
and does not guarantee that any content on such
Web sites is, or will remain, accurate or appropriate.

Contents

Preface to the third edition

As I found when preparing the second edition of this text, advances in genetics continue to be made at an ever increasing rate, which presents something of a dilemma when writing an introductory text on the subject. In the years since the second edition was published, many new applications of gene manipulation technology have been developed, covering an increasingly diverse range of disciplines and applications. The temptation in preparing this third edition, as was the case for its predecessor, was to concentrate on the applications and ignore the fundamental principles of the technology. However, in initial preparation I was convinced that a basic technical introduction to the subject should remain the major focus of the text. Thus, some of the original methods used in gene manipulation have been kept as examples of how the technology developed, even though some of these have become little used or even obsolete. From the educational point of view, this should help the reader cope with more advanced information about the subject, as a sound grasp of the basic principles is an important part of any introduction to genetic engineering. I have again been gratified by the many positive comments about the second edition, and I hope that this new edition continues to serve a useful purpose as part of the introductory literature on this fascinating subject.

In trying to strike a balance between the methodology and the applications of gene manipulation, I have retained the division of the text into three sections. **Part I** deals with an introduction to basic molecular biology, **Part II** with the methods used to manipulate genes, and **Part III** with the applications. These sections may be taken out of order if desired, depending on the level of background knowledge. Apart from a general revision of chapters retained from the second edition, there have been some additional changes made. The emerging importance of bioinformatics as a distinct discipline is recognised by a new chapter devoted to this topic. To help the student of genetic engineering, two additional features have been included. **Text boxes** highlight some of the important aspects of the topics, and **chapter summaries** have been provided, which include aims and learning outcomes along with a listing of **key words**. Along with the concept maps, I hope that these additions will help the reader to make sense of the topics and act as a support for studying the content. By using the summaries, key words, text boxes, and concept maps students should be able to tailor their study to suit their own individual learning styles. I hope that the changes have produced a balanced treatment of the field, whilst retaining the introductory nature of the text and keeping it to a reasonable length despite an overall increase in coverage.

My thanks go to my colleagues Peter Birch and John McLean for comments on various parts of the manuscript, also to Don Powell of the Wellcome Trust Sanger Institute for advice and critical comment on Chapter 9. Their help has made the book better; any errors of fact or interpretation of course remain my own responsibility. Special thanks to Katrina Halliday and her colleagues at Cambridge University Press, and to Katie Greczylo of Aptara, Inc., for their cheerful advice and patience, which helped bring the project to its conclusion. My final and biggest thank-you goes as ever to my wife, Linda, and to Charlotte, Thomas, and Anna. They have again suffered with me during the writing, and have put up with more than they should have had to. I dedicate this new edition to them, with grateful thanks.

Desmond S. T. Nicholl
Paisley 2007

An Introduction to Genetic Engineering

Third Edition

Chapter 1 summary

Aims

- To define genetic engineering as it will be described in this book
- To outline the basic features of genetic engineering
- To describe the emergence of gene manipulation technology
- To outline the structure of the book

Chapter summary/learning outcomes

When you have completed this chapter you will have knowledge of:

- The scope and nature of the subject
- The steps required to clone a gene
- The emergence and early development of the technology
- Elements of the ethical debate surrounding genetic engineering

Key words

Genetic engineering, gene manipulation, gene cloning, recombinant DNA technology, genetic modification, new genetics, molecular agriculture, genethics, DNA ligase, restriction enzyme, plasmid, extrachromosomal element, replicon, text box, aims, chapter summary, learning outcome, concept map.

Chapter 1

Introduction

1.1 | What is genetic engineering?

Progress in any scientific discipline is dependent on the availability of techniques and methods that extend the range and sophistication of experiments that may be performed. Over the past 35 years or so this has been demonstrated in a spectacular way by the emergence of genetic engineering. This field has grown rapidly to the point where, in many laboratories around the world, it is now routine practice to isolate a specific DNA fragment from the genome of an organism, determine its base sequence, and assess its function. The technology is also now used in many other applications, including forensic analysis of scene-of-crime samples, paternity disputes, medical diagnosis, genome mapping and sequencing, and the biotechnology industry. What is particularly striking about the technology of gene manipulation is that it is readily accessible by individual scientists, without the need for large-scale equipment or resources outside the scope of a reasonably well-funded research laboratory. Although the technology has become much more large-scale in recent years as genome sequencing projects have been established, it is still accessible by almost all of the bioscience community in some form or other.

The term **genetic engineering** is often thought to be rather emotive or even trivial, yet it is probably the label that most people would recognise. However, there are several other terms that can be used to describe the technology, including **gene manipulation**, **gene cloning**, **recombinant DNA technology**, **genetic modification**, and the **new genetics**. There are also legal definitions used in administering regulatory mechanisms in countries where genetic engineering is practised.

Although there are many diverse and complex techniques involved, the basic principles of genetic manipulation are reasonably simple. The premise on which the technology is based is that genetic information, encoded by DNA and arranged in the form of genes, is a resource that can be manipulated in various ways to achieve certain goals in both pure and applied science and medicine. There are

> Several terms may be used to describe the technologies involved in manipulating genes.

> Genetic material provides a rich resource in the form of information encoded by the sequence of bases in the DNA.

Fig. 1.1 The four steps in a gene cloning experiment. The term 'clone' comes from the colonies of identical host cells produced during amplification of the cloned fragments. Gene cloning is sometimes referred to as 'molecular cloning' to distinguish the process from the cloning of whole organisms.

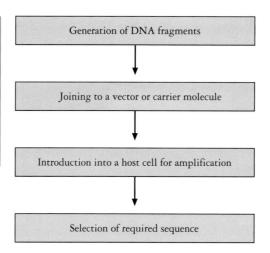

> Generation of DNA fragments
>
> ↓
>
> Joining to a vector or carrier molecule
>
> ↓
>
> Introduction into a host cell for amplification
>
> ↓
>
> Selection of required sequence

many areas in which genetic manipulation is of value, including the following:

- Basic research on gene structure and function
- Production of useful proteins by novel methods
- Generation of transgenic plants and animals
- Medical diagnosis and treatment
- Genome analysis by DNA sequencing

In later chapters we will look at some of the ways in which genetic manipulation has contributed to these areas.

Gene cloning enables isolation and identification of individual genes.

The mainstay of genetic manipulation is the ability to isolate a single DNA sequence from the genome. This is the essence of gene cloning and can be considered as a series of four steps (Fig. 1.1). Successful completion of these steps provides the genetic engineer with a specific DNA sequence, which may then be used for a variety of purposes. A useful analogy is to consider gene cloning as a form of **molecular agriculture**, enabling the production of large amounts (in genetic engineering this means micrograms or milligrams) of a particular DNA sequence. Even in the era of large-scale sequencing projects, this ability to isolate a particular gene sequence is still a major aspect of gene manipulation carried out on a day-to-day basis in research laboratories worldwide.

As well as technical and scientific challenges, modern genetics poses many moral and ethical questions.

One aspect of the new genetics that has given cause for concern is the debate surrounding the potential applications of the technology. The term **genethics** has been coined to describe the ethical problems that exist in modern genetics, which are likely to increase in both number and complexity as genetic engineering technology becomes more sophisticated. The use of transgenic plants and animals, investigation of the human genome, gene therapy, and many other topics are of concern – not just to the scientist, but to the population as a whole. Recent developments in genetically modified foods have provoked a public backlash against the technology. Additional developments in

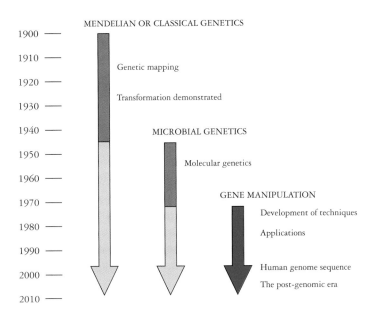

MENDELIAN OR CLASSICAL GENETICS

1900 —
1910 —
Genetic mapping
1920 —
Transformation demonstrated
1930 —
1940 —
MICROBIAL GENETICS
1950 —
Molecular genetics
1960 —
GENE MANIPULATION
1970 —
Development of techniques
1980 —
Applications
1990 —
2000 —
Human genome sequence
The post-genomic era
2010 —

Fig. 1.2 The history of genetics since 1900. Shaded areas represent the periods of major development in each branch of the subject.

the cloning of organisms, and in areas such as *in vitro* fertilisation and xenotransplantation, raise further questions. Although organismal cloning is not strictly part of gene manipulation technology, we will consider aspects of it later in this book, because this is an area of much concern and can be considered genetic engineering in its broadest sense. Research on stem cells, and the potential therapeutic benefits that this research may bring, is another area of concern that is part of the general advance in genetic technology.

Taking all the potential costs and benefits into account, it remains to be seen if we can use genetic engineering for the overall benefit of mankind and avoid the misuse of technology that often accompanies scientific achievement.

1.2 | Laying the foundations

Although the techniques used in gene manipulation are relatively new, it should be remembered that development of these techniques was dependent on the knowledge and expertise provided by microbial geneticists. We can consider the development of genetics as falling into three main eras (Fig. 1.2). The science of genetics really began with the rediscovery of Gregor Mendel's work at the turn of the century, and the next 40 years or so saw the elucidation of the principles of inheritance and genetic mapping. Microbial genetics became established in the mid 1940s, and the role of DNA as the genetic material was confirmed. During this period great advances were made in understanding the mechanisms of gene transfer between bacteria, and a broad knowledge base was established from which later developments would emerge.

Gregor Mendel is often considered the 'father' of genetics.

The discovery of the structure of DNA by James Watson and Francis Crick in 1953 provided the stimulus for the development of genetics at the molecular level, and the next few years saw a period of intense activity and excitement as the main features of the gene and its expression were determined. This work culminated with the establishment of the complete genetic code in 1966 – the stage was now set for the appearance of the new genetics.

Watson and Crick's double helix is perhaps the most 'famous' and most easily recognised molecule in the world.

1.3 | First steps

In the late 1960s there was a sense of frustration among scientists working in the field of molecular biology. Research had developed to the point where progress was being hampered by technical constraints, as the elegant experiments that had helped to decipher the genetic code could not be extended to investigate the gene in more detail. However, a number of developments provided the necessary stimulus for gene manipulation to become a reality. In 1967 the enzyme **DNA ligase** was isolated. This enzyme can join two strands of DNA together, a prerequisite for the construction of recombinant molecules, and can be regarded as a sort of molecular glue. This was followed by the isolation of the first **restriction enzyme** in 1970, a major milestone in the development of genetic engineering. Restriction enzymes are essentially molecular scissors that cut DNA at precisely defined sequences. Such enzymes can be used to produce fragments of DNA that are suitable for joining to other fragments. Thus, by 1970, the basic tools required for the construction of recombinant DNA were available.

By the end of the 1960s most of the essential requirements for the emergence of gene technology were in place.

The first recombinant DNA molecules were generated at Stanford University in 1972, utilising the cleavage properties of restriction enzymes (scissors) and the ability of DNA ligase to join DNA strands together (glue). The importance of these first tentative experiments cannot be overestimated. Scientists could now join different DNA molecules together and could link the DNA of one organism to that of a completely different organism. The methodology was extended in 1973 by joining DNA fragments to the **plasmid** pSC101, which is an **extrachromosomal element** isolated from the bacterium *Escherichia coli*. These recombinant molecules behaved as **replicons**; that is, they could replicate when introduced into *E. coli* cells. Thus, by creating recombinant molecules *in vitro*, and placing the construct in a bacterial cell where it could replicate *in vivo*, specific fragments of DNA could be isolated from bacterial colonies that formed clones (colonies formed from a single cell, in which all cells are identical) when grown on agar plates. This development marked the emergence of the technology that became known as gene cloning (Fig. 1.3).

The key to gene cloning is to ensure that the target sequence is replicated in a suitable host cell.

The discoveries in 1972 and 1973 triggered what is perhaps the biggest scientific revolution of all – the new genetics. The use of the new technology spread very quickly, and a sense of urgency and

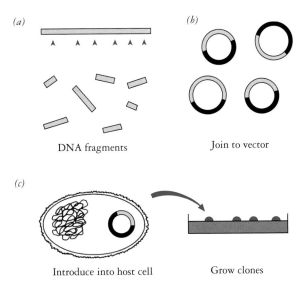

(a)

DNA fragments

(b)

Join to vector

(c)

Introduce into host cell   Grow clones

Fig. 1.3 Cloning DNA fragments. (*a*) The source DNA is isolated and fragmented into suitably sized pieces. (*b*) The fragments are then joined to a carrier molecule or vector to produce recombinant DNA molecules. In this case, a plasmid vector is shown. (*c*) The recombinant DNA molecules are then introduced into a host cell (a bacterial cell in this example) for propagation as clones.

excitement prevailed. This was dampened somewhat by the realisation that the new technology could give rise to potentially harmful organisms exhibiting undesirable characteristics. It is to the credit of the biological community that measures were adopted to regulate the use of gene manipulation and that progress in contentious areas was limited until more information became available regarding the possible consequences of the inadvertent release of organisms containing recombinant DNA. However, the development of genetically modified organisms (GMOs), particularly crop plants, has re-opened the debate about the safety of these organisms and the consequences of releasing GMOs into the environment. In addition, many of the potential medical benefits of gene manipulation, genetics, and cell biology pose ethical questions that may not be easy to answer. We will come across some of these issues later in the book.

> The development and use of genetically modified organisms (GMOs) pose some difficult ethical questions that do not arise in other areas such as gene cloning.

1.4 | What's in store?

In preparing the third edition of this book, I have retained the general organisation of the second edition. The content has been updated to better reflect the current applications of DNA technology and genetics, and several new subsections have been added to most of the chapters. However, I have again retained introductory material on molecular biology, on working with nucleic acids, and on the basic methodology of gene manipulation. I hope that this edition will therefore continue to serve as a technical introduction to the subject, whilst also giving a much broader appreciation of the applications of this exciting range of technologies.

The text is organised into three parts.

Part I (*The basis of genetic engineering*; Chapters 2–4) deals with a basic introduction to the field and the techniques underpinning the science. Chapter 2 (*Introducing molecular biology*) and Chapter 3 (*Working with nucleic acids*) provide background information about DNA and the techniques used when working with it. Chapter 4 (*The tools of the trade*) looks at the range of enzymes needed for gene manipulation.

Part II (*The methodology of gene manipulation*; Chapters 5–9) outlines the techniques and strategies needed to clone and identify genes. Chapter 5 (*Host cells and vectors*) and Chapter 6 (*Cloning strategies*) describe the various systems and protocols that may be used to clone DNA. Chapter 7 is dedicated to the polymerase chain reaction, which has now become established as a major part of modern molecular biology. Chapter 8 (*Selection, screening, and analysis of recombinants*) describes how particular DNA sequences can be selected from collections of cloned fragments. Chapter 9 (*Bioinformatics*) is a new chapter that has been added to deal with the emergence of this topic.

Part III (*Genetic engineering in action*; Chapters 10–15) deals with the applications of gene manipulation and associated technologies. Chapters include *Understanding genes, genomes, and 'otheromes'* (Chapter 10), *Genetic engineering and biotechnology* (Chapter 11), *Medical and forensic applications of gene manipulation* (Chapter 12), and *Transgenic plants and animals* (Chapter 13). Organismal cloning is examined in Chapter 14 (*The other sort of cloning*), and the moral and ethical considerations of genetic engineering are considered in Chapter 15 (*Brave new world or genetic nightmare?*).

> Each chapter is supplemented with some study guidelines to enable the student to use the text productively.

In the third edition I have expanded the range of features that should be useful as study aids where the text is used to support a particular academic course. There are now **text boxes** sprinkled throughout the chapters. The text boxes highlight key points on the way through the text and can be used as a means of summarising the content. At the start of each chapter the **aims** of the chapter are presented, along with a **chapter summary** in the form of **learning outcomes**. These have been written quite generally, so that an instructor can modify them to suit the level of detail required. A list of the **key words** in each chapter is also provided for reference. As in the first and second editions, a **concept map** is given, covering the main points of the chapter. Concept mapping is a technique that can be used to structure information and provide links between various topics. The concept maps provided here are essentially summaries of the chapters and may be examined either before or after reading the chapter. I hope that these support 'tools' continue to be a useful addition to the text for the student of genetic engineering.

Suggestions for further reading are given at the end of the book, along with tips for using the Internet and World Wide Web. No reference has been made to the primary (research) literature, as this is accessible from the books and articles mentioned in the further reading section and by searching literature databases. Many research journals are also now available online. A glossary of terms has also been provided; this may be particularly useful for readers who may be unfamiliar with the terminology used in molecular biology.

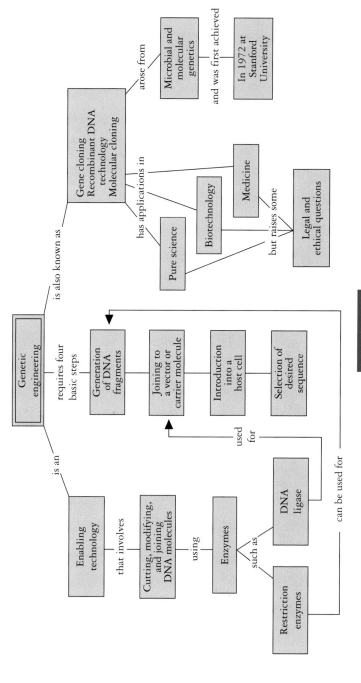

Concept map 1

Part I

The basis of genetic engineering

Chapter 2 summary

Aims

- To illustrate how living systems are organised at the genetic level
- To describe the structure and function of DNA and RNA
- To outline the organisation of genes and genomes
- To describe the processes involved in gene expression and its regulation

Chapter summary/learning outcomes

When you have completed this chapter you will have knowledge of:

- The organisation of living systems and the concept of emergent properties
- The chemistry of living systems
- The genetic code and the flow of genetic information
- The structure of DNA and RNA
- Gene structure and organisation
- Transcription and translation
- The need for regulation of gene expression
- Genome organisation
- The transcriptome and the proteome

Key words

Structure, function, emergent properties, covalent bond, atom, molecule, macromolecule, lipid, carbohydrate, protein, nucleic acid, dehydration synthesis, hydrolysis, cell membrane, plasma membrane, cell wall, genetic material, prokaryotic, eukaryotic, nucleus, enzyme, heteropolymer, codon, minimum coding requirement, genetic code, redundancy, wobble, mRNA, transcription, translation, Central Dogma, replication, reverse transcriptase, mutation, nucleotide, polynucleotide, antiparallel, purines, pyrimidines, double helix, rRNA, tRNA, ribosomes, gene, chromosome, locus, homologous pair, allele, coding strand, promoter, transcriptional unit, operon, operator, polycistronic, cistron, intron, exon, RNA processing, RNA polymerase, anticodon, adaptive regulation, differentiation, developmental regulation, housekeeping gene, constitutive gene, catabolic, inducible, repressor protein, genome, C-value, C-value paradox, inverted repeat, palindrome, foldback DNA, repetitive sequence, single-copy sequence, multigene family, transcriptome, proteome.

Chapter 2

Introducing molecular biology

This chapter presents a brief overview of the structure and function of DNA and its organisation within the genome (the total genetic complement of an organism). We will also have a look at how genes are expressed, and at the ways by which gene expression is regulated. The aim of the chapter is to provide the non-specialist reader with an introduction to the molecular biology of cells, but it should also act as a useful refresher for those who have some background knowledge of DNA. More extensive accounts of the topics presented here may be found in the sources described in Suggestions for Further Reading.

2.1 | The way that living systems are organised

Before we look at the molecular biology of the cell, it may be useful to think a little about what cells are and how living systems are organised. Two premises are useful here. First, there is a very close link between **structure** and **function** in biological systems. Second, living systems provide an excellent example of the concept of **emergent properties**. This is rather like the statement 'the whole is greater than the sum of the parts', in that living systems are organised in a hierarchical way, with each level of organisation becoming more complex. New functional features emerge as components are put together in more complicated arrangements. One often-quoted example is the reactive metal sodium and the poisonous gas chlorine, which combine to give sodium chloride (common table salt), which is of course not poisonous (although it can be harmful if taken in excess!). Thus, it is often difficult or impossible to predict the properties of a more complex system by looking at its constituent parts, which is a general difficulty with the reductionist approach to experimental science.

> Living systems are organised hierarchically, with close interdependence of structure and function.

The chemistry of living systems is based on the element carbon, which can form four **covalent bonds** with other **atoms**. By joining carbon atoms together, and incorporating other atoms, **molecules** can be built up, which in turn can be joined together to produce **macromolecules**. Biologists usually recognise four groups of macromolecules: **lipids**, **carbohydrates**, **proteins**, and **nucleic acids**. The synthesis

> Complex molecules (macromolecules) are made by joining smaller molecules together using dehydration synthesis.

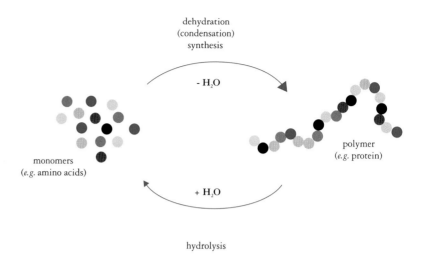

dehydration
(condensation)
synthesis

- H$_2$O

monomers
(*e.g.* amino acids)

polymer
(*e.g.* protein)

+ H$_2$O

hydrolysis

Fig. 2.1 The monomer–polymer cycle. In this example a representation of amino acids and proteins is shown. The amino acids (monomers) are joined together by removal of the elements of water (H$_2$O) during dehydration synthesis. When the protein is no longer required, it may be degraded by adding back the H$_2$O during hydrolysis. Although the cycle looks simple when presented like this, the synthesis of proteins requires many components whose functions are coordinated during the complex process of translation.

of macromolecules involves a condensation reaction between functional groups on the molecules to be joined together. This **dehydration synthesis** forms a covalent bond by removing the elements of water. In the case of the large polymeric macromolecules of the cell (polysaccharides, proteins, and nucleic acids) hundreds, thousands, or even millions of individual monomeric units may be joined together in this way. The polymers can be broken apart into their consituent monomers by adding the elements of water back to reconstitute the original groups. This is known as **hydrolysis** (literally *hydro lysis*, water breaking). The monomer/polymer cycle and dehydration/hydrolysis are illustrated in Fig. 2.1.

The cell is the basic unit of organisation in biological systems. Although there are many different types of cell, there are some features that are present in all cells. There is a **cell membrane** (the **plasma membrane**) that is the interface between the cell contents and the external environment. Some cells, such as bacteria, yeasts, and plant cells, may also have a **cell wall** that provides additional structural support. Some sort of **genetic material** (almost always DNA) is required to provide the information for cells to function, and the organisation of this genetic information provides one way of classifying cells. In **prokaryotic** cells (*e.g.* bacteria) the DNA is not compartmentalised, whereas in **eukaryotic** cells the DNA is located within a membrane-bound **nucleus**. Eukaryotic cells also utilise membranes to provide additional internal structure. Prokaryotic cells are generally smaller in size than eukaryotic cells, but all cells have a maximum upper size limit. This is largely because of the limitations of diffusion as a mechanism for gas and nutrient exchange. Typical bacterial cells have a diameter of 1–10 μm, plant and animal cells 10–100 μm. In

The cell is the basic unit of organisation in biological systems; prokaryotic cells have no nucleus, eukaryotic cells do.

multicellular eukaryotes, an increase in the size of the organism is achieved by using more cells rather than by making cells bigger.

2.2 | The flow of genetic information

To set the structure of nucleic acids in context, it is useful to think a little about what is required, in terms of genetic information, to enable a cell to carry out its various activities. It is a remarkable fact that an organism's characteristics are encoded by a four-letter alphabet, defining a language of three-letter words. The letters of this alphabet are the nitrogenous bases adenine (A), guanine (G), cytosine (C), and thymine (T). So how do these bases enable cells to function?

Life is directed by four nitrogenous bases: adenine (A), guanine (G), cytosine (C), and thymine (T).

The expression of genetic information is achieved ultimately *via* proteins, particularly the **enzymes** that catalyse the reactions of metabolism. Proteins are condensation **heteropolymers** synthesised from amino acids, of which 20 are used in natural proteins. Given that a protein may consist of several hundred amino acid residues, the number of different proteins that may be made is essentially unlimited, assuming that the correct sequence of amino acids can be specified from the genetic information. As the bases are critical informatic components, we can calculate that using the bases singly would not provide enough scope (only 4 possible arrangements) to encode 20 amino acids, as there are only 4 possible code 'combinations' (A, G, C, and T). If the bases were arranged in pairs, that would give 4^2 or 16 possible combinations – still not enough. Triplet combinations provide 4^3 or 64 possible permutations, which is more than sufficient. Thus, great diversity of protein form and function can be achieved using an elegantly simple coding system, with sets of three nucleotides (**codons**) specifying the amino acids. Thus, a protein of 300 amino acids would have a **minimum coding requirement** of 900 nucleotides on a strand of DNA. The **genetic code** or 'dictionary' is one part of molecular biology that, like the double helix, has become something of a biological icon. Although there are more possible codons that are required (64 as opposed to 20), three of these are 'STOP' codons. Several amino acids are specified by more than one codon, which accounts for the remainder, a feature that is known as **redundancy** of the code. An alternative term for this, where the first two bases in a codon are often critical with the third less so, is known as **wobble**. These features can be seen in the standard presentation of the genetic code shown in Table 2.1.

The flow of genetic information is unidirectional, from DNA to protein, with **messenger RNA (mRNA)** as an intermediate. The copying of DNA-encoded genetic information into RNA is known as **transcription** (T_C), with the further conversion into protein being termed **translation** (T_L). This concept of information flow is known as the **Central Dogma** of molecular biology and is an underlying theme in all studies of gene expression.

The flow of genetic information is from DNA to RNA to protein, via the processes of transcription (T_C) and translation (T_L). This concept is known as the Central Dogma of molecular biology.

Two further aspects of information flow may be added to this basic model to complete the picture. First, duplication of the genetic

Table 2.1. | The genetic code

First base (5′ end)	Second base				Third base (3′ end)
	U	C	A	G	
U	Phe	Ser	Tyr	Cys	U
	Phe	Ser	Tyr	Cys	C
	Leu	Ser	STOP	STOP	A
	Leu	Ser	STOP	Trp	G
C	Leu	Pro	His	Arg	U
	Leu	Pro	His	Arg	C
	Leu	Pro	Gln	Arg	A
	Leu	Pro	Gln	Arg	G
A	Ile	Thr	Asn	Ser	U
	Ile	Thr	Asn	Ser	C
	Ile	Thr	Lys	Arg	A
	Met	Thr	Lys	Arg	G
G	Val	Ala	Asp	Gly	U
	Val	Ala	Asp	Gly	C
	Val	Ala	Glu	Gly	A
	Val	Ala	Glu	Gly	G

Note: Codons read 5′→3′; thus, AUG specifies Met. The three-letter abbreviations for the amino acids are as follows: Ala, Alanine; Arg, Arginine; Asn, Asparagine; Asp, Aspartic acid; Cys, Cysteine; Gln, Glutamine; Glu, Glutamic acid; Gly, Glycine; His, Histidine; Ile, Isoleucine; Leu, Leucine; Lys, Lysine; Met, Methionine; Phe, Phenylalanine; Pro, Proline; Ser, Serine; Thr, Threonine; Trp, Tryptophan; Tyr, Tyrosine; Val, Valine. The three codons UAA, UAG, and UGA specify no amino acid and terminate translation.

material prior to cell division represents a DNA–DNA transfer, known as DNA **replication**. A second addition, with important consequences for the genetic engineer, stems from the fact that some viruses have RNA instead of DNA as their genetic material. These viruses (chiefly members of the retrovirus group) have an enzyme called **reverse transcriptase** (an RNA-dependent DNA polymerase) that produces a double-stranded DNA molecule from the single-stranded RNA genome. Thus, in these cases the flow of genetic information is reversed with respect to the normal convention. The Central Dogma is summarised in Fig. 2.2.

2.3 | The structure of DNA and RNA

In most organisms, the primary genetic material is double-stranded DNA. What is required of this molecule? First, it has to be stable, as genetic information may need to function in a living organism for up

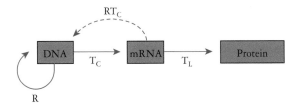

Fig. 2.2 The Central Dogma states that information flow is unidirectional, from DNA to mRNA to protein. The processes of transcription (T_C), translation (T_L), and DNA replication (R) obey this rule. An exception is found in retroviruses (RNA viruses), which have an RNA genome and carry out a process known as reverse transcription (RT_C) to produce a DNA copy of the genome following infection of the host cell.

to 100 years or more. Second, the molecule must be capable of replication, to permit dissemination of genetic information as new cells are formed during growth and development. Third, there should be the potential for limited alteration to the genetic material (**mutation**), to enable evolutionary pressures to exert their effects. The DNA molecule fulfils these criteria of stability, replicability, and mutability, and when considered with RNA provides an excellent example of the premises that we considered earlier – the very close relationship between structure and function, and the concept of emergent properties.

Nucleic acids are heteropolymers composed of monomers known as **nucleotides**; a nucleic acid chain is therefore often called a **polynucleotide**. The monomers are themselves made up of three components: a sugar, a phosphate group, and a nitrogenous base. The two types of nucleic acid (DNA and RNA) are named according to the sugar component of the nucleotide, with DNA having 2′-deoxyribose as the sugar (hence **D**eoxyribo**N**ucleic**A**cid) and RNA having ribose (hence **R**ibo**N**ucleic**A**cid). The sugar/phosphate components of a nucleotide are important in determining the structural characteristics of polynucleotides, and the nitrogenous bases determine their information storage and transmission characteristics. The structure of a nucleotide is summarised in Fig. 2.3.

> Nucleic acids are polymers composed of nucleotides; DNA is deoxyribonucleic acid, RNA is ribonucleic acid.

Nucleotides can be joined together by a 5′→3′ phosphodiester linkage, which confers directionality on the polynucleotide. Thus, the 5′ end of the molecule will have a free phosphate group, and

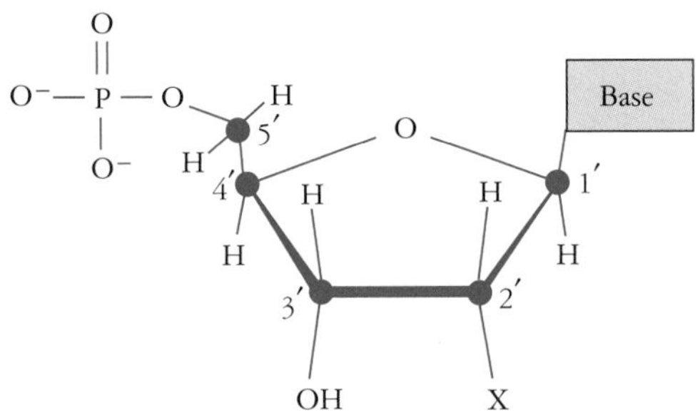

Fig. 2.3 The structure of a nucleotide. Carbon atoms are represented by solid circles, numbered 1′ to 5′. In DNA the sugar is deoxyribose, with a hydrogen atom at position X. In RNA the sugar is ribose, which has a hydroxyl group at position X. The base can be A, G, C, or T in DNA, and A, G, C, or U in RNA.

Fig. 2.4 Base-pairing arrangements in DNA. (a) An A·T base pair. The bases are linked by two hydrogen bonds (dotted lines). (b) A G·C base pair, with three hydrogen bonds.

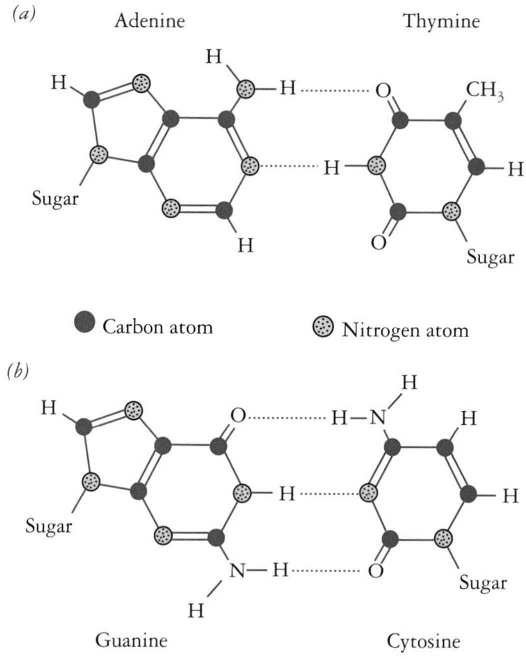

In DNA the bases pair A·T and G·C; this complementary base pairing is the key to information storage, transfer, and use.

the 3′ end a free hydroxyl group; this has important consequences for the structure, function, and manipulation of nucleic acids. In a double-stranded molecule such as DNA, the sugar–phosphate chains are found in an **antiparallel** arrangement, with the two strands running in different directions.

The nitrogenous bases are the important components of nucleic acids in terms of their coding function. In DNA the bases are as listed in Section 2.1, namely adenine (A), guanine (G), cytosine (C), and thymine (T). In RNA the base thymine is replaced by uracil (U), which is functionally equivalent. Chemically adenine and guanine are **purines**, which have a double-ring structure, whereas cytosine and thymine (and uracil) are **pyrimidines**, which have a single-ring structure. In DNA the bases are paired, A with T and G with C. This pairing is determined both by the bonding arrangements of the atoms in the bases and by the spatial constraints of the DNA molecule, the only satisfactory arrangement being a purine–pyrimidine base pair. The bases are held together by hydrogen bonds, two in the case of an A·T base pair and three in the case of a G·C base pair. The structure and base-pairing arrangement of the four DNA bases is shown in Fig. 2.4.

The DNA molecule *in vivo* usually exists as a right-handed **double helix** called the B-form. This is the structure proposed by Watson and Crick in 1953. Alternative forms of DNA include the A-form (right-handed helix) and the Z-form (left-handed helix). Although DNA structure is a complex topic, particularly when the higher-order arrangements of DNA are considered, a simple representation will suffice here, as shown in Fig. 2.5.

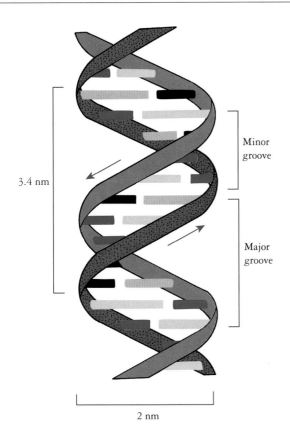

3.4 nm

Minor
groove

Major
groove

2 nm

Fig. 2.5 The double helix. This is DNA in the commonly found B-form. The right-handed helix has a diameter of 2 nm and a pitch of 3.4 nm, with 10 base pairs per turn. The sugar–phosphate 'backbones' are antiparallel (arrowed) with respect to their $5' \rightarrow 3'$ orientations. One of the sugar–phosphate chains has been shaded for clarity. The purine–pyrimidine base pairs are formed across the axis of the helix.

The structure of RNA is similar to that of DNA; the main chemical differences are the presence of ribose instead of 2′-deoxyribose and uracil instead of thymine. RNA is also most commonly single-stranded, although short stretches of double-stranded RNA may be found in self-complementary regions. There are three main types of RNA molecule found in cells: messenger RNA (mRNA), ribosomal RNA (**rRNA**), and transfer RNA (**tRNA**). Ribosomal RNA is the most abundant class of RNA molecule, making up some 85% of total cellular RNA. It is associated with **ribosomes**, which are an essential part of the translational machinery. Transfer RNAs make up about 10% of total RNA and provide the essential specificity that enables the insertion of the correct amino acid into the protein that is being synthesised. Messenger RNA, as the name suggests, acts as the carrier of genetic information from the DNA to the translational machinery and usually makes up less than 5% of total cellular RNA.

Three important types of RNA are ribosomal RNA (rRNA), messenger RNA (mRNA), and transfer RNA (tRNA).

2.4 | Gene organisation

The **gene** can be considered the basic unit of genetic information. Genes have been studied since the turn of the century, when genetics became established. Before the advent of molecular biology and the realisation that genes were made of DNA, study of the gene was largely indirect; the effects of genes were observed in phenotypes and the 'behaviour' of genes was analysed. Despite the apparent limitations of this approach, a vast amount of information about how genes functioned was obtained, and the basic tenets of transmission genetics were formulated.

As the gene was studied in greater detail, the terminology associated with this area of genetics became more extensive, and the ideas about genes were modified to take developments into account.

The term 'gene' is usually taken to represent the genetic information transcribed into a single RNA molecule, which is in turn translated into a single protein. Exceptions are genes for RNA molecules (such as rRNA and tRNA), which are not translated. In addition, the nomenclature used for prokaryotic cells is slightly different because of the way that their genes are organised. Genes are located on **chromosomes**, and the region of the chromosome where a particular gene is found is called the **locus** of that gene. In diploid organisms, which have their chromosomes arranged as **homologous pairs**, different forms of the same gene are known as **alleles**.

> The gene is the basic unit of genetic information. Genes are located on chromosomes at a particular genetic locus. Different forms of the same gene are known as alleles.

2.4.1 The anatomy of a gene

Although there is no such thing as a 'typical' gene, there are certain basic requirements for any gene to function. The most obvious is that the gene has to encode the information for the particular protein (or RNA molecule). The double-stranded DNA molecule has the potential to store genetic information in either strand, although in most organisms only one strand is used to encode any particular gene. There is the potential for confusion with the nomenclature of the two DNA strands, which may be called coding/non-coding, sense/antisense, plus/minus, transcribed/non-transcribed, or template/non-template. In some cases different authors use the same terms in different ways, which adds to the confusion. Recommendations from the International Union of Biochemistry and the International Union of Pure and Applied Chemistry favour the terms coding/non-coding, with the **coding strand** of DNA taken to be the mRNA-like strand. This convention will be used in this book where coding function is specified. The terms template and non-template will be used to describe DNA strands when there is not necessarily any coding function involved, as in the copying of DNA strands during cloning procedures. Thus, genetic information is expressed by transcription of the non-coding strand of DNA, which produces an mRNA molecule that has the same sequence as the coding strand of DNA (although the RNA has uracil substituted for thymine; see Fig. 2.9(a)). The sequence of the coding strand is usually reported when dealing with DNA sequence data, as this permits easy reference to the sequence of the RNA.

In addition to the sequence of bases that specifies the codons in a protein-coding gene, there are other important regulatory sequences associated with genes (Fig. 2.6). A site for starting transcription is required, and this encompasses a region that binds RNA polymerase, known as the **promoter** (P), and a specific start point for transcription

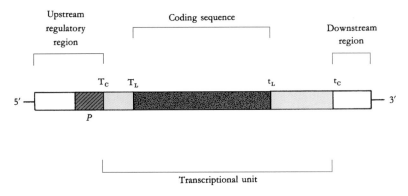

Fig. 2.6 Gene organisation. The transcriptional unit produces the RNA molecule and is defined by the transcription start site (T_C) and stop site (t_C). Within the transcriptional unit lies the coding sequence, from the translation start site (T_L) to the stop site (t_L). The upstream regulatory region may have controlling elements such as enhancers or operators in addition to the promoter (P), which is the RNA polymerase binding site.

(T_C). A stop site for transcription (t_C) is also required. From T_C start to t_C stop is sometimes called the **transcriptional unit**, that is, the DNA region that is copied into RNA. Within this transcriptional unit there may be regulatory sites for translation, namely a start site (T_L) and a stop signal (t_L). Other sequences involved in the control of gene expression may be present either upstream or downstream from the gene itself.

Genes have several important regions. A promoter is necessary for RNA polymerase binding, with the transcription start and stop sites defining the transcriptional unit.

2.4.2 Gene structure in prokaryotes

In prokaryotic cells such as bacteria, genes are usually found grouped together in **operons**. The operon is a cluster of genes that are related (often coding for enzymes in a metabolic pathway) and that are under the control of a single promoter/regulatory region. Perhaps the best known example of this arrangement is the *lac* operon (Fig. 2.7), which codes for the enzymes responsible for lactose catabolism. Within the

Genes in prokaryotes tend to be grouped together in operons, with several genes under the control of a single regulatory region.

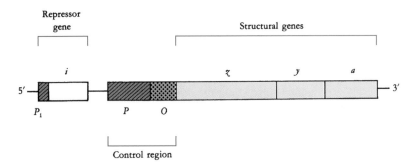

Fig. 2.7 The *lac* operon. The structural genes *lacZ*, *lacY*, and *lacA* (noted as *z*, *y*, and *a*) encode ß-galactosidase, galactoside permease, and a transacetylase, respectively. The cluster is controlled by a promoter (*P*) and an operator region (*O*). The operator is the binding site for the repressor protein, encoded by the *lacl* gene (*i*). The repressor gene lies outside the operon itself and is controlled by its own promoter, P_i.

operon there are three genes that code for proteins (termed structural genes) and an upstream control region encompassing the promoter and a regulatory site called the **operator**. In this control region there is also a site that binds a complex of cAMP (cyclic adenosine monophosphate) and CRP (cAMP receptor protein), which is important in positive regulation (stimulation) of transcription. Lying outside the operon itself is the repressor gene, which codes for a protein (the Lac repressor) that binds to the operator site and is responsible for negative control of the operon by blocking the binding of RNA polymerase.

The fact that structural genes in prokaryotes are often grouped together means that the transcribed mRNA may contain information for more than one protein. Such a molecule is known as a **polycistronic** mRNA, with the term **cistron** equating to the 'gene' as we have defined it (*i.e.* encoding one protein). Thus, much of the genetic information in bacteria is expressed *via* polycistronic mRNAs whose synthesis is regulated in accordance with the needs of the cell at any given time. This system is flexible and efficient, and it enables the cell to adapt quickly to changing environmental conditions.

2.4.3   Gene structure in eukaryotes

A major defining feature of eukaryotic cells is the presence of a membrane-bound nucleus, within which the DNA is stored in the form of chromosomes. Transcription therefore occurs within the nucleus and is separated from the site of translation, which is in the cytoplasm. The picture is complicated further by the presence of genetic information in mitochondria (plant and animal cells) and chloroplasts (plant cells only), which have their own separate genomes that specify many of the components required by these organelles. This compartmentalisation has important consequences for regulation, both genetic and metabolic, and thus gene structure and function in eukaryotes are more complex than in prokaryotes.

The most startling discovery concerning eukaryotic genes was made in 1977, when it became clear that eukaryotic genes contained 'extra' pieces of DNA that did not appear in the mRNA that the gene encoded. These sequences are known as intervening sequences or **introns**, with the sequences that will make up the mRNA being called **exons**. In many cases the number and total length of the introns exceed that of the exons, as in the chicken ovalbumin gene, which has a total of seven introns making up more than 75% of the gene. As our knowledge has developed, it has become clear that eukaryotic genes are often extremely complex, and may be very large indeed. Some examples of human gene complexity are shown in Table 2.2. This illustrates the tremendous range of sizes for human genes, the smallest of which may be only a few hundred base pairs in length. At the other end of the scale, the dystrophin gene is spread over 2.4 Mb of DNA on the X chromosome, with the 79 exons representing only 0.6% of this length of DNA.

The presence of introns obviously has important implications for the expression of genetic information in eukaryotes, in that the introns must be removed before the mRNA can be translated. This

Eukaryotic genes tend to be more complex than prokaryotic genes and often contain intervening sequences (introns). The introns form part of the primary transcript, which is converted to the mature mRNA by RNA processing.

Table 2.2. Size and structure of some human genes

Gene	Gene size (kbp)	Number of exons	% exon
Insulin	1.4	3	33
β-globin	1.6	3	38
Serum albumin	18	14	12
Blood clotting factor VIII	186	26	3
CFTR (cystic fibrosis)	230	27	2.4
Dystrophin (muscular dystrophy)	2400	79	0.6

Note: Gene sizes are given in kilobase pairs (kbp). The number of exons is shown, and the percentage of the gene that is represented by these exons is given in the final column.

is carried out in the nucleus, where the introns are spliced out of the primary transcript. Further intranuclear modification includes the addition of a 'cap' at the 5′ terminus and a 'tail' of adenine residues at the 3′ terminus. These modifications are part of what is known as **RNA processing**, and the end product is a fully functional mRNA that is ready for export to the cytoplasm for translation. The structures of the mammalian ß-globin gene and its processed mRNA are outlined in Fig. 2.8 to illustrate eukaryotic gene structure and RNA processing.

2.5 | Gene expression

As shown in Fig. 2.2, the flow of genetic information is from DNA to protein. Whilst a detailed knowledge of gene expression is not

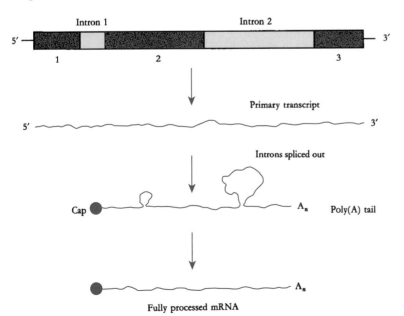

Fig. 2.8 Structure and expression of the mammalian ß-globin gene. The gene contains two intervening sequences or introns. The expressed sequences (exons) are shaded and numbered. The primary transcript is processed by capping, polyadenylation, and splicing to yield the fully functional mRNA.

required in order to understand the principles of genetic engineering, it is useful to be familiar with the main features of transcription and translation and to have some knowledge of how gene expression is controlled.

2.5.1 From genes to proteins

At this point it may be useful to introduce an analogy that I find helpful in thinking about the role of genes in determining cell structure and function. You may hear the term 'genetic blueprint' used to describe the genome. However, this is a little too simplistic, and I prefer to use the analogy of a recipe to describe how genes and proteins work. Let's consider making a cake – the recipe (gene) would be found in a particular book (chromosome), on a particular page (locus), and would contain information in the form of words (codons). One part of the recipe might read 'add 400 g of sugar and beat well', which is fairly clear and unambiguous. When put together with all the other ingredients and baked, the result is a cake in which you cannot see the sugar as an identifiable component. On the other hand, currants or blueberries would appear as identifiable parts of the cake. In a similar way many of the characteristics of an organism are determined by multiple genes, with no particular single gene product being identifiable. Conversely, in single-gene traits the effect of a particular gene may be easily identified as a phenotypic characteristic.

Mutation can also be considered in the recipe context, to give some idea of the relative severity of effect that different mutations can have. If we go back to our sugar example, what would be the effect of the last 0 of 400 being replaced by a 1, giving 401 g as opposed to 400 g? This change would almost certainly remain undetected. However, if the 4 of 400 changed to a 9, or if an additional 0 was added to 400, then things would be very different (and much sweeter!). Thus, mutations in non-critical parts of genes may be of no consequence, whereas mutation in a critical part of a gene can have extremely serious consequences. In some cases a single base insertion or substitution can have a major effect. (Think of adding a 'k' in front of the 'g' in 400 g!)

The recipe analogy is a useful one, in that it defines the role of the recipe itself (specifying the components to be put together) and also illustrates that the information is only part of the story. If the cake is not mixed or baked properly, even with the correct proportions of ingredients, it will not turn out to be a success. Genes provide the information to specify the proteins, but the whole process must be controlled and regulated if the cell is to function effectively.

2.5.2 Transcription and translation

These two processes are the critical steps involved in producing functional proteins in the cell. Transcription involves synthesis of an RNA from the DNA template provided by the non-coding strand of the transcriptional unit in question. The enzyme responsible is **RNA polymerase** (DNA-dependent RNA polymerase). In prokaryotes there is a

---

Genetic information is perhaps better thought of as a recipe than as a blueprint.

The effects of mutations can be mild or severe, depending on the type of mutation and its location in the gene sequence.

single RNA polymerase enzyme, but in eukaryotes there are three types of RNA polymerase (I, II, and III). These synthesise ribosomal, messenger, and transfer/5 S ribosomal RNAs, respectively. All RNA polymerases are large multisubunit proteins with relative molecular masses of around 500000.

Transcription has several component stages: (1) DNA/RNA polymerase binding, (2) chain initiation, (3) chain elongation, and (4) chain termination and release of the RNA. Promoter structure is important in determining the binding of RNA polymerase but will not be dealt with here. When the RNA molecule is released, it may be immediately available for translation (as in prokaryotes) or it may be processed and exported to the cytoplasm (as in eukaryotes) before translation occurs.

Translation requires an mRNA molecule, a supply of charged tRNAs (tRNA molecules with their associated amino acid residues), and ribosomes (composed of rRNA and ribosomal proteins). The ribosomes are the sites where protein synthesis occurs; in prokaryotes, ribosomes are composed of three rRNAs and some 52 different ribosomal proteins. The ribosome is a complex structure that essentially acts as a 'jig' that holds the mRNA in place so that the codons may be matched up with the appropriate **anticodon** on the tRNA, thus ensuring that the correct amino acid is inserted into the growing polypeptide chain. The mRNA molecule is translated in a $5' \rightarrow 3'$ direction, corresponding to polypeptide elongation from N terminus to C terminus.

The codon/anticodon recognition event marks the link between nucleic acid and protein.

Although transcription and translation are complex processes, the essential features (with respect to information flow) may be summarised as shown in Fig. 2.9. In conjunction with the brief descriptions presented earlier, this should provide enough background information about gene structure and expression to enable subsequent sections of the text to be linked to these processes where necessary.

2.5.3 Regulation of gene expression

Transcription and translation provide the mechanisms by which genes are expressed. However, it is vital that gene expression is controlled so that the correct gene products are produced in the cell at the right time. Why is this so important? Let's consider two types of cell – a bacterial cell and a human cell. Bacterial cells need to be able to cope with wide variations in environmental conditions and, thus, need to keep all their genetic material 'at the ready' in case particular gene products are needed. By keeping their genomes in this state of readiness, bacteria conserve energy (by not making proteins wastefully) and can respond quickly to any opportune changes in nutrient availability. This is an example of **adaptive regulation** of gene expression.

Prokaryotic genes are often regulated in response to external signals such as nutrient availability.

In contrast to bacteria, human cells (usually) experience a very different set of environmental conditions. Cells may be highly specialised and **differentiated**, and their external environment is usually stable and controlled by homeostatic mechanisms to ensure that no wide fluctuations occur. Thus, cell specialisation brings more complex

Eukaryotic genes are often regulated in response to signals generated from within the organism.

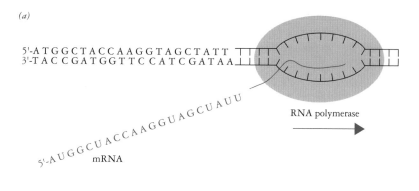

(a)

5'-A T G G C T A C C A A G G T A G C T A T T
3'-T A C C G A T G G T T C C A T C G A T A A

RNA polymerase

5'-AUGGCUACCAAGGUAGCUAUU
mRNA

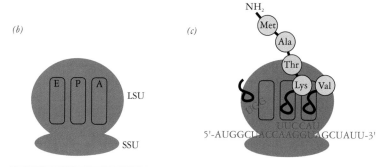

(b)

(c)

NH₂

Met

Ala

Thr

Lys   Val

E P A

LSU

SSU

5'-AUGGCUACCAAGGUAGCUAUU-3'

Fig. 2.9 Transcription and translation. (*a*) Transcription involves synthesis of mRNA by RNA polymerase. Part of the DNA/mRNA sequence is given. The mRNA has the same sequence as the coding strand in the DNA (the non-template strand), apart from U being substituted for T. (*b*) The ribosome is the site of translation and is made up of the large subunit (LSU) and the small subunit (SSU), each made up of ribosomal RNA molecules and many different proteins. There are three sites within the ribosome. The A (aminoacyl) and P (peptidyl) sites are involved in insertion of the correct tRNA–amino acid complex in the growing polypeptide chain. The E (exit) site facilitates the release of the tRNA after peptide bond formation has removed its amino acid. (*c*) The mRNA is being translated. The amino acid residue is inserted into the protein in response to the codon/anticodon recognition event in the ribosome. The first amino acid residue is encoded by AUG in the mRNA (tRNA anticodon TAC), which specifies methionine (see Table 2.1 for the genetic code). The remainder of the sequence is translated in a similar way. The ribosome translates the mRNA in a 5'→3' direction, with the polypeptide growing from its N terminus. The residues in the polypeptide chain are joined together by peptide bonds.

function but requires more controlled conditions. Differentiation is a function of development and, thus, genes in multicellular eukaryotes are often **developmentally regulated**. Gene regulation during the development and life cycle of a complex organism is, as you would expect, complex.

In addition to genes that are controlled and regulated, there are many examples of gene products that are needed at all times during a cell's life. Such genes are sometimes called **housekeeping genes** or **constitutive genes**, in that they are essentially unregulated and encode proteins that are essential at all times (such as enzymes for primary catabolic pathways).

Although a detailed discussion of the control of gene expression is outside the scope of this book, the basic principles can be illustrated by considering how bacterial operons are regulated. A bacterial cell (living outside the laboratory) will experience a wide range of environmental conditions. In particular, there will be fluctuations in the availability of nutrients. If the cell is to survive, it must conserve energy resources, which means that wasteful synthesis of non-required proteins should be prevented. Thus, bacterial cells have mechanisms that enable operons to be controlled with a high degree of sensitivity. An operon that encodes proteins involved in a **catabolic** pathway (one that breaks down materials to release energy) is often regulated by being switched 'on' only when the substance beceomes available in the extracellular medium. Thus, when the substance is absent, there are systems that keep catabolic operons switched 'off'. These are said to be **inducible** operons and are usually controlled by a negative control mechanism involving a **repressor protein** that prevents access to the promoter by RNA polymerase. The classic example of a catabolic operon is the *lac* operon (the structure of which is shown in Fig 2.7). When lactose is absent, the repressor protein binds to the operator and the system is 'off'. The system is a little 'leaky', however, and thus the proteins encoded by the operon (β-galactosidase, permease, and transacetylase) will be present in the cell at low levels. When lactose becomes available, it is transported into the cell by the permease and binds to the repressor protein, causing a conformational (shape) change so that the repressor is unable to bind to the operator. Thus, the negative control is removed, and the operon is accessible by RNA polymerase. A second level of control, based on the level of cAMP, ensures that full activity is only attained when lactose is present and energy levels are low. This dual-control mechanism is a very effective way of regulating gene expression, enabling a range of levels of expression that is a bit like a dimmer switch rather than an on/off swich. In the case of catabolic operons like the *lac* system, this ensures that the enzymes are only synthesised at maximum rate when they are really required.

> Gene regulation in bacteria enables a range of levels of gene expression to be attained, rather than a simple on/off switch.

2.6 | Genes and genomes

When techniques for the examination of DNA became established, gene structure was naturally one of the first areas where efforts were concentrated. However, genes do not exist in isolation, but as part of the **genome** of an organism. Over the past few years the emphasis in molecular biology has shifted slightly, and today we are much more likely to consider the genome as a whole – almost as a type of cellular organelle – rather than just a collection of genes. The Human Genome Project (considered in Chapter 10) is a good example of the development of the field of bioinformatics, which is one of the most active research areas in modern molecular biology.

> The genome is the total complement of DNA in the cell.

Table 2.3. | Genome size in some organisms

Organism	Genome size (Mb)
Escherichia coli (bacterium)	4.6
Saccharomyces cerevisiae (yeast)	12.1
Drosophila melanogaster (fruit fly)	150
Homo sapiens (man)	3000
Mus musculus (mouse)	3300
Nicotiana tabacum (tobacco)	4500
Triticum aestivum (wheat)	17000

Note: Genome sizes are given in megabase pairs (1 megabase = 1×10^6 bases).

2.6.1 Genome size and complexity

The amount of DNA in the haploid genome is known as the **C-value**. It would seem reasonable to assume that genome size should increase with increasing complexity of organisms, reflecting the greater number of genes required to facilitate this complexity. The data shown in Table 2.3 show that, as expected, genome size does tend to increase with organismal complexity. Thus, bacteria, yeast, fruit fly, and human genomes fit this pattern. However, mouse, tobacco, and wheat have much larger genomes than humans – this seems rather strange, as intuitively we might assume that a wheat plant is not as complex as a human being. Also, as E. coli has around 4000 genes, it appears that the tobacco plant genome has the capacity to encode 4000000 genes, and this is certainly not the case, even allowing for the increased size and complexity of eukaryotic genes. This anomaly is sometimes called the **C-value paradox**.

In addition to the size of the genome, genome complexity also tends to increase with more complex organisation. One way of studying complexity involves examining the renaturation of DNA samples. If a DNA duplex is denatured by heating the solution until the strands separate, the complementary strands will renature on cooling (Fig. 2.10). This feature can be used to provide information about the sequence complexity of the DNA in question, since sequences that are present as multiple copies in the genome will renature faster than sequences that are present as single copies only. By performing this type of analysis, eukaryotic DNA can be shown to be composed of four different abundance classes. First, some DNA will form duplex structures almost instantly, because the denatured strands have regions such as **inverted repeats** or **palindromes**, which fold back on each other to give a hairpin loop structure. This class is commonly known as **foldback DNA**. The second fastest to re-anneal are highly **repetitive sequences**, which occur many times in the genome. Following these are moderately repetitive sequences, and finally there are the **unique** or **single-copy sequences**, which rarely re-anneal under the conditions used for this type of analysis. We will consider how repetitive

Eukaryotic genomes may have a range of different types of repetitive sequences.

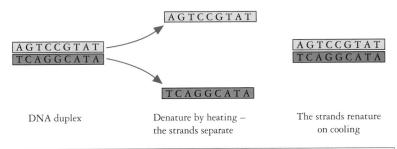

Fig. 2.10 The principle of nucleic acid hybridisation. This feature of DNA molecules is a critical part of many of the procedures involved in gene manipulation and is also an essential feature of life itself. Thus, the simple G · C and A · T base pairing (see Fig. 2.4) has profound implications for living systems and for the applications of recombinant DNA technology.

DNA sequence elements can be used in genome mapping and DNA profiling in Chapters 10 and 12.

2.6.2 Genome organisation

The C-value paradox and the sequence complexity of eukaryotic genomes raise questions about how genomes are organised. Viral and bacterial genomes tend to show very efficient use of DNA for encoding their genes, which is a consequence of (and explanation for) their small genome size. However, in the human genome, only about 3% of the total amount of DNA is actually coding sequence. Even when the introns and control sequences are added, the majority of the DNA has no obvious function. This is sometimes termed 'junk' DNA, although this is perhaps the wrong way to think about this apparently redundant DNA.

Most of the human genome is not involved in coding for proteins.

Estimating the number of genes in a particular organism is not an exact science, and a number of different methods may be used. When the full genome sequence is determined, this obviously makes gene identification much easier, although there are many cases where gene coding sequences are recognised, but the protein products are unknown in terms of their biological function.

Many genes in eukaryotes are single copy genes, and tend to be dispersed across the multiple chromosomes found in eukaryotic cell nuclei. Other genes may be part of **multigene families**, and may be grouped at a particular chromosomal location or may be dispersed. When studying gene organisation in the context of the genome itself, features such as gene density, gene size, mRNA size, intergenic distance, and intron/exon sizes are important indicators. Early analysis of human DNA indicated that the 'average' size of a coding region is around 1500 base pairs, and the average size of a gene is 10–15 kbp. Gene density is about one gene per 40–45 kbp, and the intergenic distance is around 25–30 kbp. However, as we have already seen, gene structure in eukaryotes can be very complex, and thus using 'average' estimates is a little misleading. What is clear is that genomes are now yielding much new information about gene structure and function as

Genome sequencing has greatly improved our understanding of how genomes work.

genome sequencing projects generate more data and we enter what is sometimes called the 'post-genomic era'.

2.6.3 The transcriptome and proteome

We finish this look at molecular biology by introducing two more '-omes' to complement the 'genome'. These terms have become widely used as researchers begin to delve into the bioinformatics of cells. The **transcriptome** refers to the population of transcripts at any given point in a cell's life. This expressed subset of the genomic information will be determnined by many factors affecting the status of the cell. There will be general 'housekeeping' genes for basic maintenance of cell function, but there may also be tissue-specific genes being expressed, or perhaps developmentally regulated genes will be 'on' at that particular point. Analysis of the transcriptome therefore gives a good snapshot of what the cell is engaged in at that point in time.

The **proteome** is a logical extension to the genome and transcriptome in that it represents the population of proteins in the cell. The proteome will reflect the transcriptome to a much greater extent than the transcriptome reflects the genome, although there will be some transcripts that may not be translated efficiently, and there may be proteins that persist in the cell when their transcripts have been removed from circulation. Many biologists now accept that an understanding of the proteome is critical in developing a full understanding of how cells work. Some even consider the proteome as the 'holy grail' of cell biology, comparing it with the search for the unifying theory in physics. The argument is that, if we understand how all the proteins of a cell work, then surely we have a complete understanding of cell structure and function? As with most things in biology, this is unlikely to be a simple process, although the next few years will provide much excitement for biologists as the secrets of gene expression are revealed in more detail.

Analysis of the transcriptome and proteome provides useful information about which genes a cell is expressing at any given time.

The genetic material

- is arranged as → Genes
- is usually → DNA
  - gives rise to → RNA

DNA
- arranged as a → Double helix
  - composed of → Nitrogenous bases — plus — Sugar–phosphate backbone
  - Nitrogenous bases known as:
    - Purines — such as → Adenine, Guanine
    - Pyrimidines — such as → Cytosine, Themine (Uracil–RNA)
  - which pair up →
    - A = T
    - G = C
    - (G = U)

RNA — which is of three types:
- ribosomal (rRNA) — component of → Ribosomes
- messenger (mRNA) — specifies → Codons
- transfer (tRNA) — specifies → Anticodons and amino acids
- which are involved in

RNA polymerase — which produces → RNA

Genes — two aspects:
- Gene expression — requires →
  - Transcription (T_C)
  - Translation (T_l) — which produces → Proteins
- Gene structure
  - in prokaryotes → Relatively simple — genes grouped in → Operons
  - in eukaryotes → Complex — genes have → Introns, Exons, 5'–Cap, 3'–Poly(A)

Concept map 2

Chapter 3 summary

Aims

- To outline the requirements for working with nucleic acids
- To illustrate the range of techniques for isolation, handling, and processing of nucleic acids
- To describe the principles of nucleic acid hybridisation, gel electrophoresis, and DNA sequencing

Chapter summary/learning outcomes

When you have completed this chapter you will have knowledge of:

- Basic laboratory requirements for gene manipulation
- Methods used to isolate nucleic acids
- Handling and quantifying DNA and RNA
- Labelling nucleic acids
- Separating nucleic acids by gel electrophoresis
- Principles of DNA sequencing
- DNA sequencing methods
- Requirements for large-scale DNA sequencing

Key words

Physical containment, biological containment, deproteinisation, ribonuclease, affinity chromatography, gradient centrifugation, microgram, nanogram, picogram, ethidium bromide, precipitation, isopropanol, ethanol, radiolabelling, scintillation counter, radioactive probe, specific activity, end labelling, polynucleotide kinase, nick translation, DNA polymerase I, DNase I, primer extension, oligonucleotide primer, Klenow fragment, nucleic acid hybridisation, autoradiograph, gel electrophoresis, polyanionic, agarose, polyacrylamide, SAGE, PAGE, DNA sequencing, restriction mapping, nested fragments, ordered sequencing strategy, shotgun sequencing, dideoxynucleoside triphosphates.

Chapter 3

Working with nucleic acids

Before examining some of the specific techniques used in gene manipulation, it is useful to consider the basic methods required for handling, quantifying, and analysing nucleic acid molecules. It is often difficult to make the link between theoretical and practical aspects of a subject, and an appreciation of the methods used in routine work with nucleic acids may be of help when the more detailed techniques of gene cloning and analysis are described.

3.1 | Laboratory requirements

One of the striking aspects of gene manipulation technology is that many of the procedures can be carried out with a fairly basic laboratory setup. Although applications such as large-scale DNA sequencing and production-scale biotechnology require major facilities and investment, it is still possible to do high-quality work within a 'normal' research laboratory. The requirements can be summarised under three headings:

Gene manipulation can be carried out with relatively modest laboratory facilities.

- General laboratory facilities
- Cell culture and containment
- Processing and analysis

General facilities include aspects such as laboratory layout and furnishings, and provision of essential services such as water (including distilled and/or deionised water sources), electrical power, gas, compressed air, vacuum lines, drainage, and so forth. Most of the normal services would be provided as part of any laboratory establishment and present no particular difficulty and expense beyond the norm.

Cell culture and containment facilities are essential for growing the cell lines and organisms required for the research, with the precise requirements depending on the type of work being carried out. Most labs will require facilities for growing bacterial cells, with the need for equipment such as autoclaves, incubators (static and rotary), centrifuges, and protective cabinets in which manipulation can be performed. Mammalian cell culture requires slightly more

The biology of gene manipulation requires facilities for the growth, containment, and processing of different types of cells and organisms.

sophisticated facilities, and plant or algal culture usually requires integration of lighting into the culture cabinets. In many cases some form of **physical containment** is required to prevent the escape of organisms during manipulation. The overall level of containment required depends on the type of host and vector being used, with the combination providing (usually) a level of **biological containment** in that the host is usually disabled and does not survive beyond the laboratory. The overall containment requirements will usually be specified by national bodies that regulate gene manipulation, and these may apply to bacterial and mammalian cell culture facilities. Thus, a simple cloning experiment with *E. coli* may require only normal microbiological procedures, whereas an experiment to clone human genes using viral vectors in mammalian cell lines may require the use of more stringent safety systems.

For processing and analysis of cells and cell components such as DNA, there is a bewildering choice of different types of equipment. At the most basic level, the type of automatic pipette and microcentrifuge tube can be important. A researcher struggling with pipettes that don't work properly, or with tubes that have caps that are very hard to open, will soon get frustrated! At the other end of the scale, equipment such as ultracentrifuges and automated DNA sequencers may represent a major investment for the lab and need to be chosen carefully. Much of the other equipment is (relatively) small and low-cost, with researchers perhaps having a particular brand preference.

Progress in science is dependent on the creativity of good minds, advances in technology, and the state of current knowledge; it is often constrained by availability of funds.

In addition to what might be termed infrastructure and equipment, the running costs of a laboratory need to be taken into account as this is likely to be a major part of the expenditure in any given year after start-up. Funding for research is a major issue for anyone embarking on a career in science, and the ability to attract significant research funds is a major part of the whole process of science. A mid-sized research team in a university (say three academic staff, five postdoctoral research assistants, six PhD/MSc students, and four technical staff) might easily cost in the region of a million pounds a year to run when salary, overhead, equipment, and consumables are considered. Thus, a significant part of a senior research scientist's time may be taken up with securing grants for various projects, often with no guarantee of available long-term funding.

3.2 | Isolation of DNA and RNA

Every gene manipulation experiment requires a source of nucleic acid, in the form of either DNA or RNA. It is therefore important that reliable methods are available for isolating these components from cells. There are three basic requirements: (1) opening the cells in the sample to expose the nucleic acids for further processing, (2) separation of the nucleic acids from other cell components, and (3) recovery of the nucleic acid in purified form. A variety of techniques may be used, ranging from simple procedures with few steps up to more complex purifications involving several different stages. These days most molecular biology supply companies sell kits that enable purification of nucleic acids from a range of sources.

The first step in any isolation protocol is disruption of the starting material, which may be viral, bacterial, plant, or animal. The method used to open cells should be as gentle as possible, preferably utilising enzymatic degradation of cell wall material (if present) and detergent lysis of cell membranes. If more vigorous methods of cell disruption are required (as is the case with some types of plant cell material), there is the danger of shearing large DNA molecules, and this can hamper the production of representative recombinant molecules during subsequent processing.

Following cell disruption, most methods involve a **deproteinisation** stage. This can be achieved by one or more extractions using phenol or phenol/chloroform mixtures. On the formation of an emulsion and subsequent centrifugation to separate the phases, protein molecules partition into the phenol phase and accumulate at the interface. The nucleic acids remain mostly in the upper aqueous phase and may be precipitated from solution using isopropanol or ethanol (see Section 3.3). Some techniques do not require the use of phenolic mixtures and are safer and more pleasant to use than phenol-based extraction media.

If a DNA preparation is required, the enzyme **ribonuclease** (RNase) can be used to digest the RNA in the preparation. If mRNA is needed for cDNA synthesis, a further purification can be performed by **affinity chromatography** using oligo(dT)-cellulose to bind the poly(A) tails of eukaryotic mRNAs (Fig. 3.1). This gives substantial enrichment for mRNA and enables most of the contaminating DNA, rRNA, and tRNA to be removed.

The technique of **gradient centrifugation** is often used to prepare DNA, particularly plasmid DNA (pDNA). In this technique a caesium chloride (CsCl) solution containing the DNA preparation is spun at high speed in an ultracentrifuge. Over a long period (up to 48 h in some cases) a density gradient is formed and the pDNA forms a band at one position in the centrifuge tube. The band may be taken off and the CsCl removed by dialysis to give a pure preparation of pDNA. As an alternative to gradient centrifugation, size exclusion chromatography (gel filtration) or similar techniques may be used.

> Cells have to be opened to enable nucleic acids to be isolated; opening cells should be done as gently as possible to avoid shearing large DNA molecules.

> Once broken open, cell preparations can be deproteinised and the nucleic acids purified by a range of techniques. Some applications require highly purified nucleic acid preparations; some may be able to use partially purified DNA or RNA.

3.3 | Handling and quantification of nucleic acids

It is often necessary to use very small amounts of nucleic acid (typically **micro-**, **nano-**, or **picograms**) during a cloning experiment. It is obviously impossible to handle these amounts directly, so most of the measurements that are done involve the use of aqueous solutions of DNA and RNA. The concentration of a solution of nucleic acid can be determined by measuring the absorbance at 260 nm, using a spectrophotometer. An A_{260} of 1.0 is equivalent to a concentration of

> Solutions of nucleic acids are used to enable very small amounts to be handled easily, measured, and dispensed.

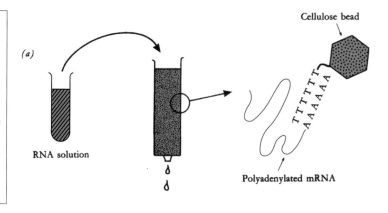

Fig. 3.1 Preparation of mRNA by affinity chromatography using oligo(dT)-cellulose. (*a*) Total RNA in solution is passed through the column in a high-salt buffer, and the oligo(dT) tracts bind the poly(A) tails of the mRNA. (*b*) Residual RNA is washed away with a high-salt buffer, and (*c*) the mRNA is eluted by washing with a low salt buffer. (*d*) The mRNA is then precipitated under ethanol and collected by centrifugation.

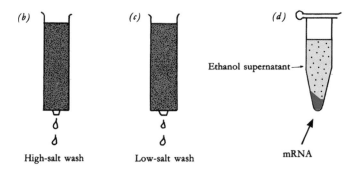

$50\ \mu g\ ml^{-1}$ for double-stranded DNA, or $40\ \mu g\ ml^{-1}$ for single-stranded DNA or RNA. If the A_{280} is also determined, the A_{260}/A_{280} ratio indicates if there are contaminants present, such as residual phenol or protein. The A_{260}/A_{280} ratio should be around 1.8 for pure DNA and 2.0 for pure RNA preparations.

In addition to spectrophotometric methods, the concentration of DNA may be estimated by monitoring the fluorescence of bound **ethidium bromide**. This dye binds between the DNA bases (intercalates) and fluoresces orange when illuminated with ultraviolet (uv) light. By comparing the fluorescence of the sample with that of a series of standards, an estimate of the concentration may be obtained. This method can detect as little as 1–5 ng of DNA and may be used when uv-absorbing contaminants make spectrophotometric measurements impossible. Having determined the concentration of a solution of nucleic acid, any amount (in theory) may be dispensed by taking the appropriate volume of solution. In this way nanogram or picogram amounts may be dispensed with reasonable accuracy.

Precipitation of nucleic acids is an essential technique that is used in a variety of applications. The two most commonly used precipitants are **isopropanol** and **ethanol**, ethanol being the preferred choice for most applications. When added to a DNA solution in a ratio, by volume, of 2:1 in the presence of 0.2 M salt, ethanol causes the nucleic acids to come out of solution. Although it used to be thought that low temperatures (-20°C or -70°C) were necessary, this is not an

Nucleic acids can be concentrated by using alcohol to precipitate the DNA or RNA from solution; the precipitate is recovered by centrifugation and can then be processed as required.

absolute requirement, and 0°C appears to be adequate. After precipitation the nucleic acid can be recovered by centrifugation, which causes a pellet of nucleic acid material to form at the bottom of the tube. The pellet can be dried and the nucleic acid resuspended in the buffer appropriate to the next stage of the experiment.

3.4 | Labelling of nucleic acids

A major problem encountered in many cloning procedures is that of keeping track of the small amounts of nucleic acid involved. This problem is magnified at each stage of the process, because losses mean that the amount of material usually diminishes after each step. One way of tracing the material is to label the nucleic acid with a marker of some sort, so that the material can be identified at each stage of the procedure. So what can be used as the label?

3.4.1 Types of label – radioactive or not?

Radioactive tracers have been used extensively in biochemistry and molecular biology for a long time, and procedures are now well established. The most common isotopes used are tritium (^{3}H), carbon-14 (^{14}C), sulphur-35 (^{35}S), and phosphorus-32 (^{32}P). Tritium and ^{14}C are low-energy emitters, with ^{35}S being a 'medium'-energy emitter and ^{32}P being a high-energy emitter. Thus, ^{32}P is more hazardous than the other isotopes and requires particular care in use. There are also strict statutory requirements for the storage and disposal of radioactive waste materials. Partly because of the inherent dangers of working with high-energy isotopes, the use of alternative technologies such as fluorescent dyes, or enzyme-linked labels, has become popular in recent years. Although these methods do offer advantages for particular applications (such as DNA sequencing), for routine tracing experiments a radioactive label is still often the preferred choice. In this case the term **radiolabelling** is often used to describe the technique.

> Radioactive isotopes are often used to label nucleic acids, although they are more hazardous than non-radioactive labelling methods.

One way of tracing DNA and RNA samples is to label the nucleic acid with a radioactive molecule (usually a deoxynucleoside triphosphate (dNTP), labelled with ^{3}H or ^{32}P), so that portions of each reaction may be counted in a **scintillation counter** to determine the amount of nucleic acid present. This is usually done by calculation, taking into account the amount of radioactivity present in the sample.

A second application of radiolabelling is in the production of highly radioactive nucleic acid molecules for use in hybridisation experiments. Such molecules are known as **radioactive probes** and have a variety of uses (see Sections 3.5 and 8.2). The difference between labelling for tracing purposes and labelling for probes is largely one of **specific activity**, that is, the measure of how radioactive the molecule is. For tracing purposes, a low specific activity will suffice, but for probes a high specific activity is necessary. In probe preparation the radioactive label is usually the high-energy ß-emitter ^{32}P. Some common methods of labelling nucleic acid molecules are described next.

> Radioactive probes are very useful for identifying specific DNA or RNA sequences.

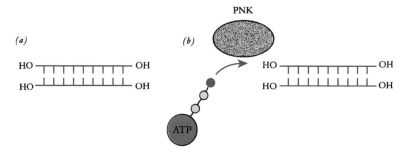

Fig. 3.2 End labelling DNA using polynucleotide kinase (PNK). (*a*) DNA is dephosphorylated using phosphatase to generate 5'-OH groups. (*b*) The terminal phosphate of [γ-^{32}P]ATP (solid circle) is then transferred to the 5' terminus by PNK. The reaction can also occur as an exchange reaction with 5'-phosphate termini.

3.4.2 End labelling

In the **end labelling** technique, the enzyme **polynucleotide kinase** is used to transfer the terminal phosphate group of ATP onto 5'-hydroxyl termini of nucleic acid molecules. If the ATP donor is radioactively labelled, this produces a labelled nucleic acid of relatively low specific activity, as only the termini of each molecule become radioactive (Fig. 3.2).

3.4.3 Nick translation

Nick translation relies on the ability of the enzyme **DNA polymerase I** (see Section 4.2.2) to translate (move along the DNA) a nick created in the phosphodiester backbone of the DNA double helix. Nicks may occur naturally or may be caused by a low concentration of the nuclease **DNase I** in the reaction mixture. DNA polymerase I catalyses a strand-replacement reaction that incorporates new dNTPs into the DNA chain. If one of the dNTPs supplied is radioactive, the result is a highly labelled DNA molecule (Fig. 3.3).

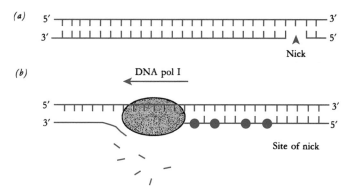

Fig. 3.3 Labelling DNA by nick translation. (*a*) A single-strand nick is introduced into the phosphodiester backbone of a DNA fragment using DNase I. (*b*) DNA polymerase I then synthesises a copy of the template strand, degrading the non-template strand with its 5'→3' exonuclease activity. If [α-^{32}P]dNTP is supplied this will be incorporated into the newly synthesised strand (solid circles).

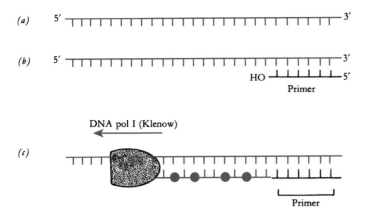

Fig. 3.4 Labelling DNA by primer extension (oligolabelling). (*a*) DNA is denatured to give single-stranded molecules. (*b*) An oligonucleotide primer is then added to give a short double-stranded region with a free 3′-OH group. (*c*) The Klenow fragment of DNA polymerase I can then synthesise a copy of the template strand from the primer, incorporating [α-^{32}P]dNTP (solid circles) to produce a labelled molecule with a very high specific activity.

3.4.4 Labelling by primer extension

Labelling by **primer extension** refers to a technique that uses random oligonucleotides (usually hexadeoxyribonucleotide molecules – sequences of six deoxynucleotides) to prime synthesis of a DNA strand by DNA polymerase. The DNA to be labelled is denatured by heating, and the **oligonucleotide primers** annealed to the single-stranded DNAs. The **Klenow fragment** of DNA polymerase (see Section 4.2.2) can then synthesise a copy of the template, primed from the 3′-hydroxyl group of the oligonucleotide. If a labelled dNTP is incorporated, DNA of very high specific activity is produced (Fig. 3.4).

In most labelling reactions not all the radioactive dNTP is incorporated into the target sequence, and non-incorporated isotope is usually removed before using the probe.

In a radiolabelling reaction it is often desirable to separate the labelled DNA from the unincorporated nucleotides present in the reaction mixture. A simple way of doing this is to carry out a small-scale gel filtration step using a suitable medium. The whole process can be carried out in a Pasteur pipette, with the labelled DNA coming through the column first, followed by the free nucleotides. Fractions can be collected and monitored for radioactivity, and the data used to calculate total activity of the DNA, specific activity, and percentage incorporation of the isotope.

3.5 | Nucleic acid hybridisation

In addition to providing information about sequence complexity (as discussed in Section 2.6.1), **nucleic acid hybridisation** can be used as an extremely sensitive detection method, capable of picking out specific DNA sequences from complex mixtures. Usually a single pure sequence is labelled with ^{32}P and used as a probe. The probe is denatured before use so that the strands are free to base-pair with their

The simple base-pairing relationship between complementary sequences has very far-reaching consequences – both for the cell and its functioning and for the scientist who wishes to exploit this feature.

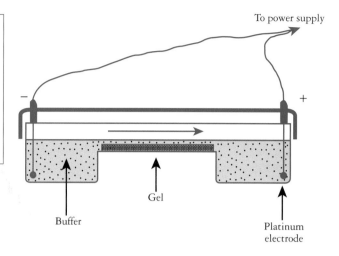

Fig. 3.5 A typical system used for agarose gel electrophoresis. The gel is just covered with buffer; therefore, the technique is sometimes called *submerged agarose gel electrophoresis* (SAGE). Nucleic acid samples placed in the gel will migrate towards the positive electrode as indicated by the horizontal arrow.

complements. The DNA to be probed is also denatured and is usually fixed to a supporting membrane made from nitrocellulose or nylon. Hybridisation is carried out in a sealed plastic bag or tube at 65–68°C for several hours to allow the duplexes to form. The excess probe is then washed off and the degree of hybridisation can be monitored by counting the sample in a scintillation spectrometer or by preparing an **autoradiograph**, where the sample is exposed to X-ray film. Some of the applications of nucleic acid hybridisation as a method for identifying cloned DNA fragments will be discussed in Chapter 8.

3.6 | Gel electrophoresis

The technique of **gel electrophoresis** is vital to the genetic engineer, as it represents the main way by which nucleic acid fragments may be visualised directly. The method relies on the fact that nucleic acids are **polyanionic** at neutral pH; that is, they carry multiple negative charges because of the phosphate groups on the phosphodiester backbone of the nucleic acid strands. This means that the molecules will migrate towards the positive electrode when placed in an electric field. As the negative charges are distributed evenly along the DNA molecule, the charge/mass ratio is constant; thus, mobility depends on fragment length. The technique is carried out using a gel matrix, which separates the nucleic acid molecules according to size. A typical nucleic acid electrophoresis setup is shown in Fig. 3.5.

The type of matrix used for electrophoresis has important consequences for the degree of separation achieved, which is dependent on the porosity of the matrix. Two gel types are commonly used: **agarose** and **polyacrylamide**. Agarose is extracted from seaweed and can be purchased as a dry powder that is melted in buffer at an appropriate concentration, normally in the range 0.3–2.0% (w/v). On cooling, the agarose sets to form the gel. Agarose gels are usually run

Separation of biomolecules by gel electrophoresis is one of the most powerful techniques in molecular biology.

Table 3.1.	Separation characteristics for agarose and polyacrylamide gels	
Gel type	Separation range (base pairs)
0.3% agarose	50 000 to 1 000
0.7% agarose	20 000 to 300
1.4% agarose	6 000 to 300
4% polyacrylamide	1 000 to 100
10% polyacrylamide	500 to 25
20% polyacrylamide	50 to 1

Source: From Schleif and Wensimk (1981), *Practical Methods in Molecular Biology*, Springer-Verlag, New York. Reproduced with permission.

in the apparatus shown in Fig. 3.5, using the submerged agarose gel electrophoresis (**SAGE**) technique. Polyacrylamide-based gel electrophoresis (**PAGE**) is sometimes used to separate small nucleic acid molecules; in applications such as **DNA sequencing** (see Section 3.7), as the pore size is smaller than that achieved with agarose. The useful separation ranges of agarose and polyacrylamide gels are shown in Table 3.1.

Electrophoresis is carried out by placing the nucleic acid samples in the gel and applying a potential difference across it. This potential difference is maintained until a marker dye (usually bromophenol blue, added to the sample prior to loading) reaches the end of the gel. The nucleic acids in the gel are usually visualised by staining with the intercalating dye ethidium bromide and examining under uv light. Nucleic acids show up as orange bands, which can be photographed to provide a record (Fig. 3.6). The data can be used to estimate the sizes of unknown fragments by construction of a calibration curve using standards of known size, as migration is inversely proportional to the $\log_{10}$ of the number of base pairs. This is particularly useful in the technique of **restriction mapping** (see Section 4.1.3).

In addition to its use in the analysis of nucleic acids, PAGE is used extensively for the analysis of proteins. The methodology is different from that used for nucleic acids, but the basic principles are similar. One common technique is SDS-PAGE, in which the detergent SDS (sodium dodecyl sulphate) is used to denature multisubunit proteins and cover the protein molecules with negative charges. In this way the inherent charge of the protein is masked, and the charge/mass ratio becomes constant. Thus, proteins can be separated according to their size in a similar way to DNA molecules.

3.7 | DNA sequencing

The ability to determine the sequence of bases in DNA is a central part of modern molecular biology and provides what might be considered

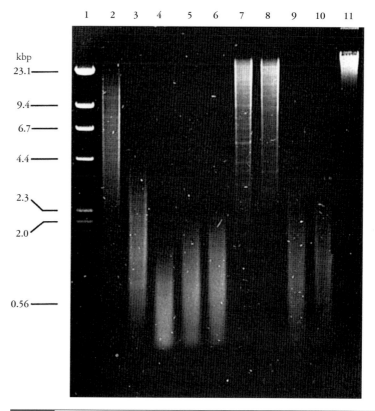

Fig. 3.6 Black-and-white photograph of an agarose gel, stained with ethidium bromide, under uv irradiation. The DNA samples show up as orange smears or as orange bands on a purple background. Individual bands (lane 1) indicate discrete fragments of DNA – in this case, the fragments are of phage λ DNA cut with the restriction enzyme *Hind*III. The sizes of the fragments (in kbp) are indicated. The remaining lanes contain samples of DNA from an alga, cut with various restriction enzymes. Because of the heterogeneous nature of these samples, the fragments merge into one another and show up as a smear on the gel. Samples that have migrated farthest (lanes 3, 4, 5, 9, and 10) are made up of smaller fragments than those that have remained near the top of the gel (lanes 2, 7, 8, and 11). Photograph courtesy of Dr N. Urwin.

the ultimate structural information. Rapid methods for sequence analysis were developed in the late 1970s, and the technique is now used in laboratories worldwide. In recent years the basic techniques have been revolutionised by automation, which has improved the efficiency of sequencing to the point where genome sequencing is possible.

3.7.1 Principles of DNA sequencing

By definition, the determination of a DNA sequence requires that the bases are identified in a sequential technique that enables the processive identification of each base in turn. There are three main requirements for this to be achieved:

- DNA fragments need to be prepared in a form suitable for sequencing.

- The technique used must achieve the aim of presenting each base in turn in a form suitable for identification.
- The detection method must permit rapid and accurate identification of the bases.

Generation and preparation of DNA fragments is fairly simple on a purely technical level. The fragments are often cloned sequences that are presented for sequencing in a suitable vector; with careful attention to detail, this can be achieved quite easily. Much more difficult is the 'informatic problem' of knowing where the fragment is within the genome; there are basically two approaches to solving this problem, which will be described in Section 3.7.2.

The sequencing protocol is essentially a technical procedure rather than an experimental one. There are several variants of the basic procedure, but the most widely used techniques are based on the enzymatic method (described in more detail in Section 3.7.4). Whatever the method, the desired result is to generate a set of overlapping fragments that terminate at different bases and differ in length by one nucleotide. This is known as a set of **nested fragments**.

Assuming that the technique has generated a set of nested fragments, the detection step is the final stage of the sequencing procedure. This usually involves separation of the fragments on a polyacrylamide gel. Slab gels, in which fragments are radioactively labelled, generate an autoradiograph. Automated sequencing procedures tend to use fluorescent labels and a continuous electrophoresis to separate the fragments, which are identified as they pass a detector.

There are two main methods for sequencing DNA. In one method, developed by Allan Maxam and Walter Gilbert, chemicals are used to cleave the DNA at certain positions, generating a set of fragments that differ by one nucleotide. The same result is achieved in a different way in the second method, developed by Fred Sanger and Alan Coulson, which involves enzymatic synthesis of DNA strands that terminate in a modified nucleotide. Analysis of fragments is similar for both methods and involves gel electrophoresis and autoradiography (assuming that a radioactive label has been used). The enzymatic method (and variants of the basic technique) has now almost completely replaced the chemical method as the technique of choice, although there are some situations where chemical sequencing can provide useful data to confirm information generated by the enzymatic method.

As already mentioned, fluorographic detection methods can be used in place of radioactive isotopes. This is particularly important in DNA sequencing, as it speeds up the process and enables the technique to be automated. We will look at this in more detail in Chapter 10 when we consider genome sequencing.

3.7.2 Preparation of DNA fragments

The difficulty of preparing a fragment for sequencing is largely dependent on the scale of the sequencing project. If the aim is to sequence part of a gene that has already been isolated and identified, the process is relatively straightforward and usually requires that the

> The principles on which DNA sequencing is based are fairly simple; the procedures required to achieve the desired result are rather more complex.

> Two rapid sequencing techniques were developed in the 1970s: the chemical method and the enzymatic method. Modern DNA-sequencing technologies are based on the enzymatic method.

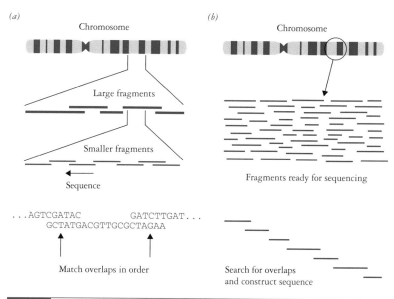

Fig. 3.7 Two strategies for sequencing large stretches of DNA. (*a*) An ordered approach is shown, where the relative positions of the cloned fragments is known. This approach uses data from physical maps to enable clones to be assigned to particular parts of the chromosome. By extending this approach as cloned fragments become smaller (and ultimately reach sequencing size), the sequence data can be assembled as a set of ordered fragment sequences. This approach is sometimes called the clone contig method. (*b*) The shotgun approach is shown. Here there is no attempt to determine the relative positional order of the sequenced fragments. Instead, the fragments are generated and sequenced randomly, with the resulting sequence being assembled by matching up regions of sequence overlap. The two approaches have their own merits; the ordered approach is more difficult at the stage of cloning and sequencing, but assembling the sequence is a little simpler than in the shotgun method.

The strategy employed to tackle large-scale sequencing projects depends on a number of factors, but essentially comes down to a choice between the 'ordered' and 'shotgun' approaches.

fragment is of a suitable length and in a suitable form for the sequencing procedure in use. If, however, the aim is to sequence a much larger piece of DNA (such as an entire chromosome), the problem is much greater. In this case the sequencing strategy is important, and there are two approaches to this. The first is an **ordered sequencing strategy** (often called the 'clone-by-clone' or 'clone contig' method; contig is an abbreviation of *contiguous*). The fragments are tracked as part of the strategy, and their relative order is noted as the project progresses. The sequence is put together by reference to the order of the fragments. The second approach is the so-called **shotgun sequencing** method, where the fragments are generated and processed at random. Assembly of the sequence is then carried out by searching for sequence overlaps using a computer. The two strategies are shown in Fig. 3.7.

3.7.3   Maxam–Gilbert (chemical) sequencing

A defined fragment of DNA is required as the starting material. This need not be cloned in a plasmid vector, so the technique is applicable to any DNA fragment. The DNA is radiolabelled with ^{32}P at the 5′ ends

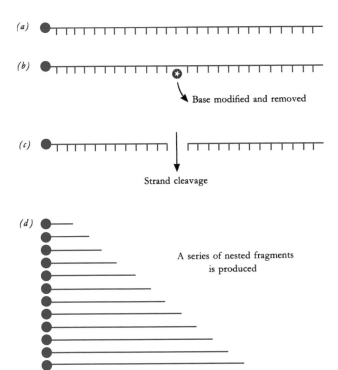

(a)

(b)

Base modified and removed

(c)

Strand cleavage

(d)

A series of nested fragments
is produced

Fig. 3.8 DNA sequencing using the chemical (Maxam–Gilbert) method.
(*a*) Radiolabelled single-stranded DNA is produced. (*b*) The bases in the DNA are chemically modified and removed, with, on average, one base being affected per molecule. (*c*) The phosphodiester backbone is then cleaved using piperidine. (*d*) The process produces a set of fragments differing in length by one nucleotide, labelled at their 5′ termini.

of each strand, and the strands are denatured, separated, and purified to give a population of labelled strands for the sequencing reactions (Fig. 3.8). The next step is a chemical modification of the bases in the DNA strand. This is done in a series of four or five reactions with different specificities, and the reaction conditions are chosen so that, on average, only one modification will be introduced into each copy of the DNA molecule. The modified bases are then removed from their sugar groups and the strands cleaved at these positions using the chemical piperidine. The theory is that, given the large number of molecules and the different reactions, this process will produce a set of nested fragments.

3.7.4 Sanger–Coulson (dideoxy or enzymatic) sequencing

Although the end result is similar to that attained by the chemical method, the Sanger–Coulson procedure is totally different from that of Maxam and Gilbert. In this case a copy of the DNA to be sequenced is made by the Klenow fragment of DNA polymerase (see Section 4.2.4). The template for this reaction is single-stranded DNA, and a primer must be used to provide the 3′ terminus for DNA polymerase to begin synthesising the copy (Fig. 3.9). The production of nested fragments is achieved by the incorporation of a modified dNTP in each reaction. These dNTPs lack a hydroxyl group at the 3′ position of deoxyribose, which is necessary for chain elongation to proceed. Such modified dNTPs are known as **dideoxynucleoside triphosphates** (ddNTPs). The

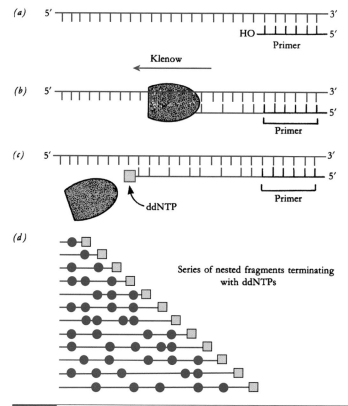

Fig. 3.9 DNA sequencing using the dideoxy chain termination (Sanger–Coulson) method. (*a*) A primer is annealed to a single-stranded template and (*b*) the Klenow fragment of DNA polymerase I is used to synthesise a copy of the DNA. A radiolabelled dNTP (often [α-^{35}S]dNTP, solid circles) is incorporated into the DNA. (*c*) Chain termination occurs when a dideoxynucleoside triphosphate (ddNTP) is incorporated. (*d*) A series of four reactions, each containing one ddNTP in addition to the four dNTPs required for chain elongation, generates a set of radiolabelled nested fragments.

four ddNTPs (A, G, T, and C forms) are included in a series of four reactions, each of which contains the four normal dNTPs. The concentration of the dideoxy form is such that it will be incorporated into the growing DNA chain infrequently. Each reaction, therefore, produces a series of fragments terminating at a specific nucleotide, and the four reactions together provide a set of nested fragments. The DNA chain is labelled by including a radioactive dNTP in the reaction mixture. This is usually [α-^{35}S]dATP, which enables more sequence to be read from a single gel than the ^{32}P-labelled dNTPs that were used previously.

The generation of fragments for dideoxy sequencing is more complicated than for chemical sequencing and usually involves subcloning into different vectors. Many plasmid vectors are now available (see Section 5.2), and some types can be used directly for DNA-sequencing experiments. Another method is to clone the DNA into a

(a)

(b)

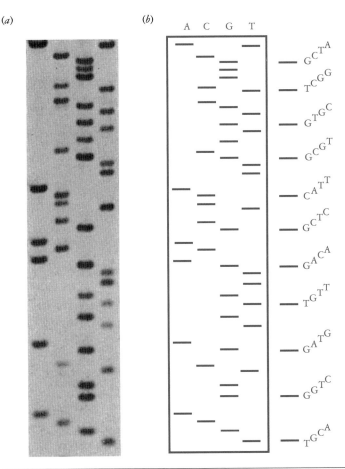

Fig. 3.10 Reading a DNA sequence. (a) An autoradiograph of part of a sequencing gel, and (b) a tracing of the autoradiograph. Each lane corresponds to a reaction containing one of the four ddNTPs used in the chain-termination technique for DNA sequencing. The sequence is read from the bottom of the gel, each successive fragment being one nucleotide longer than the preceding one. Several hundred bases can be read from a good autoradiograph, although there may be regions that are not clear and thus need to be reprocessed. Usually any given piece of DNA is sequenced several times to ensure that the best data possible are obtained. Photograph courtesy of Dr N. Urwin.

vector such as the bacteriophage M13 (see Section 5.3.3), which produces single-stranded DNA during infection. This provides a suitable substrate for the sequencing reactions.

3.7.5 Electrophoresis and reading of sequences

Separation of the DNA fragments created in sequencing reactions is achieved by PAGE. For the standard lab procedure (small-scale non-automated), a single gel system is used. The gels usually contain 6–20% polyacrylamide and 7 M urea, which acts as a denaturant to reduce the effects of DNA secondary structure. This is important because fragments that differ in length by only one base are being separated. The gels are very thin (0.5 mm or less) and are run at high-power settings, which causes them to heat up to 60–70°C. This also helps to maintain denaturing conditions. Sometimes two lots of samples are loaded onto the same gel at different times to maximise the amount of sequence information obtained.

After the gel has been run, it is removed from the apparatus and may be dried onto a paper sheet to facilitate handling. It is then exposed to X-ray film. The emissions from the radioactive label sensitise the silver grains, which turn black when the film is developed and fixed. The result is known as an autoradiograph (Fig. 3.10(*a*)). Reading the autoradiograph is straightforward – the sequence is read from the smallest fragment upwards, as shown in Fig. 3.10(*b*). Using this method, sequences of up to several hundred bases may be read from single gels. The sequence data are then compiled and studied using a computer, which can perform analyses such as translation into amino acid sequences and identification of restriction sites, regions of sequence homology, and other structural motifs such as promoters and control regions.

3.7.6 Automation of DNA sequencing

One of the major advances in technology that enabled sequencing to move from single-gel lab-based systems up to large-scale 'production line' sequencing was the automation of many parts of the process. Whereas a good lab scientist or technician could sequence maybe a few hundred bases per day, this was not going to solve the problem of determining *genome* sequences as opposed to *gene* sequences. Improving the technology by orders of magnitude was required. This was achieved by improvements in sample preparation and handling, with robotic processing enabling high-volume throughput. In a similar way the automation of the sequencing reactions, and linear continuous capillary electrophoresis techniques, enabled scale-up of the sequence-determination stage of the process.

In addition to the challenges of sequence determination, a parallel challenge lay in the need to develop sufficient computing power to deal with the vast amounts of data generated by the newly improved technologies employed in genome sequencing. In fact, it could be argued that the most critical part of the whole process is the data analysis side of things – without the ability to interrogate sequence data, the sequence remains essentially silent. These aspects will be dealt with in more detail when we look at bioinformatics in Chapter 9 and genome sequencing in Chapter 10.

> Sequencing technology had to be improved by orders of magnitude in order to tackle genome-sequencing projects within realistic time frames.

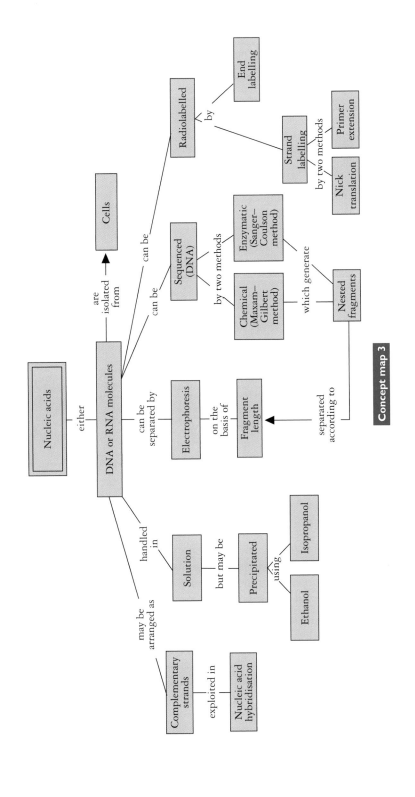

Concept map 3

Chapter 4 summary

Aims

- To outline the range of enzymes used in gene manipulation
- To describe the mode of action and uses of type II restriction enzymes
- To describe the mode of action and uses of a range of DNA modifying enzymes
- To describe the mode of action and use of DNA ligase

Chapter summary/learning outcomes

When you have completed this chapter you will have knowledge of:

- The various groups of enzymes used in gene manipulation
- Type II restriction endonucleases
- Restriction mapping
- Endonucleases, exonucleases, polymerases, and end-modifying enzymes
- Reverse transcriptase
- DNA ligase

Key words

Enzyme, restriction enzyme, restriction-modification system, methylation, recognition sequence, nuclease, endonucleases, *Escherichia coli*, *Bacillus amyloliquefaciens*, blunt ends, sticky ends, restriction mapping, DNA modifying enzymes, exonucleases, Bal 31, exonuclease III, deoxyribonuclease I, S_1-nuclease, ribonuclease, DNA polymerase I, Klenow fragment, reverse transcriptase, complementary DNA, alkaline phosphatase, polynucleotide kinase, terminal transferase, DNA ligase, polymerase chain reaction.

Chapter 4

The tools of the trade

As well as having what might be termed a good 'infrastructure' (a good laboratory setup, with access to various essential items of equipment), the genetic engineer needs to be able to cut and join DNA from different sources. This is the essence of creating recombinant DNA in the test tube. In addition, certain modifications to the DNA may have to be carried out during the various steps required to produce, clone, and identify recombinant DNA molecules. The tools that enable these manipulations to be performed are **enzymes**, which are purified from a wide range of organisms and can be bought from various suppliers. In this chapter we will have a look at some of the important classes of enzymes that make up the genetic engineer's toolkit.

4.1 | Restriction enzymes – cutting DNA

The **restriction enzymes**, which cut DNA at defined sites, represent one of the most important groups of enzymes for the manipulation of DNA. These enzymes are found in bacterial cells, where they function as part of a protective mechanism called the **restriction-modification system**. In this system the restriction enzyme hydrolyses any exogenous DNA that appears in the cell. To prevent the enzyme acting on the host cell DNA, the modification enzyme of the system (a methylase) modifies the host DNA by **methylation** of particular bases in the restriction enzyme's **recognition sequence**, which prevents the enzyme from cutting the DNA.

> Restriction enzymes act as a 'protection' system for bacteria in that they hydrolyse exogenous DNA that is not methylated by the host modification enzyme.

Restriction enzymes are of three types (I, II, or III). Most of the enzymes commonly used today are type II enzymes, which have the simplest mode of action. These enzymes are **nucleases** (see Section 4.2.1), and as they cut at an internal position in a DNA strand (as opposed to beginning degradation at one end) they are known as **endonucleases**. Thus, the correct designation of such enzymes is that they are type II restriction endonucleases, although they are often simply called restriction enzymes. In essence they may be thought of as molecular scissors.

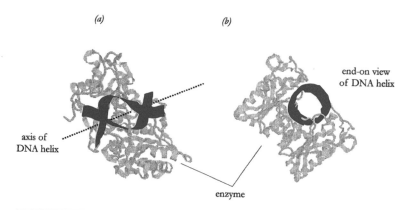

Fig. 4.1 Binding of the restriction enzyme *Bam*HI to the DNA helix. This shows how the enzyme wraps around the helix to facilitate hydrolysis of the phosphodiester linkages. (*a, b*) Different views with respect to the axis of the helix. This illustrates the very close relationship between structure and function in biology. Generated using *RasMol* molecular modelling software (Roger Sayle, Public Domain). From Nicholl (2000), *Cell & Molecular Biology*, Advanced Higher Monograph Series, Learning and Teaching Scotland. Reproduced with permission.

4.1.1   Type II restriction endonucleases

Restriction enzyme nomenclature is based on a number of conventions. The generic and specific names of the organism in which the enzyme is found are used to provide the first part of the designation, which comprises the first letter of the generic name and the first two letters of the specific name. Thus, an enzyme from a strain of *Escherichia coli* is termed *Eco*, one from *Bacillus amyloliquefaciens* is *Bam*, and so on. Further descriptors may be added, depending on the bacterial strain involved and on the presence or absence of extrachromosomal elements. Two widely used enzymes from the aforementioned bacteria are *Eco*RI and *Bam*HI. The binding of *Bam*HI to its recognition sequence is shown in Fig. 4.1.

> Restriction enzymes are named according to the bacterium from which they are purified.

The value of restriction endonucleases lies in their specificity. Each particular enzyme recognises a specific sequence of bases in the DNA, the most common recognition sequences being four, five, or six base pairs in length. Thus, given that there are four bases in the DNA, and assuming a random distribution of bases, the expected frequency of any particular sequence can be calculated as 4^n, where n is the length of the recognition sequence. This predicts that tetranucleotide sites will occur every 256 base pairs, pentanucleotide sites every 1024 base pairs, and hexanucleotide sites every 4096 base pairs. There is, as one might expect, considerable variation from these values, but generally the fragment lengths produced will lie around the calculated value. Thus, an enzyme recognising a tetranucleotide sequence (sometimes called a 'four-cutter') will produce shorter DNA fragments than a six-cutter. Some of the most commonly used restriction enzymes are listed in Table 4.1, with their recognition sequences and cutting sites.

> Different restriction enzymes generate different ranges of DNA fragment sizes; the size of fragment is linked to the frequency of occurrence of the recognition sequence.

Table 4.1.	Recognition sequences and cutting sites for some restriction endonucleases		
Enzyme	Recognition sequence	Cutting sites	Ends
*Bam*HI	5′-GGATCC-3′	G▾GATCC CCTAG▴G	5′
*Eco*RI	5′-GAATTC-3′	G▾AATTC CTTAA▴G	5′
*Hae*III	5′-GGCC-3′	GG▾CC CC▴GG	Blunt
*Hpa*I	5′-GTTAAC-3′	GTT▾AAC CAA▴TTG	Blunt
*Pst*I	5′-CTGCAG-3′	CTGCA▾G G▴ACGTC	3′
*Sau*3A	5′-GATC-3′	▾GATC CTAG▴	5′
*Sma*I	5′-CCCGGG-3′	CCC▾GGG GGG▴CCC	Blunt
*Sst*I	5′-GAGCTC-3′	GAGCT▾C C▴TCGAG	3′
*Xma*I	5′-CCCGGG-3′	C▾CCGGG GGGCC▴C	5′

Note: The recognition sequences are given in single-strand form, written 5′→3′. Cutting sites are given in double-stranded form to illustrate the type of ends produced by a particular enzyme; 5′ and 3′ refer to 5′- and 3′-protruding termini, respectively. The point at which the phosphodiester bonds are broken is shown by the arrow on each strand of the recognition sequence.

4.1.2 Use of restriction endonucleases

Restriction enzymes are very simple to use – an appropriate amount of enzyme is added to the target DNA in a buffer solution, and the reaction is incubated at the optimum temperature (usually 37°C) for a suitable length of time. Enzyme activity is expressed in units, with one unit being the amount of enzyme that will cleave one microgram of DNA in one hour at 37°C. Although most experiments require complete digestion of the target DNA, there are some cases where various combinations of enzyme concentration and incubation time may be used to achieve only partial digestion (see Section 6.3.2).

The type of DNA fragment that a particular enzyme produces depends on the recognition sequence and on the location of the cutting site within this sequence. As we have already seen, fragment length is dependent on the frequency of occurrence of the recognition sequence. The actual cutting site of the enzyme will determine the type of ends that the cut fragment has, which is important with regard to further manipulation of the DNA. Three types of fragment

One very useful feature of restriction enzymes is that they can generate cohesive or 'sticky' ends that can be used to join DNA from two different sources together to generate recombinant DNA molecules.

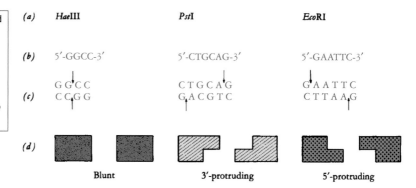

Fig. 4.2 Types of ends generated by different restriction enzymes. (*a*) The enzymes are listed, with their (*b*) recognition sequences and (*c*) cutting sites, respectively. (*d*) A schematic representation of the types of ends generated is also shown.

may be produced: (1) **blunt ends** (sometimes known as flush-ended fragments), (2) fragments with protruding 3′ ends, and (3) fragments with protruding 5′ ends. An example of each type is shown in Fig. 4.2.

Enzymes such as *Pst*I and *Eco*RI generate DNA fragments with cohesive or '**sticky' ends**, as the protruding sequences can base pair with complementary sequences generated by the same enzyme. Thus, by cutting two different DNA samples with the same enzyme and mixing the fragments together, recombinant DNA can be produced, as shown in Fig. 4.3. This is one of the most useful applications of restriction enzymes and is a vital part of many manipulations in genetic engineering.

Fig. 4.3 Generation of recombinant DNA. DNA fragments from different sources can be joined together if they have cohesive ('sticky') ends, as produced by many restriction enzymes. On annealing the complementary regions, the phosphodiester backbone is sealed using DNA ligase.

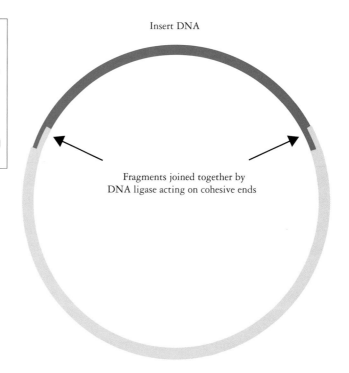

Table 4.2.	Digestion of a 15 kbp DNA fragment with three restriction enzymes

BamHI	EcoRI	PstI	BamHI + EcoRI	BamHI + PstI	EcoRI + PstI	BamHI + EcoRI + PstI
14	12	8	11	8	7	6
1	3	7	3	6	5	5
			1	1	3	3
						1

Note: Data shown are lengths (in kbp) of fragments that are produced on digestion of a 15 kbp DNA fragment with the enzymes *Bam*HI, *Eco*RI, and *Pst*I. Single, double, and triple digests were carried out as indicated. Fragments produced by each digest are listed in order of length.

4.1.3 Restriction mapping

Most pieces of DNA will have recognition sites for various restriction enzymes, and it is often beneficial to know the relative locations of some of these sites. The technique used to obtain this information is known as **restriction mapping**. This involves cutting a DNA fragment with a selection of restriction enzymes, singly and in various combinations. The fragments produced are run on an agarose gel and their sizes determined. From the data obtained, the relative locations of the cutting sites can be worked out. A fairly simple example can be used to illustrate the technique, as outlined in the following.

Let us say that we wish to map the cutting sites for the restriction enzymes *Bam*HI, *Eco*RI, and *Pst*I, and that the DNA fragment of interest is 15 kb in length. Various digestions are carried out, and the fragments arising from these are analysed and their sizes determined. The results obtained are shown in Table 4.2. As each of the single enzyme reactions produces two DNA fragments, we can conclude that the DNA has a single cutting site for each enzyme. The double digests enable a map to be drawn up, and the triple digest confirms this. Construction of the map is outlined in Fig. 4.4.

A physical map of a piece of DNA can be assembled by determining where the restriction enzyme recognition sequences are relative to each other; this is known as restriction mapping.

4.2 | DNA modifying enzymes

Restriction enzymes (described earlier) and DNA ligase (Section 4.3) provide the cutting and joining functions that are essential for the production of recombinant DNA molecules. Other enzymes used in genetic engineering may be loosely termed **DNA modifying enzymes**, with the term used here to include degradation, synthesis, and alteration of DNA. Some of the most commonly used enzymes are described next.

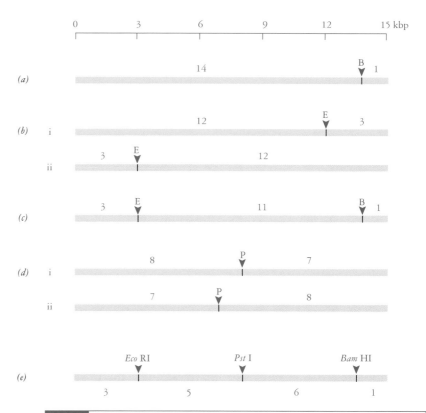

Fig. 4.4 Restriction mapping. (*a*) The 15 kb fragment yields two fragments of 14 and 1 kb when cut with *Bam*HI. (*b*) The *Eco*RI fragments of 12 and 3 kb can be orientated in two ways with respect to the *Bam*HI site, as shown in (*b*)i and (*b*)ii. The *Bam*HI/*Eco*RI double digest gives fragments of 11, 3, and 1 kb, and therefore the relative positions of the *Bam*HI and *Eco*RI sites are as shown in (*c*). Similar reasoning with the orientation of the 8 and 7 kb *Pst*I fragments (*d*) gives the final map (*e*).

4.2.1   Nucleases

In addition to restriction endonucleases, there are several other types of nuclease enzymes that are important in the manipulation of DNA.

Nuclease enzymes degrade nucleic acids by breaking the phosphodiester bond that holds the nucleotides together. Restriction enzymes are good examples of endonucleases, which cut within a DNA strand. A second group of nucleases, which degrade DNA from the termini of the molecule, are known as **exonucleases**.

Apart from restriction enzymes, there are four useful nucleases that are often used in genetic engineering. These are **Bal 31** and **exonuclease III** (exonucleases), and **deoxyribonuclease I** (DNase I) and **S$_1$-nuclease** (endonucleases). These enzymes differ in their precise mode of action and provide the genetic engineer with a variety of strategies for attacking DNA. Their features are summarised in Fig. 4.5.

In addition to DNA-specific nucleases, there are **ribonucleases** (RNases), which act on RNA. These may be required for many of the stages in the preparation and analysis of recombinants and are usually used to get rid of unwanted RNA in the preparation. However, as well as being useful, ribonucleases can pose some unwanted

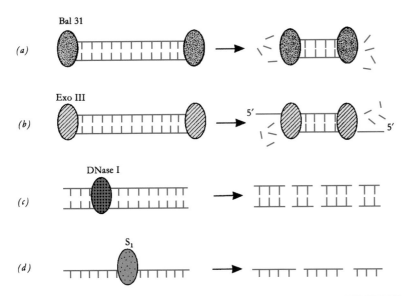

Fig. 4.5 Mode of action of various nucleases. (*a*) Nuclease Bal 31 is a complex enzyme. Its primary activity is a fast-acting 3′ exonuclease, which is coupled with a slow-acting endonuclease. When Bal 31 is present at a high concentration these activities effectively shorten DNA molecules from both termini. (*b*) Exonuclease III is a 3′ exonuclease that generates molecules with protruding 5′ termini. (*c*) DNase I cuts either single-stranded or double-stranded DNA at essentially random sites. (*d*) Nuclease S₁ is specific for single-stranded RNA or DNA. Modified from Brown (1990), *Gene Cloning*, Chapman and Hall; and Williams and Patient (1988), *Genetic Engineering*, IRL Press. Reproduced with permission.

problems. They are remarkably difficult to inactivate and can be secreted in sweat. Thus, contamination with RNases can be a problem in preparing recombinant DNA, particularly where cDNA is prepared from an mRNA template. In this case it is vital to avoid RNase contamination by wearing gloves and ensuring that all glass and plastic equipment is treated to avoid ribonuclease contamination.

Not all nucleases are helpful! Ribonucleases can be a problem when working with purified preparations of RNA, and care must be taken to remove or inactivate RNase activity.

4.2.2 Polymerases

Polymerase enzymes synthesise copies of nucleic acid molecules and are used in many genetic engineering procedures. When describing a polymerase enzyme, the terms 'DNA-dependent' or 'RNA-dependent' may be used to indicate the type of nucleic acid template that the enzyme uses. Thus, a DNA-dependent DNA polymerase copies DNA into DNA, an RNA-dependent DNA polymerase copies RNA into DNA, and a DNA-dependent RNA polymerase transcribes DNA into RNA. These enzymes synthesise nucleic acids by joining together nucleotides whose bases are complementary to the template strand bases. The synthesis proceeds in a 5′→3′ direction, as each subsequent nucleotide addition requires a free 3′-OH group for the formation of the phosphodiester bond. This requirement also means that a short double-stranded region with an exposed 3′-OH (a primer) is necessary for synthesis to begin.

Polymerases are the copying enzymes of the cell; they are also essential parts of the genetic engineer's armoury. These enzymes are template-dependent and can be used to copy long stretches of DNA or RNA.

The enzyme **DNA polymerase I** has, in addition to its polymerase function, $5'\rightarrow3'$ and $3'\rightarrow5'$ exonuclease activities. The enzyme catalyses a strand-replacement reaction, where the $5'\rightarrow3'$ exonuclease function degrades the non-template strand as the polymerase synthesises the new copy. A major use of this enzyme is in the nick translation procedure for radiolabelling DNA (outlined in Section 3.4.3).

The $5'\rightarrow3'$ exonuclease function of DNA polymerase I can be removed by cleaving the enzyme to produce what is known as the **Klenow fragment**. This retains the polymerase and $3'\rightarrow5'$ exonuclease activities. The Klenow fragment is used where a single-stranded DNA molecule needs to be copied; because the $5'\rightarrow3'$ exonuclease function is missing, the enzyme cannot degrade the non-template strand of dsDNA during synthesis of the new DNA. The $3'\rightarrow5'$ exonuclease activity is supressed under the conditions normally used for the reaction. Major uses for the Klenow fragment include radiolabelling by primed synthesis and DNA sequencing by the dideoxy method (see Sections 3.4.4 and 3.7.4) in addition to the copying of single-stranded DNAs during the production of recombinants.

> A modified form of DNA polymerase I called the Klenow fragment is a useful polymerase that is used widely in a number of applications.

Reverse transcriptase (RTase) is an RNA-dependent DNA polymerase, and therefore produces a DNA strand from an RNA template. It has no associated exonuclease activity. The enzyme is used mainly for copying mRNA molecules in the preparation of cDNA (**complementary** or **copy DNA**) for cloning (see Section 6.2.1), although it will also act on DNA templates.

> Reverse transcriptase is a key enzyme in the generation of cDNA; the enzyme is an RNA-dependent DNA polymerase, which produces a DNA copy of an mRNA molecule.

4.2.3 Enzymes that modify the ends of DNA molecules

The enzymes **alkaline phosphatase**, **polynucleotide kinase**, and **terminal transferase** act on the termini of DNA molecules and provide important functions that are used in a variety of ways. The phosphatase and kinase enzymes, as their names suggest, are involved in the removal or addition of phosphate groups. Bacterial alkaline phosphatase (there is also a similar enzyme, calf intestinal alkaline phosphatase) removes phosphate groups from the $5'$ ends of DNA, leaving a $5'$-OH group. The enzyme is used to prevent unwanted ligation of DNA molecules, which can be a problem in certain cloning procedures. It is also used prior to the addition of radioactive phosphate to the $5'$ ends of DNAs by polynucleotide kinase (see Section 3.4.2).

> In many applications it is often necessary to modify the ends of DNA molecules using enzymes such as phosphatases, kinases, and transferases.

Terminal transferase (terminal deoxynucleotidyl transferase) repeatedly adds nucleotides to any available $3'$ terminus. Although it works best on protruding $3'$ ends, conditions can be adjusted so that blunt-ended or $3'$-recessed molecules may be utilised. The enzyme is mainly used to add homopolymer tails to DNA molecules prior to the construction of recombinants (see Section 6.2.2).

4.3 | DNA ligase – joining DNA molecules

DNA ligase is an important cellular enzyme, as its function is to repair broken phosphodiester bonds that may occur at random or as a consequence of DNA replication or recombination. In genetic engineering it is used to seal discontinuities in the sugar–phosphate chains that arise when recombinant DNA is made by joining DNA molecules from different sources. It can therefore be thought of as molecular glue, which is used to stick pieces of DNA together. This function is crucial to the success of many experiments, and DNA ligase is therefore a key enzyme in genetic engineering.

DNA ligase is essentially 'molecular glue'; with restriction enzymes, it provides the tools for cutting and joining DNA molecules.

The enzyme used most often in experiments is T4 DNA ligase, which is purified from *E. coli* cells infected with bacteriophage T4. Although the enzyme is most efficient when sealing gaps in fragments that are held together by cohesive ends, it will also join blunt-ended DNA molecules together under appropriate conditions. The enzyme works best at 37°C, but is often used at much lower temperatures (4–15°C) to prevent thermal denaturation of the short base-paired regions that hold the cohesive ends of DNA molecules together.

The ability to cut, modify, and join DNA molecules gives the genetic engineer the freedom to create recombinant DNA molecules. The technology involved is a test-tube technology, with no requirement for a living system. However, once a recombinant DNA fragment has been generated *in vitro*, it usually has to be amplified so that enough material is available for subsequent manipulation and analysis. Amplification usually requires a biological system, unless the **polymerase chain reaction (PCR)** is used (see Chapter 7 for detailed discussion of PCR). We must, therefore, examine the types of living systems that can be used for the propagation of recombinant DNA molecules. These systems are described in the next chapter.

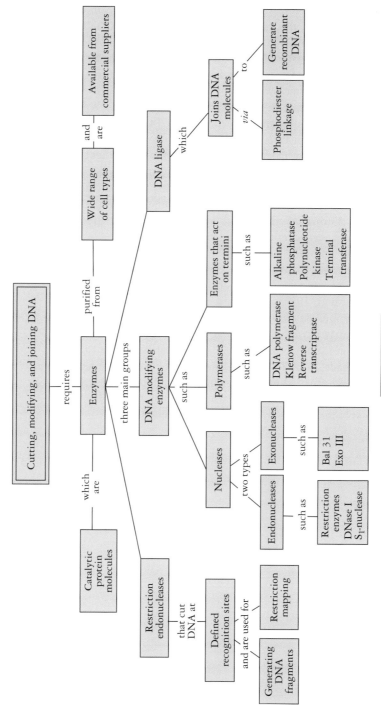

Concept map 4

Part II

The methodology of gene manipulation

Chapter 5 summary

Aims

- To outline the types of host cell used in gene manipulation
- To describe the features of plasmid and bacteriophage vectors
- To describe vectors for use in eukaryotic hosts
- To outline the range of methods available to get recombinant DNA into host cells

Chapter summary/learning outcomes

When you have completed this chapter you will have knowledge of:

- The host and vector requirements for gene cloning
- Plasmid-based vectors for use in prokaryotic hosts
- Vectors based on bacteriophages
- Vectors for use in eukaryotic host cells
- Artificial chromosomes
- Transformation and transfection
- Packaging recombinant DNA *in vitro*
- Microinjection and biolistic DNA delivery methods

Key words

Recombinant DNA, gene cloning, transgenic, vector, host, primary host, prokaryotic, nucleoid, shuttle vector, nucleus, organelle, transgenic, origin of replication, selectable marker, plasmid, conjugative, non-conjugative, copy number, stringent, relaxed, deletion derivative, ampicillin, tetracycline, insertional inactivation, polylinker, multiple cloning site, genomic library, bacteriophage, capsid, virulent, temperate, lytic, lysogenic, *cos* site, adsorption, multiplicity of infection, prophage, packaged, bacterial lawn, plaque, plaque forming unit, replicative form, insertion vector, replacement vector, substitution vector, stuffer fragment, polystuffer, packaging *in vitro*, cosmid, phagemid, yeast episomal plasmid, yeast integrative plasmid, yeast replicative plasmid, yeast centromere plasmid, artificial chromosome, YACs, BACs, transformation, transfection, growth transformation, competent, transformant, concatemer, packaging extract, protoplast, electroporation, microinjection, biolistic, microprojectile, macroprojectile.

Chapter 5

Host cells and vectors

In Chapter 4 we looked at the tools required to construct **recombinant DNA**. This process is carried out in the test tube and produces recombinant molecules that must be processed further to enable selection of the required sequence. In some experiments hundreds of thousands of different DNA fragments may be produced, and the isolation of a particular sequence would seem to be an almost impossible task. It is a bit like looking for the proverbial needle in a haystack – with the added complication that the needle is made of the same material as the haystack! Fortunately the methods available provide a relatively simple way to isolate specific gene sequences. To achieve this, we need to move away from a purely *in vitro* process and begin to use the properties and characteristics of living systems.

Three things have to be done to isolate a gene from a collection of recombinant DNA sequences:

- The individual recombinant molecules have to be physically separated from each other.
- The recombinant sequences have to be amplified to provide enough material for further analysis.
- The specific fragment of interest has to be selected by some sort of sequence-dependent method.

In this chapter we will look at the first two of these requirements, which in essence represent the systems and techniques involved in **gene cloning**. This is an essential part of most genetic manipulation programmes. Even if the desired result is a **transgenic** organism, the gene to be used must first be isolated and characterised, and therefore cloning systems are required. Some of the methods used for selecting specific sequences will be described later, in Chapter 8.

The biology of gene cloning is concerned with the selection and use of a suitable carrier molecule or **vector**, and a living system or **host** in which the vector can be propagated. In this chapter the various types of host cell will be described first, followed by vector systems and methods for getting DNA into cells.

> Gene cloning utilises the characteristics of living systems to propagate recombinant DNA molecules; in essence this can be considered as a form of molecular agriculture.

> Gene cloning is achieved by using a vector (carrier) to propagate the desired sequence in a host cell. Choosing the right vector/host combination is one of the critical stages of a cloning procedure.

Table 5.1.	Types of host used for genetic engineering		
Major group	Prokaryotic/eukaryotic	Type	Examples
Bacteria	Prokaryotic	Gram −	*Escherichia coli*
		Gram +	*Bacillus subtilis*
			Streptomyces spp.
Fungi	Eukaryotic	Microbial	*Saccharomyces cerevisiae*
		Filamentous	*Aspergillus nidulans*
Plants	Eukaryotic	Protoplasts	Various types
		Intact cells	Various types
		Whole organism	Various types
Animals	Eukaryotic	Insect cells	*Drosophila melanogaster*
		Mammalian cells	Various types
		Oocytes	Various types
		Whole organism	Various types

Note: Bacteria and fungi are generally cultured in liquid media and/or agar plates, using relatively simple growth media. Plant and animal cells may be subjected to manipulation either in tissue culture or as cells in the whole (or developing) organism. Growth requirements for these cells are often more exacting than for the microbial host cells.

5.1 | Host cell types

The type of host cell used for a particular application will depend mainly on the purpose of the cloning procedure. If the aim is to isolate a gene for structural analysis, the requirements may call for a simple system that is easy to use. If the aim is to express the genetic information in a higher eukaryote such as a plant, a more specific system will be required. These two aims are not necessarily mutually exclusive; often a simple **primary host** is used to isolate a sequence that is then introduced into a more complex system for expression. The main types of host cell are shown in Table 5.1, and are described in the following sections.

5.1.1 Prokaryotic hosts

An ideal host cell should be easy to handle and propagate, should be available as a wide variety of genetically defined strains, and should accept a range of vectors. The bacterium *Escherichia coli* fulfils these requirements and is used in many cloning protocols. *E. coli* has been studied in great detail, and many different strains were isolated by microbial geneticists as they investigated the genetic mechanisms of this **prokaryotic** organism. Such studies provided the essential background information on which genetic engineering is based.

E. coli is a gram-negative bacterium with a single chromosome packed into a compact structure known as the **nucleoid**. The genome size is some 4.6×10^6 base pairs, and the complete sequence is now known. The processes of gene expression (transcription and translation) are coupled, with the newly synthesised mRNA being immediately available for translation. There is no post-transcriptional modification of the primary transcript as is commonly found in eukaryotic cells. *E. coli* can therefore be considered as one of the simplest host cells. Much of the gene cloning that is carried out routinely in laboratories involves the use of *E. coli* hosts, with many genetically different strains available for specific applications.

In addition to *E. coli*, other bacteria may be used as hosts for gene cloning experiments, with examples including species of *Bacillus*, *Pseudomonas*, and *Streptomyces*. There are, however, certain drawbacks with most of these. Often there are fewer suitable vectors available for use in such cells than is the case for *E. coli*, and getting recombinant DNA into the cell can cause problems. This is particularly troublesome when primary cloning experiments are envisaged, such as the direct introduction of ligated recombinant DNA into the host cell. For this type of application, reliable and efficient procedures are required to maximise the yield of cloned fragments. It is, therefore, often more sensible to perform an initial cloning procedure in *E. coli*, to isolate the required sequence, and then to introduce the purified DNA into the target host. Many of the drawbacks can be overcome by using this approach, particularly when vectors that can function in the target host and in *E. coli* (**shuttle vectors**) are used. Use of bacteria other than *E. coli* will not be discussed further in this book; details may be found in some of the texts mentioned in the section Suggestions for Further Reading.

> The bacterium *Escherichia coli* is the most commonly used prokaryotic host cell, with a wide variety of different strains available for particular applications.

5.1.2 Eukaryotic hosts

One disadvantage of using an organism such as *E. coli* as a host for cloning is that it is a prokaryote, and therefore lacks the membrane-bound **nucleus** (and other **organelles**) found in eukaryotic cells. This means that certain eukaryotic genes may not function in *E. coli* as they would in their normal environment, which can hamper their isolation by selection mechanisms that depend on gene expression. Also, if the production of a eukaryotic protein is the desired outcome of a cloning experiment, it may not be easy to ensure that a prokaryotic host produces a fully functional protein.

> Prokaryotic host cells have certain limitations when the cloning and expression of genes from eukaryotes is the aim of the procedure.

Eukaryotic cells range from microbes, such as yeast and algae, to cells from complex multicellular organisms, such as ourselves. The microbial cells have many of the characteristics of bacteria with regard to ease of growth and availability of mutants. Higher eukaryotes present a different set of problems to the genetic engineer, many of which require specialised solutions. Often the aim of a gene manipulation experiment in a higher plant or animal is to alter the genetic makeup of the organism by creating a **transgenic**, rather than to isolate a gene for further analysis or to produce large amounts of a particular protein. Transgenesis is discussed further in Chapter 13.

The yeast *Saccharomyces cerevisiae* is one of the most commonly used eukaryotic microbes in genetic engineering. It has been used for centuries in the production of bread and beer and has been studied extensively. The organism is amenable to classical genetic analysis, and a range of mutant cell types is available. In terms of genome complexity, *S. cerevisiae* has about 3.5 times more DNA than *E. coli*. The complete genome sequence is now known. Other fungi that may be used for gene cloning experiments include *Aspergillus nidulans* and *Neurospora crassa*.

Plant and animal cells may also be used as hosts for gene manipulation experiments. Unicellular algae such as *Chlamydomonas reinhardtii* have all the advantages of microorganisms plus the structural and functional organisation of plant cells, and their use in genetic manipulation will increase as they become more widely studied. Other plant (and animal) cells are usually grown as cell cultures, which are much easier to manipulate than cells in a whole organism. Mammalian cell lines in particular are very important sources of cells from gene manipulation procedures. Some aspects of genetic engineering in plant and animal cells are discussed in the final section of this book.

> Microbes (such as yeast) and mammalian cell lines are two examples of eukaryotic host cells that have become widely used in gene manipulation.

5.2 | Plasmid vectors for use in *E. coli*

There are certain essential features that vectors must posess. Ideally they should be fairly small DNA molecules, to facilitate isolation and handling. There must be an **origin of replication**, so that their DNA can be copied and thus maintained in the cell population as the host organism grows and divides. It is desirable to have some sort of **selectable marker** that will enable the vector to be detected, and the vector must also have at least one unique restriction endonuclease recognition site to enable DNA to be inserted during the production of recombinants. **Plasmids** have these features and are extensively used as vectors in cloning experiments. Some features of plasmid vectors are described next.

5.2.1 What are plasmids?

Many types of plasmids are found in nature, in bacteria and some yeasts. They are circular DNA molecules, relatively small when compared to the host cell chromosome, that are maintained mostly in an extrachromosomal state. Although plasmids are generally dispensable (*i.e.* not essential for cell growth and division), they often confer traits (such as antibiotic resistance) on the host organism, which can be a selective advantage under certain conditions. The antibiotic resistance genes encoded by plasmid DNA (pDNA) are often used in the construction of vectors for genetic engineering, as they provide a convenient means of selecting cells containing the plasmid. When plated on growth medium that contains the appropriate antibiotic, only the plasmid-containing cells will survive. This is a very simple and powerful selection method.

> Plasmids are extrachromosomal genetic elements that are not essential for bacteria to survive but often confer advantageous traits (such as antibiotic resistance) on the host cell.

Table 5.2.	Properties of some naturally occurring plasmids			
Plasmid	Size (kb)	Conjugative?	Copy number	Selectable markers
ColE1	7.0	No	10–15	E1^{imm}
RSF1030	9.3	No	20–40	Ap^r
CloDF13	10.0	No	10	DF13^{imm}
pSC101	9.7	No	1–2	Tc^r
RK6	42	Yes	10–40	Ap^r, Sm^r
F	103	Yes	1–2	–
R1	108	Yes	1–2	Ap^r, Cm^r, Sm^r, Sn^r, Km^r
RK2	56.4	Yes	3–5	Ap^r, Km^r, Tc^r

Note: Antibiotic abbreviations are as follows: Ap, ampicillin; Cm, chloramphenicol; Km, kanamycin; Sm, streptomycin; Sn, sulphonamide; Tc, tetracycline. E1^{imm} and DF13^{imm} represent immunity to the homologous but not to the heterologous colicin. Thus, plasmid ColE1 is resistant to the effects of its own colicin (E1) but not to colicin DF13. Copy number is the number of plasmids per chromosome equivalent.

Source: After Winnacker (1987), *From Genes to Clones*, VCH. Data from Helsinki (1979), *Critical Reviews in Biochemistry* **7**, 83–101, copyright (1979) CRC Press Inc., Boca Raton, Florida; Kahn *et al.* (1979), *Methods in Enzymology* **68**, 268–280, copyright (1979) Academic Press; Thomas (1981), *Plasmid* **5**, 10–19, copyright (1981) Academic Press. Reproduced with permission.

Plasmids can be classified into two groups, **conjugative** and **non-conjugative** plasmids. Conjugative plasmids can mediate their own transfer between bacteria by the process of conjugation, which requires functions specified by the *tra* (transfer) and *mob* (mobilising) regions carried on the plasmid. Non-conjugative plasmids are not self-transmissible but may be mobilised by a conjugation-proficient plasmid if their *mob* region is functional. A further classification is based on the number of copies of the plasmid found in the host cell, a feature known as the **copy number**. Low-copy-number plasmids tend to exhibit **stringent** control of DNA replication, with replication of the pDNA closely tied to host cell chromosomal DNA replication. High-copy-number plasmids are termed **relaxed** plasmids, with DNA replication not dependent on host cell chromosomal DNA replication. In general terms, conjugative plasmids are large, show stringent control of DNA replication, and are present at low copy numbers, whilst non-conjugative plasmids are small, show relaxed DNA replication, and are present at high copy numbers. Some examples of plasmids are shown in Table 5.2.

5.2.2 Basic cloning plasmids

For genetic engineering, naturally occurring plasmids have been extensively modified to produce vectors that have the desired characteristics. In naming plasmids, p is used to designate plasmid, and this is usually followed by the initials of the worker(s) who isolated or constructed the plasmid. Numbers may be used to classify the particular isolate. An important plasmid in the history of gene manipulation is pBR322, which was developed by Francisco Bolivar and his

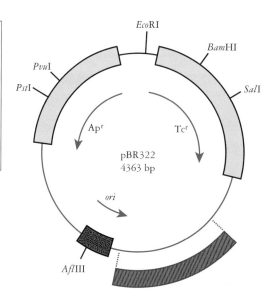

Fig. 5.1 Map of plasmid pBR322. Important regions indicated are the genes for ampicillin and tetracycline resistance (Ap^r and Tc^r) and the origin of replication (*ori*). Some unique restriction sites are given. The hatched region shows the two fragments that were removed from pBR322 to generate pAT153.

pBR322 is a very 'famous' plasmid and has all the essential requirements for a cloning vector – relatively small size, useful restriction enzyme sites, an origin of replication, and antibiotic resistance genes.

New plasmid vectors can be constructed by re-arranging various parts of the plasmid. This can involve addition or deletion of DNA to change the characteristics of the vector. In this way a wide range of vectors can be constructed with relative ease.

colleagues. Construction of pBR322 involved a series of manipulations to get the right pieces of DNA together, with the final result containing DNA from three sources. The plasmid has all the features of a good vector, such as low molecular weight, antibiotic resistance genes, an origin of replication, and several single-cut restriction endonuclease recognition sites. A map of pBR322 is shown in Fig. 5.1.

Two aspects of plasmid vector development are worthy of note at this point, as exemplified by pBR322. First, there are several plasmids in the pBR series, each with slightly different features. Second, the pBR series has been the basis for the development of many more plasmid vectors, often by subcloning parts of the vector and joining with other DNA sequences, perhaps taken from a different vector. These two aspects can be traced through other 'families' of plasmid-based vectors. In the early days of plasmid vector development, scientists were usually willing to share their vectors freely. Whilst this is still the case in many applications, there are sometimes issues involving intellectual property rights and trademarks where plasmids have been developed by companies and marketed on a commercial basis.

One variant of pBR322 is worth describing to illustrate how a relatively simple change can affect plasmid properties and perhaps improve some aspects of the original. The plasmid is pAT153, which is still widely used and has some advantages over its progenitor. The vector pAT153 is a **deletion derivative** of pBR322 (see Fig. 5.1). The plasmid was isolated by removal of two fragments of DNA from pBR322, using the restriction enzyme *Hae*II. The amount of DNA removed was small (705 base-pairs), but the effect was to increase the copy number some threefold, and to remove sequences necessary for mobilisation. Thus, pAT153 is in some respects a 'better' vector than pBR322, as it is present as more copies per cell and has a greater degree of biological containment because it is not mobilisable.

In vectors such as pBR322 and pAT153, the presence of two antibiotic resistance genes (Ap^r and Tc^r) enables selection for cells harbouring the plasmid, as such cells will be resistant to both **ampicillin** and **tetracycline**. An added advantage is that the unique restriction sites within the antibiotic resistance genes permit selection of cloned DNA by what is known as **insertional inactivation**, where the inserted DNA interrupts the coding sequence of the resistance gene and so alters the phenotype of the cell carrying the recombinant. This is discussed further in Section 8.1.2.

5.2.3 Slightly more exotic plasmid vectors

Although plasmids pBR322 and pAT153 are still often used for many applications in gene cloning, there are situations where other plasmid vectors may be more suitable. Generally these have been constructed so that they have particular characteristics not found in the simpler vectors, such as a wider range of restriction sites for cloning DNA fragments. They may contain specific promoters for the expression of inserted genes, or they may offer other advantages such as direct selection for recombinants. Despite these advantages, the well-tried vectors such as pBR322 and pAT153 are often more than sufficient if a relatively simple procedure is being used.

One series of plasmid vectors that has proved popular is the pUC family. These plasmids have a region that contains several unique restriction endonuclease sites in a short stretch of the DNA. This region is known as a **polylinker** or **multiple cloning site** and is useful because of the choice of site available for insertion of DNA fragments during recombinant production. A map of one of the pUC vectors, with the restriction sites in its polylinker region, is shown in Fig. 5.2. In addition to the multiple cloning sites in the polylinker region, the pUC plasmids have a region of the β-galactosidase gene that codes for what is known as the α-peptide. This sequence contains the polylinker region, and insertion of a DNA fragment into one of the cloning sites results in a non-functional α-peptide. This forms the basis for a powerful direct recombinant screening method using the chromogenic substrate X-gal, as outlined in Sections 8.1.1 and 8.1.2.

Over the past few years, many different types of plasmid vectors have been derived from the basic cloning plasmids. Today there are many different plasmids available for specific purposes, often from commercial sources. These vectors are sometimes provided as part of a 'cloning kit' that contains all the essential components to conduct a cloning experiment. This has made the technology much more accessible to a greater number of scientists, although it has not yet become totally foolproof! Some commercially available plasmids are listed in Table 5.3.

Although plasmid vectors have many useful properties and are essential for gene manipulation, they do have a number of disadvantages. One of the major drawbacks is the size of DNA fragment that can be inserted into plasmids; the maximum is around 5 kb of DNA for many plasmids before cloning efficiency or insert stability

Multiple cloning sites (polylinkers) increase the flexibility of vectors by providing a range of restriction sites for cloning.

Plasmid vectors have an upper size limit for efficient cloning, which can sometimes restrict their use where a large number of clones is required. In this case it makes sense to clone longer DNA fragments, and a different vector system is needed.

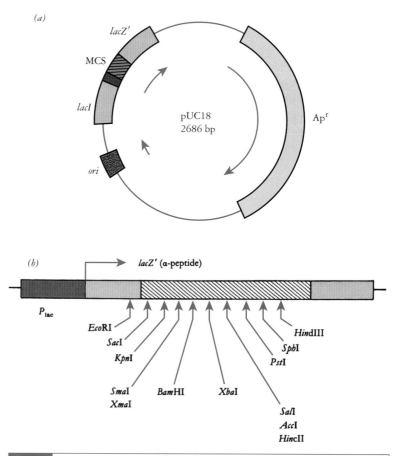

Fig. 5.2 Map of plasmid pUC18. (*a*) The physical map, with the positions of the origin of replication (*ori*) and the ampicillin resistance gene (Ap^r) indicated. The *lacI* gene (lac repressor), multiple cloning site (MCS) or polylinker, and the *lacZ'* gene (α-peptide fragment of β-galactosidase) are also shown. (*b*) The polylinker region. This has multiple restriction sites immediately downstream from the lac promoter (P_{lac}). The in-frame insert used to create the MCS is hatched. Plasmid pUC19 is identical with pUC18 apart from the orientation of the polylinker region, which is reversed.

are affected. In many cases this is not a problem, but in some applications it is important to maximise the size of fragments that may be cloned. Such a case is the generation of a **genomic library**, in which all the sequences present in the genome of an organism are represented. For this type of approach, vectors that can accept larger pieces of DNA are required. Examples of suitable vectors are those based on **bacteriophage** lambda (λ); these are considered in the next section.

5.3 | Bacteriophage vectors for use in *E. coli*

Although bacteriophage-based vectors are in many ways more specialised than plasmid vectors, they fulfil essentially the same function (*i.e.* they act as carrier molecules for fragments of DNA). Two types

Table 5.3. Some commercially available plasmid vectors

Vector	Features	Applications	Supplier
pBR322	Ap^r Tc^r Single cloning sites	General cloning and subcloning in *E. coli*	Various
pAT153	Ap^r Tc^r Single cloning sites	General cloning and subcloning in *E. coli*	Various
pGEM®-3Z	Ap^r MCS SP6/T7 promoters *lacZ* α-peptide	General cloning and *in vitro* transcription in *E. coli* and single-stranded DNA production	Promega
pCI®	Ap^r MCS T7 promoter CMV enhancer/promoter	Expression of genes in mammalian cells	Promega
pET-3	Ap^r MCS T7 promoter	Expression of genes in bacterial cells	Stratagene
pCMV-Script®	Neo^r Large MCS CMV enhancer/promoter	High level expression of genes in mammalian cells and cloning of PCR products	Stratagene

Note: There are hundreds of variants available from many different suppliers. A good source of information is the supplier's catalogue or website. Ap^r, ampicillin resistance; Tc^r, tetracycline resistance; Neo^r, neomycin resistance (selection using kanamycin in bacteria, G418 in mammalian cells); MCS, multiple cloning site; SP6/T7 are promoters for *in vitro* transcription; *lacZ*, β-galactosidase gene; CMV, human cytomegalovirus. The terms marked ® are registered trademarks of the suppliers.

of bacteriophage (λ and M13) have been extensively developed for cloning purposes; these will be described to illustrate the features of bacteriophages and the vectors derived from them.

5.3.1 What are bacteriophages?

In the 1940s Max Delbrück, and the 'Phage Group' that he brought into existence, laid the foundations of modern molecular biology by studying bacteriophages. These are literally 'eaters of bacteria' – viruses that are dependent on bacteria for their propagation. The term 'bacteriophage' is often shortened to 'phage' and can be used to describe either one or many particles of the same type. Thus, we might say that a test tube contained one λ phage or 2×10^6 λ phage particles. The plural term 'phages' is used when different types of phage are being considered; we therefore talk of T4, M13, and λ as being phages.

Structurally, phages fall into three main groups: (1) tailless, (2) head with tail, and (3) filamentous. The genetic material may be single- or double-stranded DNA or RNA, with double-stranded DNA (dsDNA) found most often. In tailless and tailed phages the genome

Bacteriophages are essentially bacterial viruses and usually consist of a DNA genome enclosed in a protein head (capsid). As with other viruses, they depend on the host cell for their propagation and do not exist as free-living organisms.

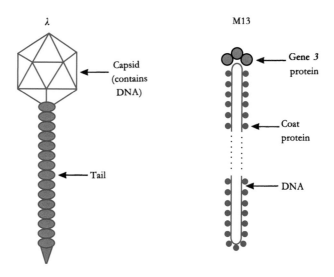

Fig. 5.3 Structure of bacteriophages λ and M13. Phage λ has a capsid or head that encloses the double-stranded DNA genome. The tail region is required for adsorption to the host cell. M13 has a simpler structure, with the single-stranded DNA genome being enclosed in a protein coat. The gene 3 product is important in both adsorption and extrusion of the phage. M13 is not drawn to scale; in reality it is a long thin structure.

is encapsulated in an icosahedral protein shell called a **capsid** (sometimes known as a phage coat or head). In typical dsDNA phages, the genome makes up about 50% of the mass of the phage particle. Thus, phages represent relatively simple systems when compared to bacteria, and for this reason they have been extensively used as models for the study of gene expression. The structure of phages λ and M13 is shown in Fig. 5.3.

Phages may be classified as either **virulent** or **temperate**, depending on their life cycles. When a phage enters a bacterial cell it can produce more phage and kill the cell (this is called the **lytic** growth cycle), or it can integrate into the chromosome and remain in a quiescent state without killing the cell (this is the **lysogenic** cycle). Virulent phages are those that exhibit a lytic life cycle only. Temperate phages exhibit lysogenic life cycles, but most can also undergo the lytic response when conditions are suitable. The best-known example of a temperate phage is λ, which has been the subject of intense research effort and is now more or less fully characterised in terms of its structure and mode of action.

Bacteriophage λ has played a major role in the development of bacterial genetics and molecular biology. In addition to fundamental aspects of gene regulation, λ has been used as the basis for a wide variety of cloning vectors.

The genome of phage λ is 48.5 kb in length, and encodes some 46 genes (Fig. 5.4). The entire genome has been sequenced (this was the first major sequencing project to be completed, and represents one of the milestones of molecular genetics), and all the regulatory sites are known. At the ends of the linear genome there are short (12 bp) single-stranded regions that are complementary. These act as cohesive or 'sticky' ends, which enable circularisation of the genome

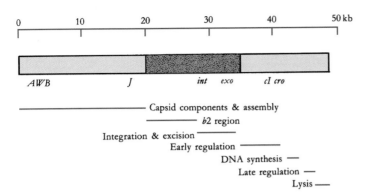

Fig. 5.4 Map of the phage λ genome. Some of the genes are indicated. Functional regions are shown by horizontal lines and annotated. The non-essential region that may be manipulated in vector construction is shaded.

following infection. The region of the genome that is generated by the association of the cohesive ends is known as the *cos* **site**.

Phage infection begins with **adsorption**, which involves the phage particle binding to receptors on the bacterial surface (Fig. 5.5). When the phage has adsorbed, the DNA is injected into the cell and the life cycle can begin. The genome circularises and the phage initiates

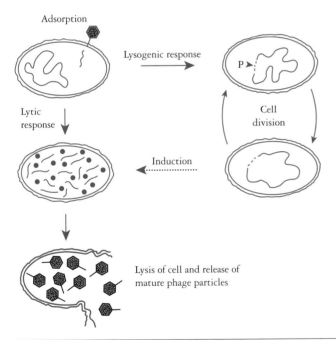

Fig. 5.5 Life cycle of bacteriophage λ. Infection occurs when a phage particle is adsorbed and the DNA injected into the host cell. In the lytic response, the phage utilises the host cell replication mechanism and produces copies of the phage genome and structural proteins. Mature phage particles are then assembled and released by lysis of the host cell. In the lysogenic response the phage DNA integrates into the host genome as a prophage (P), which can be maintained through successive cell divisions. The lytic response can be induced in a lysogenic bacterium in response to a stimulus such as ultraviolet light.

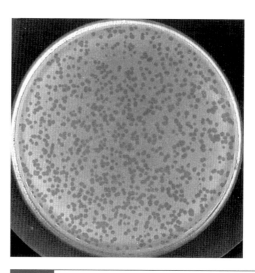

Fig. 5.6 Bacteriophage plaques. Particles of phage λ were mixed with a strain of *E. coli* and plated using a soft agar overlay. After overnight incubation the bacterial cells grow to form a lawn, in which regions of phage infection appear as cleared areas or plaques. The plaques are areas where lysis of the bacterial cells has occurred. Photograph courtesy of Dr M. Stronach.

either the lytic or lysogenic cycle, depending on a number of factors that include the nutritional and metabolic state of the host cell and the **multiplicity of infection** (MOI – the ratio of phage to bacteria during adsorption). If the lysogenic cycle is initiated, the phage genome integrates into the host chromosome and is maintained as a **prophage**. It is then replicated with the chromosomal DNA and passed on to daughter cells in a stable form. If the lytic cycle is initiated, a complex sequence of transcriptional events essentially enables the phage to take over the host cell and produce multiple copies of the genome and the structural proteins. These components are then assembled or **packaged** into mature phage, which are released following lysis of the host cell.

To determine the number of bacteriophage present in a suspension, serial dilutions of the phage stock are mixed with an excess of indicator bacteria (MOI is very low) and plated onto agar using a soft agar overlay. On incubation, the bacteria will grow to form what is termed a **bacterial lawn**. Phage that grow in this lawn will cause lysis of the cells that the phage infects, and as this growth spreads a cleared area or **plaque** will develop (Fig. 5.6). Plaques can then be counted to determine the number of **plaque forming units** in the stock suspension and may be picked from the plate for further growth and analysis. Phage may be propagated in liquid culture by infecting a growing culture of the host cell and incubating until cell lysis is complete; the yield of phage particles depends on the MOI and the stage in the bacterial growth cycle at which infection occurs.

The filamentous phage M13 differs from λ both structurally (Fig. 5.3) and in its life cycle. The M13 genome is a single-stranded circular DNA molecule 6407 bp in length. The phage will infect only *E. coli* that have F-pili (threadlike protein 'appendages' found on conjugation-proficient cells). When the DNA enters the cell, it is converted to a double-stranded molecule known as the **replicative form** (RF), which replicates until there are about 100 copies in the cell. At this point DNA replication becomes asymmetric, and single-stranded copies of the genome are produced and extruded from the cell as M13 particles. The bacterium is not lysed and remains viable during this process, although growth and division are slower than in non-infected cells.

5.3.2 Vectors based on bacteriophage λ

The utility of phage λ as a cloning vector depends on the fact that not all of the λ genome is essential for the phage to function. Thus, there is scope for the introduction of exogenous DNA, although certain requirements have had to be met during the development of cloning vectors based on phage λ. First, the arrangement of genes on the λ genome will determine which parts can be removed or replaced for the addition of exogenous DNA. It is fortunate that the central region of the λ genome (between positions 20 and 35 on the map shown in Fig. 5.4) is largely dispensable, so no complex rearrangement of the genome *in vitro* is required. The central region controls mainly the lysogenic properties of the phage, and much of this region can be deleted without impairing the functions required for the lytic infection cycle. Second, wild-type λ phage will generally have multiple recognition sites for the restriction enzymes commonly used in cloning procedures. This can be a major problem, as it limits the choice of sites for the insertion of DNA. In practice, it is relatively easy to select for phage that have reduced numbers of sites for particular restriction enzymes, and the technique of mutagenesis *in vitro* may be used to modify remaining sites that are not required. Thus, it is possible to construct phage that have the desired combination of restriction enzyme recognition sites.

The λ genome has to be modified and re-arranged to produce the right combination of features for a cloning vector.

One of the major drawbacks of λ vectors is that the capsid places a physical constraint on the amount of DNA that can be incorporated during phage assembly, which limits the size of exogenous DNA fragments that can be cloned. During packaging, viable phage particles can be produced from DNA that is between approximately 38 and 51 kb in length. Thus, a wild-type phage genome could accommodate only around 2.5 kb of cloned DNA before becoming too large for viable phage production. This limitation has been minimised by careful construction of vectors to accept pieces of DNA that are close to the theoretical maximum for the particular construct. Such vectors fall into two main classes: (1) **insertion vectors** and (2) **replacement** or **substitution vectors**. The difference between these two types of vector is outlined in Fig. 5.7.

As with plasmids, there is now a bewildering variety of λ vectors available for use in cloning experiments, each with slightly different characteristics. The choice of vector has to be made carefully,

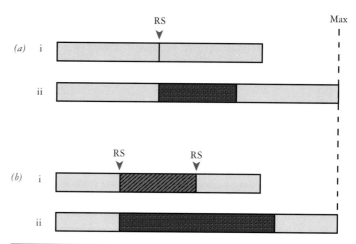

Fig. 5.7 Insertion and replacement phage vectors. (*a*) An insertion vector is shown in part i. Such vectors have a single restriction site (RS) for cloning into. To generate a recombinant, the target DNA is inserted into this site. The size of fragment that may be cloned is therefore determined by the difference between the vector size and the maximum packagable fragment size (Max). Insert DNA is shaded in part ii. (*b*) A replacement vector is shown in part i. These vectors have two restriction sites (RS), which flank a region known as the stuffer fragment (hatched). Thus, a section of the phage genome is replaced during cloning into this site, as shown in part ii. This approach enables larger fragments to be cloned than is possible with insertion vectors.

with aspects such as the size of DNA fragments to be cloned and the preferred selection/screening method being taken into account. To illustrate the structural characteristics of λ vectors, two insertion and two replacement vectors are described briefly. Although not as widely used as some other λ vectors today, these illustrate some of the important aspects of vector design. Functional aspects of some λ vectors are discussed in Chapter 8 when selection and screening methods are considered.

Insertion vectors have a single recognition site for one or more restriction enzymes, which enables DNA fragments to be inserted into the λ genome. Examples of insertion vectors include λgt10 and Charon 16A. The latter is one of a series of vectors named after the ferryman of Greek mythology, who conveyed the spirits of the dead across the river Styx – a rather apt example of what we might call 'bacteriophage culture'! These two insertion vectors are illustrated in Fig. 5.8. Each has a single *Eco*RI site into which DNA can be inserted. In λgt10 (43.3 kb) this generates left and right 'arms' of 32.7 and 10.6 kb, respectively, which can in theory accept insert DNA fragments up to approximately 7.6 kb in length. The *Eco*RI site lies within the *cI* gene (λ repressor), and this forms the basis of a selection/screening method based on plaque formation and morphology (see Section 8.1.2). In Charon 16A (41.8 kb), the arms generated by *Eco*RI digestion are 19.9 kb (left arm) and 21.9 kb (right arm), and fragments of up to approximately 9 kb may be cloned. The *Eco*RI site in Charon 16A lies within

Insertion vectors, as the name suggests, are vectors into which DNA fragments are inserted without removal of part of the vector DNA.

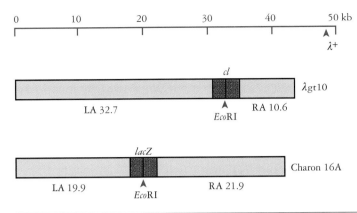

Fig. 5.8 Bacteriophage λ insertion vectors λgt10 and Charon 16A. The *cl* and *lacZ* genes in λgt10 and Charon 16A, respectively, are shaded. Within these genes there is an *Eco*RI site for cloning into. The lengths of the left and right arms (LA and RA, in kb) are given. The size of the wild-type λ genome is marked on the scale bar as λ⁺. Redrawn from Winnacker (1987), *From Genes to Clones*, VCH. Reproduced with permission.

the β-galactosidase gene (*lacZ*), which enables the detection of recombinants using X-gal (see Section 8.1.2).

Insertion vectors offer limited scope for cloning large pieces of DNA; thus, replacement vectors were developed in which a central **'stuffer' fragment** is removed and replaced with the insert DNA. Two examples of λ replacement vectors are EMBL4 and Charon 40 (Fig. 5.9). EMBL4 (41.9 kb) has a central 13.2 kb stuffer fragment flanked by inverted polylinker sequences containing sites for the restriction enzymes *Eco*RI, *Bam*HI, and *Sal*I. Two *Sal*I sites are also present in the stuffer fragment. DNA may be inserted into any of the cloning sites; the choice depends on the method of preparation of the fragments. Often a partial *Sau*3A or *Mbo*I digest is used in the preparation of a

> Replacement vectors have restriction sites flanking a region of non-essential DNA that can be removed and replaced with a DNA fragment for cloning. This increases the size of insert that can be accepted by phage-based vectors.

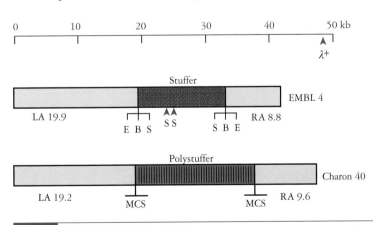

Fig. 5.9 Bacteriophage λ replacement vectors EMBL4 and Charon 40. The stuffer fragment in EMBL4 is 13.2 kb and is flanked by inverted polylinkers containing the sites for *Eco*RI (E), *Bam*HI (B), and *Sal*I (S). In Charon 40 the polystuffer is composed of short repeated regions that are cleaved by *Nae*I. The multiple cloning site (MCS) in Charon 40 carries a wider range of restriction sites than that in EMBL4.

genomic library (see Section 3.3.2), which enables insertion into the *Bam*HI site. Such inserts may be released from the recombinant by digestion with *Eco*RI. During preparation of the vector for cloning, the *Bam*HI digestion (which generates sticky ends for accepting the insert DNA) is often followed by a *Sal*I digestion. This cleaves the stuffer fragment at the two internal *Sal*I sites and also releases short *Bam*HI/*Sal*I fragments from the polylinker region. This is helpful because it prevents the stuffer fragment from re-annealing with the left and right arms and generating a viable phage that is non-recombinant.

DNA fragments between approximately 9 and 22 kb may be cloned in EMBL4; the lower limit represents the minimum size required to form viable phage particles (left arm + insert + right arm must be greater than 38 kb) and the upper the maximum packagable size of around 51 kb. These size constraints can act as a useful initial selection method for recombinants, although an additional genetic selection mechanism can be employed with EMBL4 (the Spi⁻ phenotype; see Section 8.1.4).

Charon 40 is a replacement vector in which the stuffer fragment is composed of multiple repeats of a short piece of DNA. This is known as a **polystuffer**, and it has the advantage that the restriction enzyme *Nae*I will cut the polystuffer into its component parts. This enables efficient removal of the polystuffer during vector preparation, and most of the surviving phage will be recombinant. The polystuffer is flanked by polylinkers with a more extensive range of restriction sites than that found in EMBL4, which increases the choice of restriction enzymes that may be used to prepare the insert DNA. The size range of fragments that may be cloned in Charon 40 is similar to that for EMBL4.

5.3.3 Vectors based on bacteriophage M13

Two aspects of M13 infection are of value to the genetic engineer. First, the RF is essentially similar to a plasmid and can be isolated and manipulated using the same techniques. A second advantage is that the single-stranded DNA produced during the infection is useful in techniques such as DNA sequencing by the dideoxy method (see Section 3.7.4). This aspect alone made M13 immediately attractive as a potential vector.

Unlike phage λ, M13 does not have any non-essential genes. The 6407 bp genome is also used very efficiently in that most of it is taken up by gene sequences, so that the only part available for manipulation is a 507 bp intergenic region. This has been used to construct the M13mp series of vectors, by inserting a polylinker/*lacZ* α-peptide sequence into this region (Fig. 5.10). This enables the X-gal screening system to be used for the detection of recombinants, as is the case with the pUC plasmids. When M13 is grown on a bacterial lawn, 'plaques' appear because of the reduction in growth of the host cells (which are not lysed), and these may be picked for further analysis.

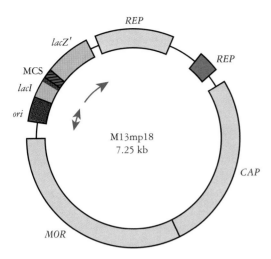

Fig. 5.10 Map of the filamentous phage vector M13mp18. The double-stranded replicative form is shown. The polylinker region (MCS) is the same as that found in plasmid pUC18 (Fig. 5.2). Genes in the *REP* region encode proteins that are important for DNA replication. The *CAP* and *MOR* regions contain genes that specify functions associated with capsid formation and phage morphogenesis, respectively. The vector M13mp19 is identical except for the orientation of the polylinker region.

A second disadvantage of M13 vectors is the fact that they do not function efficiently when long DNA fragments are inserted into the vector. Although in theory there should be no limit to the size of clonable fragments, as the capsid structure is determined by the genome size (unlike phage λ), there is a marked reduction in cloning efficiency with fragments longer than about 1.5 kb. In practice this was not a major problem, as the main use of the early M13 vectors was in subcloning small DNA fragments for sequencing. In this application single-stranded DNA production, coupled with ease of purification of the DNA from the cell culture, outweighs any size limitation, although this has also been alleviated by the construction of hybrid plasmid/M13 vectors (see Section 5.4.1).

5.4 | Other vectors

So far we have concentrated on what we might call 'basic' plasmid and bacteriophage vectors for use in *E. coli* hosts. Although these vectors still represent a major part of the technology of gene manipulation, there has been continued development of more sophisticated bacterial vectors, as well as vectors for other organisms. One driving force in this has been the need to clone and analyse ever larger pieces of DNA, as the emphasis in molecular biology has shifted towards the analysis of genomes rather than simply genes in isolation. In addition, the commercial development of integrated approaches to cloning procedures has required new vectors. Such kit-based products are often

As cloning methodology developed, the limitations of plasmid- and phage-based vectors placed some constraints on what could be achieved, and there was an increasing need for vectors with increased cloning capacities.

marketed as 'cloning technologies'. Cloning kits have been a successful addition to the gene manipulator's armoury, often reducing the time taken to achieve a particular outcome. In this section we will look at the features of some additional bacterial vectors and some vectors for use in other organisms.

5.4.1 Hybrid plasmid/phage vectors

One feature of phage vectors is that the technique of **packaging *in vitro*** (see Section 5.5.2 for details) is sequence independent, apart from the requirement of having the *cos* sites separated by DNA of packagable size (38–51 kb). This has been exploited in the construction of vectors that are made up of plasmid sequences joined to the *cos* sites of phage λ. Such vectors are known as **cosmids**. They are small (4–6 kb) and can therefore accommodate cloned DNA fragments up to some 47 kb in length. As they lack phage genes, they behave as plasmids when introduced into *E. coli* by the packaging/infection mechanism of λ. Cosmid vectors therefore offer an apparently ideal system – a highly efficient and specific method of introducing the recombinant DNA into the host cell, and a cloning capacity some twofold greater than the best λ replacement vectors. However, they are not without disadvantages, and often the gains of using cosmids instead of phage vectors are offset by losses in terms of ease of use and further processing of cloned sequences.

> Cosmids and phagemids incorporate some of the characteristics of both plasmid- and phage-based vectors.

Hybrid plasmid/phage vectors in which the phage functions are expressed and utilised in some way are known as **phagemids**. One such series of vectors is the λZAP family, produced by Stratagene. Features of these phagemids include the potential to excise cloned DNA fragments *in vivo* as part of a plasmid. This automatic excision is useful in that it removes the need to subclone inserts from λ into plasmid vectors for further manipulation.

Hybrid plasmid/phage vectors have been developed to overcome the size limitation of the M13 cloning system and are now widely used for applications such as DNA sequencing and the production of probes for use in hybridisation studies. These vectors are essentially plasmids that contain the f1 (M13) phage origin of replication. When cells containing the plasmid are superinfected with phage, they produce single-stranded copies of the plasmid DNA and secrete these into the medium as M13-like particles. Vectors such as pEMBL9 or pBluescript can accept DNA fragments of up to 10 kb. Some commercially available vectors based on bacteriophages are listed in Table 5.4.

5.4.2 Vectors for use in eukaryotic cells

When eukaryotic host cells are considered, vector requirements become a little more complex than is the case for prokaryotic hosts. Bacteria are relatively simple in genetic terms, whereas eukaryotic cells have multiple chromosomes that are held within the membrane-bound nucleus. Given the wide variety of eukaryotes, it is not

Table 5.4. | Some commercially available bacteriophage-based vectors

Vector	Features	Applications	Supplier
λGT11	λ insertion vector Insert capacity 7.2 kbp *lacZ* gene	cDNA library construction Expression of inserts	Various
λEMBL3/4	λ replacement vectors Insert capacity 9–23 kbp	Genomic library construction	Various
λZAP Express®	λ-based insertion vector Capacity of 12 kbp *In vivo* excision of inserts Expression of inserts	cDNA library construction Also genomic/PCR cloning	Stratagene
λFIX®II	λ-based replacement vector, capacity 9–23 kbp Spi^+/P2 selection system to reduce non-recombinant background	Genomic library construction	Stratagene
pBluescript®II	Phagemid vector Produces single-stranded DNA	*In vitro* transcription DNA sequencing	Stratagene
SuperCos 1	Cosmid vector with Ap^r and Neo^r markers, plus T3 and T7 promoters Capacity 30–42 kbp	Generation of cosmid-based genomic DNA libraries T3/7 promoters allow end-specific transcripts to be generated for chromosome walking techniques	Stratagene

Note: As with plasmid vectors, there are many variants available from a range of different suppliers. A good source of information is the supplier's catalogue or website. Ap^r, ampicillin resistance; Neo^r, neomycin resistance (selection using kanamycin in bacteria, G418 in mammalian cells); T3/7 are promoters for *in vitro* transcription; *lacZ*, β-galactosidase gene; SV40, promoter for expression in eukaryotic cells. Terms marked ® are registered trademarks of Stratagene.

surprising that vectors tend to be highly specialised and designed for specific purposes.

The unicellular yeast *S. cerevisiae* has had a major impact on eukaryotic gene manipulation technology. A range of vectors for use in yeast cells has been developed, with the choice of vector depending on the particular application. **Yeast episomal plasmids** (YEps) are based on the naturally occurring yeast 2 μm plasmid and may replicate autonomously or integrate into a chromosomal location. **Yeast integrative plasmids** are designed to integrate into the chromosome in a similar way to the YEps, and **yeast replicative plasmids** remain as independent plasmids and do not integrate. Plasmids that contain sequences from around the centromeric region of chromosomes are known as **yeast centromere plasmids**, and these behave essentially as mini-chromosomes.

> A range of plasmid-based vectors for the yeast *Saccharomyces cerevisiae* was developed from the naturally occurring yeast 2μm plasmid.

| Table 5.5. | Some possible vectors for plant and animal cells |

Cell type	Vector type	Genome	Examples
Plant cells	Plasmid	DNA	Ti plasmids of *Agrobacterium tumefaciens* – these are well-established vector systems
	Viral	DNA	Cauliflower mosaic viruses, geminiviruses – these are not highly developed
Animal cells	Plasmid	DNA	Many vectors available commercially, often using sequences/promoters/origins of replication from Simian virus 40 (SV40) and/or cytomegalovirus (CMV)
	Viral	DNA	Baculoviruses for insect cells Papilloma viruses Adenovirus SV40 Vaccinia virus
	Viral	RNA	Retroviruses
	Transposon	DNA	P elements in *Drosophila melanogaster*

Note: Certain aspects of 'vectorology' are more advanced than others. Often a particular type of system will be developed as a vector, becoming extensively modified in the process as different versions are generated. In areas where vector technology is not well developed, techniques such as PCR can sometimes be used to overcome any limitations of the system.

When dealing with higher eukaryotes that are multicellular, such as plants and animals, the problems of introducing recombinant DNA into the organism become slightly different than those that apply to microbial eukaryotes such as yeast. The aims of genetic engineering in higher eukaryotes can be considered as broadly twofold:

- To express cloned genes in plant and animal cells in tissue culture, for basic research on gene expression or for the production of useful proteins
- To alter the genetic makeup of the organism and produce a transgenic, in which all the cells will carry the genetic modification

The latter aim in particular can pose technical difficulties, as the recombinant DNA has to be introduced very early in development or in some sort of vector that will promote the spread of the recombinant sequence throughout the organism.

Vectors used for plant and animal cells may be introduced into cells directly by techniques such as those described in Section 5.5.3, or they may have a biological entry mechanism if based on viruses or other infectious agents such as *Agrobacteria*. Some examples of the types of system that have been used in the development of vectors for plant and animal cells are given in Table 5.5. The use of specific vectors is described further in Part III of this book when considering some of the applications of gene manipulation technology in eukaryotes.

Vectors for use in plant and animal cells have properties that enable them to function in these cell types; they are often more specialised than the basic primary cloning vectors such as λ.

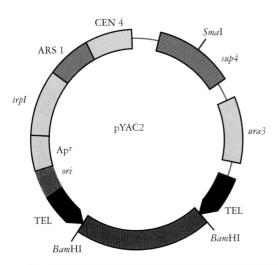

Fig. 5.11 Map of the yeast artificial chromosome vector pYAC2. This carries the origin of replication (*ori*; shaded) and ampicillin resistance gene (Apr) from pBR322, and yeast sequences for replication (ARS1) and chromosome structure (centromere, CEN4; and telomeres, TEL). The TEL sequences are separated by a fragment flanked by two *Bam*HI sites. The genes *trp1* and *ura3* may be used as selectable markers in yeast. The cloning site *Sma*I lies within the *sup4* gene (hatched). From Kingsman and Kingsman (1988), *Genetic Engineering*, Blackwell. Reproduced with permission.

5.4.3 Artificial chromosomes

The development of vectors for cloning very large pieces of DNA was essential to enable large genome sequencing projects to proceed at a reasonable rate, although genomes such as *S. cerevisiae* have been sequenced mainly by using cosmid vectors to construct the genomic libraries required. However, even insert sizes of 40–50 kb are too small to cope with projects such as the Human Genome Project (see Chapter 10 for a more detailed treatment of genome analysis). The development of yeast **artificial chromosomes** (YACs, see Fig. 5.11) has enabled DNA fragments in the megabase range to be cloned, although there have been some problems of insert instability. YACs are the most sophisticated yeast vectors and to date represent the largest-capacity vectors available. They have centromeric and telomeric regions, and the recombinant DNA is therefore maintained essentially as a yeast chromosome.

A further development of artificial chromosome technology came with the construction of bacterial artificial chromosomes (**BACs**). These are based on the F plasmid, which is much larger than the standard plasmid cloning vectors and, therefore, offers the potential of cloning larger fragments. BACs can accept inserts of around 300 kb, and many of the instability problems of YACs can be avoided by using the bacterial version. Much of the sequencing of the human genome has been accomplished using a library of BAC recombinants. Vectors based on the phage P1 have also been developed, both as phage vectors and also as P1-based artificial chromosomes.

Artificial chromosomes are elegantly simple vectors that mimic the natural construction of chromosomal DNA, with telomeres, a centromere, and an origin of replication in addition to features designed for ease of use, such as selectable markers.

5.5 | Getting DNA into cells

Manipulation of vector and insert DNAs to produce recombinant molecules is carried out in the test tube, and we are then faced with the task of getting the recombinant DNA into the host cell for propagation. The efficiency of this step is often a crucial factor in determining the success of a given cloning experiment, particularly when a large number of recombinants is required. Efficiency may not be an issue where a subcloning procedure is used, as the target sequence is likely to have been cloned (or perhaps generated using the polymerase chain reaction; see Chapter 7). Therefore, the target sequence will be available in relatively large amounts, so that efficiency of the cloning protocol is not often a major concern. The methods available for getting recombinant DNA into cells depend on the type of host/vector system and range from very simple procedures to much more complicated and esoteric ones. In this section we will consider some of these methods.

5.5.1 Transformation and transfection

The techniques of **transformation** and **transfection** represent the simplest methods available for getting recombinant DNA into cells. In the context of cloning in *E. coli* cells, transformation refers to the uptake of plasmid DNA, and transfection to the uptake of phage DNA. Transformation is also used more generally to describe uptake of any DNA by any cell and can also be used in a different context when talking about a **growth transformation** such as occurs in the production of a cancerous cell.

Transformation of *E. coli* cells with recombinant plasmid DNA is one of the classic techniques of gene manipulation.

Transformation in bacteria was first demonstrated in 1928 by Frederick Griffith, in his famous 'transforming principle' experiment that paved the way for the discoveries that eventually showed that genes were made of DNA. However, not all bacteria can be transformed easily, and it was not until the early 1970s that transformation was demonstrated in *E. coli*, the mainstay of gene manipulation technology. To effect transformation of *E. coli*, the cells need to be made **competent**. This is achieved by soaking the cells in an ice-cold solution of calcium chloride, which induces competence in a way that is still not fully understood. Transformation of competent cells is carried out by mixing the plasmid DNA with the cells, incubating on ice for 20–30 min, and then giving a brief heat shock (2 min at 42°C is often used), which appears to enable the DNA to enter the cells. The transformed cells are usually incubated in a nutrient broth at 37°C for 60–90 min to enable the plasmids to become established and permit phenotypic expression of their traits. The cells can then be plated out onto selective media for propagation of cells harbouring the plasmid.

Transformation is an inefficient process in that only a very small percentage of competent cells become transformed, representing

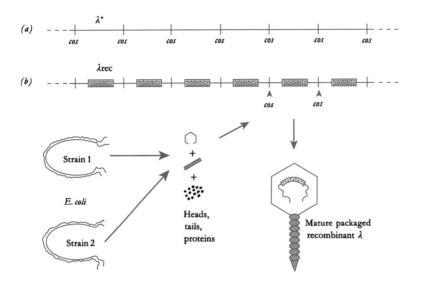

Fig. 5.12 Packaging bacteriophage DNA. (*a*) A concatemeric DNA molecule composed of wild-type phage DNA ($\lambda+$). The individual genomes are joined at the *cos* sites. (*b*) Recombinant genomes (λrec) are shown being packaged *in vitro*. A mixed lysate from two bacterial strains supplies the head and tail precursors and the proteins required for the formation of mature λ particles. On adding this mixture to the concatemer, the DNA is cleaved at the *cos* sites (arrowed) and packaged into individual phage particles, each containing a recombinant genome.

uptake of a fraction of the plasmid DNA that is available. Thus, the process can become the critical step in a cloning experiment where a large number of individual recombinants is required, or when the starting material is limiting. Despite these potential disadvantages, transformation is an essential technique, and with care can yield up to 10^9 transformed cells (**transformants**) per microgram of input DNA, although transformation frequencies of around 10^6 or 10^7 transformants per microgram are more often achieved in practice. Many biological supply companies offer a variety of competent cell strains that have been pre-treated to yield high transformation frequencies. Whilst more expensive than 'home-made' competent cells, these ready-to-go cells have become popular as they do save on preparation time. Transfection is a similar process to transformation, the difference being that phage DNA is used instead of plasmid DNA. It is again a somewhat inefficient process, and it has largely been superseded by packaging *in vitro* for applications that require the introduction of phage DNA into *E. coli* cells.

> Transformation efficiency is often a limiting factor in using the technique, and this may be critical if the aim of the procedure is to prepare a representative clone bank.

5.5.2 Packaging phage DNA *in vitro*

During the lytic cycle of phage λ, the phage DNA is replicated to form what is known as a **concatemer**. This is a very long DNA molecule composed of many copies of the λ genome, linked together by the *cos* sites (Fig. 5.12(*a*)). When the phage particles are assembled the

DNA is packaged into the capsid, which involves cutting the DNA at the *cos* sites using a phage-encoded endonuclease. Mature phage particles are thus produced, ready to be released on lysis of the cell, and capable of infecting other cells. This process normally occurs *in vivo*, the particular functions being encoded by the phage genes. However, it is possible to carry out the process in the test tube, which enables recombinant DNA that is generated as a concatemer to be packaged into phage particles.

To enable packaging *in vitro*, the components of the λ capsid, and the endonuclease, must be available. In practice, two strains of bacteria are used to produce a lysate known as a **packaging extract**. Each strain is mutant in one function of phage morphogenesis, so that the packaging extracts will not work in isolation. When the two are mixed with the concatemeric recombinant DNA under suitable conditions, all the components are available and phage particles are produced. These particles can then be used to infect *E. coli* cells, which are plated out to obtain plaques. The process of packaging *in vitro* is summarised in Fig. 5.12(*b*).

> Packaging recombinant bacteriophage DNA *in vitro* mimics the normal process that occurs during phage maturation and assembly and has proved to be a very useful method for the construction of genomic libraries.

5.5.3 Alternative DNA delivery methods

The methods available for introducing DNA into bacterial cells are not easily transferred to other cell types. The phage-specific packaging system is not available for other systems, and transformation by normal methods may prove impossible or too inefficient to be a realistic option. However, there are alternative methods for introducing DNA into cells. Often these are more technically demanding and less efficient than the bacterial methods, but reliable results have been achieved in many situations where there appeared to be no hope of getting recombinant DNA molecules into the desired cell.

Most of the problems associated with getting DNA into non-bacterial cells have involved plant cells. Animal cells are relatively flimsy and can be transformed readily. However, plant cells pose the problem of a rigid cell wall, which is a barrier to DNA uptake. This can be alleviated by the production of **protoplasts**, in which the cell wall is removed enzymatically. The protoplasts can then be transformed using a technique such as **electroporation**, where an electrical pulse is used to create transient holes in the cell membrane through which the DNA can pass. The protoplasts can then be regenerated. In addition to this application, protoplasts also have an important role to play in the generation of hybrid plant cells by fusing protoplasts together.

> Introducing recombinant DNA into eukaryotic cells can involve biological methods or one of a range of techniques such as electroporation, microinjection, or biolistics.

An alternative to transformation procedures is to introduce DNA into the cell by some sort of physical method. One way of doing this is to use a very fine needle and inject the DNA directly into the nucleus. This technique is called **microinjection** (Fig. 5.13) and has been used successfully with both plant and animal cells. The cell is held on a glass tube by mild suction and the needle used to pierce the membrane. The technique requires a mechanical micromanipulator

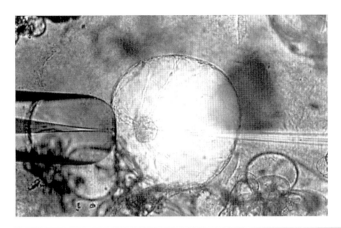

Fig. 5.13 Microinjection of a protoplast-derived potato cell. The cell is held on a glass capillary (on the left of the photograph) by gentle suction. The microinjection needle is made by drawing a heated glass capillary out to a fine point. Using a micromanipulator (a mechanical device for fine control of the capillary) the needle has been inserted into the cell (on the right of the photograph), where its tip can be seen approaching the cell nucleus. Photograph courtesy of Dr K. Ward.

and a microscope, and plenty of practice! One obvious disadvantage is that this technique is labour-intensive and not suitable for primary cloning procedures where large numbers of recombinants are required. However, in certain specialised cases it is an excellent method for targetting DNA delivery once a suitable recombinant has been identified and developed to the point where microinjection is feasible.

An ingenious and somewhat bizarre development has proved extremely useful in transformation of plant cells. The technique, which is called **biolistic** DNA delivery, involves literally shooting DNA into cells (Fig. 5.14). The DNA is used to coat microscopic tungsten particles known as **microprojectiles**, which are then accelerated on

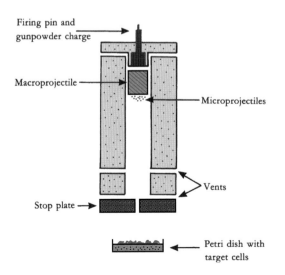

Fig. 5.14 Biolistic apparatus. The DNA is coated onto microprojectiles, which are accelerated by the macroprojectile on firing the gun. At the stop plate the macroprojectile is retained in the chamber and the microprojectiles carry on to the target tissue. Other versions of the apparatus, driven by compressed gas instead of a gunpowder charge, are available.

a **macroprojectile** by firing a gunpowder charge or by using compressed gas to drive the macroprojectile. At one end of the 'gun' there is a small aperture that stops the macroprojectile but allows the microprojectiles to pass through. When directed at cells, these microprojectiles carry the DNA into the cell and, in some cases, stable transformation will occur.

```
                                    Cosmids
                                    Phagemids
                         other      Yeast artificial
                         types      chromsomes
                         include    Viruses
                            │
Gene cloning ── uses a ──┐  │
                          │  │
                    Vector or              Replacement ── such as ── EMBL4
                    carrier molecule         vectors                 Charon 40
                          │
           ┌── two main types ──┐
           │                    │
        Plasmids              Phages              λ ── two types ──┐
           │                    │                 │                │
      which are            such as ──┐      Insertion ── such as ── λgt10
           │                    │    │       vectors                Charon 16A
   Circular DNA              M13   │
   molecules                       λ
           │
   and have been
           │
   Specially
   constructed

                    which
                      │
          ┌── may have ──┐    ┌── should have ──┐
          │              │    │                 │
   Multiple                An origin of
   cloning                 replication
   site                    Selectable
   Expression              marker(s)
   signals                 Single
          │                restriction
          │                site
          └── and can be ──┘
                  │
            Inserted into
            a host cell
                  │
                  by
                  │
                  ▼
   requires a
       │
   Host cell
       │
   including both
       │
   Prokaryotic ── and ── Eukaryotic
       │                     │
   such as               such as
       │                     │
   E. coli              S. cerevisiae
   B. subtilis          Plant cells
       │                Animal cells
       │                     │
       └── need method for ──┘
                  │
            Introduction
            of cloned
            DNA
                  │
               such as
                  │
            Transformation
            Transfection
            Packaging in vitro
            Microinjection
            Biolistics
```

Concept map 5

Chapter 6 summary

Aims

- To outline the range of strategies that may be employed to clone a DNA sequence
- To describe the rationale and methodology involved in cloning cDNA prepared from messenger RNA
- To describe the rationale and methodology involved in cloning from genomic DNA
- To illustrate some aspects of advanced cloning strategies
- To describe the use of artificial chromosomes for cloning large DNA fragments

Chapter summary/learning outcomes

When you have completed this chapter you will have knowledge of:

- Possible routes for cloning cDNA and genomic DNA
- The methods used to prepare cDNA
- Cloning cDNA in plasmid and bacteriophage vectors
- The use of linkers, adaptors, and homopolymer tailing
- The preparation of genomic DNA fragments for cloning
- Genomic library construction and amplification
- Expression of cloned DNA molecules
- The use of artificial chromosome vectors for cloning large fragments of DNA

Key words

Generation, joining, propagation, selection, polymerase chain reaction, messenger RNA, genomic DNA, transcriptome, housekeeping genes, tissue-specific gene expression, proteome, genome, mRNA diversity, abundance class, complementary DNA, copy DNA, cDNA, blunt-end ligation, linker molecule, homopolymer tailing, concatemer, bimolecular recombinant, adaptor, annealing, ligation, gene, clone bank, clone library, genomic library, mechanical shearing, random fragmentation, partial restriction enzyme digestion, primary library, amplification, skewing, priming, promoter, strong promoter, weak promoter, consensus sequence, Pribnow box, TATA box, Hogness box, CAAT box, inducible, repressible, BACs, YACs, artificial chromosome.

Chapter 6

Cloning strategies

In the previous two chapters we have examined the two essential components of genetic engineering: (1) the ability to cut, modify, and join DNA molecules *in vitro*, and (2) the host/vector systems that allow recombinant DNA molecules to be propagated. With these components at his or her disposal, the genetic engineer has to devise a cloning strategy that will enable efficient use of the technology to achieve the aims of the experiment. As we saw in Chapter 1, there are basically four stages to any cloning experiment (Fig. 1.1). These are the **generation** of DNA fragments, **joining** the fragments to a suitable vector to produce recombinant DNA, **propagation** of the recombinants in a host cell, and (finally!) the identification and **selection** of the required sequence. In this chapter we will look at some of the strategies that are available for completing the first three of these stages by the traditional methods of gene cloning, largely restricting the discussion to cloning eukaryotic DNA in *E. coli*. The use of the **polymerase chain reaction** (PCR) in amplification and cloning of sequences is discussed in Chapter 7, as this is now a widely used protocol that in some cases bypasses standard cloning techniques. Selection of cloned sequences is discussed in Chapter 8, although the type of selection method that will be used does have to be considered when choosing host/vector combinations for a particular cloning exercise.

> To complete the four key stages, and achieve a successful outcome to a cloning procedure, a clear overall strategy is required at the outset.

6.1 | Which approach is best?

The complexity of any cloning experiment depends largely on two factors: (1) the overall aims of the procedure and (2) the type of source material from which the nucleic acids will be isolated for cloning. Thus, a strategy to isolate and sequence a relatively small DNA fragment from *E. coli* will be different (and will probably involve fewer stages) than a strategy to produce a recombinant protein in a transgenic eukaryotic organism. There is no single cloning strategy that will cover all requirements. Each project will, therefore, be unique and will present its own set of problems that have to be addressed by choosing the appropriate path through the maze of possibilities

> Each cloning procedure is unique and presents a set of challenges that must be overcome by selection of the appropriate techniques – this is often made easier by using an optimised kit from a single supplier.

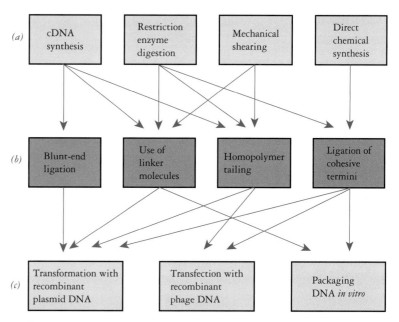

Fig. 6.1 Routes available for cloning by what might now be considered 'traditional' methods. Possible methods available for three key stages of a cloning procedure are shown as follows: (*a*) the generation of DNA fragments, (*b*) joining to a vector, and (*c*) introducing the recombinant DNA into a host cell. Commonly used or preferred routes are indicated by arrows. Although cloning technology has become more sophisticated in recent years, the basic premises on which the procedures are based have not changed substantially, and routes such as those indicated are still used in many applications. Redrawn from Old and Primrose (1989), *Principles of Gene Manipulation*, Blackwell. Reproduced with permission.

(see Fig. 6.1). Fortunately, most of the confusion can be eliminated by careful design of the experimental procedures and rigorous interpretation of results at each stage of the process. These days, the popularity of 'kit cloning', using a cloning technology from a particular supplier, has overcome some of the pitfalls of the 'home-made' approach. This is largely due to the fact that kit components can be optimised and batch-tested to ensure that they work well together. Despite this, cloning can have its troublespots, and the technology has not yet become foolproof – care, patience, and attention to detail are essential if a successful outcome is to be achieved.

When dealing with eukaryotic organisms, the first major decision is whether to begin with **messenger RNA (mRNA)** or **genomic DNA**. Although the DNA represents the complete genome of the organism, it may contain non-coding DNA such as introns, control regions, and repetitive sequences. This can sometimes present problems, particularly if the genome is large and the aim is to isolate a single-copy gene. However, if the primary interest is in the control of gene expression, it is obviously necessary to isolate the control sequences, so genomic DNA is the only alternative.

The choice of starting material is the first critical step in any cloning experiment and is influenced by the overall aims of the procedure.

Messenger RNA has two advantages over genomic DNA as a source material. First, it represents the **transcriptome** (*i.e.* the genetic information that is being expressed by the particular cell type from which it is prepared). This can be a very powerful preliminary selection mechanism, as not all the genomic DNA will be represented in the mRNA population. Also, if the gene of interest is highly expressed, this may be reflected in the abundance of its mRNA, and this can make isolation of the clones easier. A second advantage of mRNA is that it, by definition, represents the coding sequence of the gene, with any introns having been removed during RNA processing. Thus, production of recombinant protein is much more straightforward if a clone of the mRNA is available.

Although genomic DNA and mRNA are the two main sources of nucleic acid molecules for cloning, it is possible to synthesise DNA *in vitro* if the amino acid sequence of the protein is known. Whilst this is a laborious task for long stretches of DNA, it is a useful technique in some cases, particularly if only short sections of a gene need to be synthesised to complete a sequence prior to cloning. The PCR (see Chapter 7) can also be used to generate DNA fragments that may then be cloned for further processing.

Having decided on the source material, the next step is to choose the type of host/vector system. Even when cloning in *E. coli* hosts there is still a wide range of strains available, and care must be taken to ensure that the optimum host/vector combination is chosen. When choosing a vector, the method of joining the DNA fragments to the vector and the means of getting the recombinant molecules into the host cell are two main considerations. In practice the host/vector systems in *E. coli* are well defined, so it is a relatively straightforward task to choose the best combination, given the type of fragments to be cloned and the desired outcome of the experiment. However, the great variety of vectors, host cells, and cloning kits available from suppliers can be confusing to the first-time gene manipulator, and often a recommendation from an experienced colleague is the best way to proceed. Often a particular laboratory will have a set of favoured procedures that have become established and work well, and the maxim 'if it ain't broke, don't fix it' is a good one to bear in mind!

In devising a cloning strategy all the aforementioned points have to be considered. Often there will be no ideal solution to a particular problem, and a compromise will have to be accepted. By keeping the overall aim of the experiments in mind, the researcher can minimise the effects of such compromises and choose the most efficient cloning route.

> Although genomic DNA and mRNA are two major sources of nucleic acid for cloning, techniques such as the polymerase chain reaction (PCR) may provide the starting material in some cases.

6.2 | Cloning from mRNA

Each type of cell in a multicellular organism will produce a range of mRNA molecules. In addition to the expression of general **housekeeping genes** whose products are required for basic cellular metabolism,

		Number of	Abundance
Source		different mRNAs	(molecules/cell)
Mouse liver cytoplasmic poly(A)$^+$ RNA		9	12000
		700	300
		11500	15
Chick oviduct polysomal poly(A)$^+$ RNA		1	100000
		7	4000
		12500	5

Table 6.1. mRNA abundance classes

Note: The diversity of mRNAs is indicated by the number of different mRNA molecules. There is one mRNA that is present in chick oviduct cells at a very high level (100 000 molecules per cell). This mRNA encodes ovalbumin, the major egg white protein.
Source: After Old and Primrose (1989), *Principles of Gene Manipulation*, 4th edition, Blackwell. Mouse data from Young *et al.* (1976), Biochemistry 15, 2823–2828, copyright (1976) American Chemical Society. Chick data from Axel *et al.* (1976), Cell **11**, 247–254, copyright (1976) Cell Press. Reproduced with permission.

cells exhibit **tissue-specific gene expression**. Thus, liver cells, kidney cells, skin cells, *etc.* will each synthesise a different spectrum of tissue-specific proteins (the **proteome**). This requires expression of a particular subset of genes in the **genome**, achieved by synthesis of a set of mRNAs (the transcriptome). In addition to the **diversity of mRNAs** produced by each cell type, there may well be different **abundance classes** of particular mRNAs. This has important consequences for cloning from mRNA, as it is easier to isolate a specific cloned sequence if it is present as a high proportion of the starting mRNA population. Some examples of mRNA abundance classes are shown in Table 6.1.

> As mRNA represents the expressed part of the genome at any given time, it is sensible to use mRNA as the starting point if the coding sequence of a gene is the target.

6.2.1 Synthesis of cDNA

It is not possible to clone mRNA directly, so it has to be converted into DNA before being inserted into a suitable vector. This is achieved using the enzyme reverse transcriptase (RTase; see Section 4.2.2) to produce **complementary DNA** (also known as **copy DNA** or **cDNA**). The classic early method of cDNA synthesis utilises the poly(A) tract at the 3' end of the mRNA to bind an oligo(dT) primer, which provides the 3'-OH group required by RTase (Fig. 6.2). Given the four dNTPs and suitable conditions, RTase will synthesise a copy of the mRNA to produce a cDNA · mRNA hybrid. The mRNA can be removed by alkaline hydrolysis and the single-stranded (ss) cDNA converted into double-stranded (ds) cDNA by using a DNA polymerase. In this second-strand synthesis the priming 3'-OH is generated by short hairpin loop regions that form at the end of the ss cDNA. After second-strand synthesis, the ds cDNA can be trimmed by S$_1$ nuclease to give a flush-ended molecule, which can then be cloned in a suitable vector.

> The first step in cloning from mRNA is to convert the mRNA into double-stranded complementary DNA (cDNA; also known as copy DNA) using the enzymes reverse transcriptase and DNA polymerase.

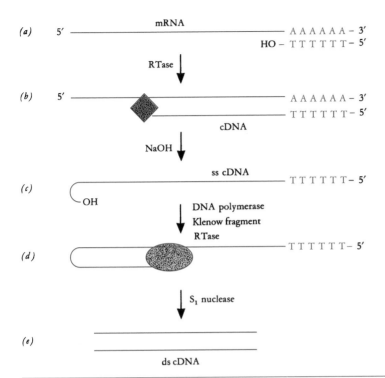

(a) 5′ ————————— mRNA —————————— A A A A A A – 3′
                                   HO – T T T T T – 5′

RTase ↓

(b) 5′ ————————————◆———————————— A A A A A A – 3′
                        ————————— T T T T T – 5′
                          cDNA

NaOH ↓

(c)   ss cDNA
                              ————————— T T T T T – 5′
     ⌒
     ⌐OH   DNA polymerase ↓
                    Klenow fragment
                    RTase ↓
                              ————————— T T T T T – 5′

(d) ⌐——————————⬤————————————— T T T T T – 5′

S₁ nuclease ↓

(e) ⌐———————————————————————
     —————————————————————————
                ds cDNA

Fig. 6.2 Synthesis of cDNA. Poly(A)$^+$ RNA (mRNA) is used as the starting material.
(a) A short oligo(dT) primer is annealed to the poly(A) tail on the mRNA, which provides
the 3′-OH group for reverse transcriptase to begin copying the mRNA (b). The mRNA is
removed by alkaline hydrolysis to give an ss cDNA molecule (c). This has a short ds
hairpin loop structure that provides a 3′-OH terminus for (d) second-strand synthesis by
a DNA polymerase (T4 DNA polymerase, Klenow fragment, or RTase). (e) The
double-stranded cDNA is trimmed with S₁ nuclease to produce a blunt-ended ds cDNA
molecule. An alternative to the alkaline hydrolysis step is to use RNase H, which creates
nicks in the mRNA strand of the mRNA–cDNA hybrid. By using this in conjunction with
DNA polymerase I, a nick translation reaction synthesises the second cDNA strand.

Several problems are often encountered in synthesising cDNA
using the method just outlined. First, synthesis of full-length cDNAs
may be inefficient, particularly if the mRNA is relatively long. This is
a serious problem if expression of the cDNA is required, as it may not
contain the entire coding sequence of the gene. Such inefficient full-
length cDNA synthesis also means that the 3′ regions of the mRNA
tend to be over-represented in the cDNA population. Second, problems can arise from the use of S₁ nuclease, which may remove some
important 5′ sequences when it is used to trim the ds cDNA.

More recent methods for cDNA synthesis overcome the aforementioned problems to a great extent, and the original method is
now rarely used. One of the simplest adaptations involves the use
of oligo(dC) tailing to permit oligo(dG)-primed second-strand cDNA
synthesis (Fig. 6.3). The dC tails are added to the 3′ termini of the
cDNA using the enzyme terminal transferase. This functions most

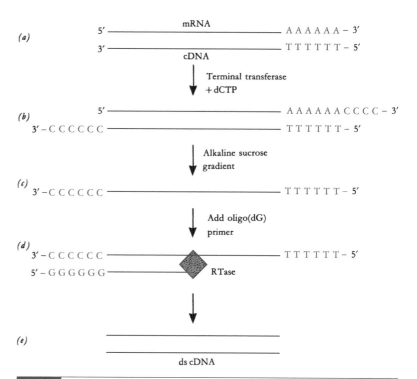

Fig. 6.3 Oligo(dG)-primed second-strand cDNA synthesis. (*a*) First-strand synthesis is as shown in Fig. 6.2, generating an mRNA–cDNA hybrid. (*b*) This is tailed with C residues using terminal transferase. (*c*) Fractionation through an alkaline sucrose gradient hydrolyses the mRNA and permits recovery of full-length cDNA molecules. (*d*) An oligo(dG) primer is annealed to the C tails, and reverse transcriptase is used to synthesise the second strand. (*e*) This generates a double-stranded full-length cDNA molecule. From Old and Primrose (1989), *Principles of Gene Manipulation*, Blackwell. Reproduced with permission.

efficiently on accessible 3′ termini, and the tailing reaction therefore favours full-length cDNAs in which the 3′ terminus is not 'hidden' by the mRNA template. The method also obviates the need for S$_1$ nuclease treatment and, thus, full-length cDNA production is enhanced further.

As we have already mentioned in a number of contexts, many suppliers now produce kits for cDNA synthesis and cloning. Often these have been optimised for a particular application, and the number of steps involved is usually reduced to a minimum. In many ways the mystique that surrounded cDNA synthesis in the early days has now gone, and the techniques available make full-length cDNA synthesis a relatively straightforward business. The key to success is to obtain good-quality mRNA preparations and to take great care in handling these. In particular, contamination with nucleases must be avoided.

Although the poly(A) tract of eukaryotic mRNAs is often used for priming cDNA synthesis, there may be cases where this is not appropriate. Where the mRNA is not polyadenylated, random oligonucleotide primers may be used to initiate cDNA synthesis. Or, if all or

> The synthesis of cDNA is now routine in many laboratories and is straightforward if care is taken to ensure that good-quality mRNA is prepared and nuclease contamination is avoided.

part of the amino acid sequence of the desired protein is known, a specific oligonucleotide primer can be synthesised and used to initiate cDNA synthesis. This can be of great benefit in that specific mRNAs may be copied into cDNA, which simplifies the screening procedure when the clones are obtained. An additional possibility with this approach is to use the PCR to amplify the desired sequence selectively.

Having generated the cDNA fragments, the cloning procedure can continue. The choice of vector system – plasmid or phage, or perhaps cosmid or phagemid – will probably have been made before beginning the procedure or will have been determined by the manufacturer of the cloning kit. Examples of cloning strategies based on the use of plasmid and phage vectors are given next.

6.2.2 Cloning cDNA in plasmid vectors

Although many workers prefer to clone cDNA using a bacteriophage vector system, plasmids are still often used, particularly where isolation of the desired cDNA sequence involves screening a relatively small number of clones. Joining the cDNA fragments to the vector is usually achieved by one of the three methods outlined in Fig. 6.1 for cDNA cloning: **blunt-end ligation**, the use of **linker molecules**, and **homopolymer tailing**. Although favoured for cDNA cloning, these methods may also be used with genomic DNA (see Section 6.3). Each of the three methods will be described briefly.

Blunt-end ligation is exactly what it says – the joining of DNA molecules with blunt (flush) ends, using DNA ligase (see Section 4.3). In cDNA cloning, the blunt ends may arise as a consequence of the use of S_1 nuclease, or they may be generated by filling in the protruding ends with DNA polymerase. The main disadvantage of blunt-end ligation is that it is an inefficient process, as there is no specific intermolecular association to hold the DNA strands together whilst DNA ligase generates the phosphodiester linkages required to produce the recombinant DNA. Thus, high concentrations of the participating DNAs must be used, so that the chances of two ends coming together are increased. The effective concentration of DNA molecules in cloning reactions is usually expressed as the concentration of termini; thus, one talks about 'picomoles of ends', which can seem rather strange terminology to the uninitiated.

The conditions for end ligation must be chosen carefully. In theory, when vector DNA and cDNA are mixed, there are several possible outcomes. The desired result is for one cDNA molecule to join with one vector molecule, thus generating a recombinant with one insert. However, if concentrations are not optimal, the insert or vector DNAs may self-ligate to produce circular molecules, or the insert/vector DNAs may form **concatemers** instead of **bimolecular recombinants**. In practice, the vector is often treated with a phosphatase (either bacterial alkaline phosphatase or calf intestinal alkaline phophatase; see Section 4.2.3) to prevent self-ligation, and the concentrations of the vector and insert DNAs are chosen to favour the production of recombinants.

Inserting the cDNA into the vector can be achieved in a number of ways – the aim is always to generate bimolecular recombinants for cloning.

Ligation reactions are driven by the concentrations of the termini that are to be ligated – this is effectively the 'number of ends' that are available to react with each other.

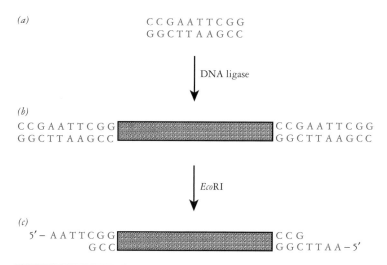

(a)

CCGAATTCGG
GGCTTAAGCC

↓ DNA ligase

(b)

CCGAATTCGG [linker] CCGAATTCGG
GGCTTAAGCC [linker] GGCTTAAGCC

↓ EcoRI

(c)

5′ – AATTCGG [linker] CCG
      GCC [linker] GGCTTAA – 5′

Fig. 6.4 Use of linkers. (a) The 10-mer 5′-CCGAATTCGG-3′ contains the recognition site for *Eco*RI. (b) The linker is added to blunt-ended DNA using DNA ligase. (c) The construct is then digested with *Eco*RI, which cleaves the linker to generate protruding 5′ termini. Redrawn from Winnacker (1987), *From Genes to Clones*, VCH. Reproduced with permission.

One potential disadvantage of blunt-end ligation is that it may not generate restriction enzyme recognition sequences at the cloning site, thus hampering excision of the insert from the recombinant. This is usually not a major problem, as many vectors now have a series of restriction sites clustered around the cloning site. Thus, DNA inserted by blunt-end ligation can often be excised by using one of the restriction sites in the cluster. Another approach involves the use of **linkers**, which are self-complementary oligomers that contain a recognition sequence for a particular restriction enzyme. One such sequence would be 5′-CCGAATTCGG-3′, which in ds form will contain the recognition sequence for *Eco*RI (GAATTC). Linkers are synthesised chemically and can be added to cDNA by blunt-end ligation (Fig. 6.4). When they have been added, the cDNA/linker is cleaved with the linker-specific restriction enzyme, thus generating sticky ends prior to cloning. This can pose problems if the cDNA contains sites for the restriction enzyme used to cleave the linker, but these may be overcome by using a methylase to protect any internal recognition sites from digestion by the enzyme.

A second approach to cloning by addition of sequences to the ends of DNA molecules involves the use of **adaptors** (Fig. 6.5). These are ss non-complementary oligomers that may be used in conjunction with linkers. When annealed together, a linker/adaptor with one blunt end and one sticky end is produced, which can be added to the cDNA to provide sticky-end cloning without digestion of the linkers.

The use of homopolymer tailing has proved to be a popular and effective means of cloning cDNA. In this technique, the enzyme terminal transferase (see Section 4.2.3) is used to add homopolymers of

It is usually beneficial to try to ensure that restriction endonuclease recognition sites are generated when cDNA is inserted into the vector, so that the cloned fragment can be excised easily.

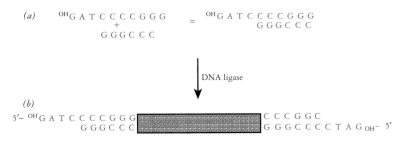

(a)   ᴼᴴG A T C C C C G G G   =   ᴼᴴG A T C C C C G G G
                 +   G G G C C C
            G G G C C C

DNA ligase

(b)

5′– ᴼᴴG A T C C C C G G G ▓▓▓▓▓▓▓▓▓▓▓▓▓▓ C C C G G C
              G G G C C C ▓▓▓▓▓▓▓▓▓▓▓▓▓▓ G G G C C C C T A G ₒₕ– 5′

Fig. 6.5 Use of adaptors. In this example a *Bam*HI adaptor (5′-GATCCCCGGG-3′) is annealed with a single-stranded *Hpa*II linker (3′-GGGCCC-5′) to generate a double-stranded sticky-ended molecule, as shown in (*a*). This is added to blunt-ended DNA using DNA ligase. The DNA therefore gains protruding 5′ termini without the need for digestion with a restriction enzyme, as shown in (*b*). The 5′ terminus of the adaptor can be dephosphorylated to prevent self-ligation. Redrawn from Winnacker (1987), *From Genes to Clones*, VCH. Reproduced with permission.

dA, dT, dG, or dC to a DNA molecule. Early experiments in recombinant production used dA tails on one molecule and dT tails on the other, although the technique is now most often used to clone cDNA into the *Pst*I site of a plasmid vector by dG · dC tailing. Homopolymers have two main advantages over other methods of joining DNAs from different sources. First, they provide longer regions for **annealing** DNAs together than, for example, cohesive termini produced by restriction enzyme digestion. This means that **ligation** need not be carried out *in vitro*, as the cDNA–vector hybrid is stable enough to survive introduction into the host cell, where it is ligated *in vivo*. A second advantage is specificity. As the vector and insert cDNAs have different but complementary 'tails', there is little chance of self-annealing, and the generation of bimolecular recombinants is favoured over a wider range of effective concentrations that is the case for other annealing/ligation reactions.

An example of the use of homopolymer tailing is shown in Fig. 6.6. The vector is cut with *Pst*I and tailed by terminal transferase in the presence of dGTP. This produces dG tails. The insert DNA is tailed with dC in a similar way, and the two can then be annealed. This regenerates the original *Pst*I site, which enables the insert to be cut out of the recombinant using this enzyme.

Introduction of cDNA–plasmid recombinants into suitable *E. coli* hosts is achieved by transformation (Section 5.5.1), and the desired transformants can then be selected by the various methods available (see Chapter 8). The transformation step is often the critical point with respect to cloning efficiency and the size of insert that can be cloned, and these factors must be taken into account when considering the vector/host combination and the size of the intended cDNA clone.

The homopolymer tailing procedure is a classic example of the development of a technique aimed at achieving a particular result whilst minimising the risk of unwanted outcomes.

6.2.3 Cloning cDNA in bacteriophage vectors
Although plasmid vectors have been used extensively in cDNA cloning protocols, there are situations where they may not be appropriate. If

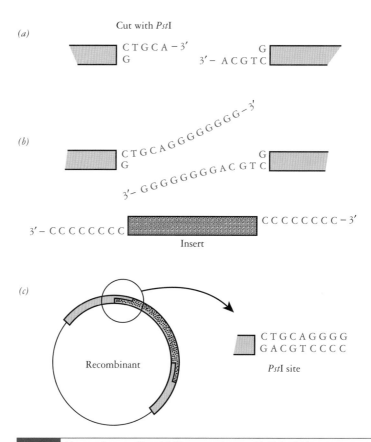

Fig. 6.6 Homopolymer tailing. (*a*) The vector is cut with *Pst*I, which generates protruding 3′-OH termini. (*b*) The vector is then tailed with dG residues using terminal transferase. The insert DNA is tailed with dC residues in a similar way. (*c*) The dC and dG tails are complementary and the insert can therefore be annealed with the vector to generate a recombinant. The *Pst*I sites are regenerated at the ends of the insert DNA, as shown.

Bacteriophage vectors offer certain advantages over plasmid vectors for cloning cDNA. However, for some applications plasmids are still the vector of choice, particularly if expression of the cDNA is the desired outcome.

a large number of recombinants is required, as might be the case if a low-abundance mRNA was to be cloned, phage vectors may be more suitable. The chief advantage here is that packaging *in vitro* may be used to generate the recombinant phage, which greatly increases the efficiency of the cloning process. In addition, it is much easier to store and handle large numbers of phage clones than is the case for bacterial colonies carrying plasmids. Given that isolation of a cDNA clone of a rare mRNA species may require screening hundreds of thousands of independent clones, ease of handling becomes a major consideration.

Cloning cDNA in phage λ vectors is, in principle, no different than cloning any other piece of DNA. However, the vector has to be chosen carefully, as cDNA cloning has slightly different requirements than genomic DNA cloning in λ vectors (see Section 6.3). Generally cDNAs will be much shorter than genomic DNA fragments, so an

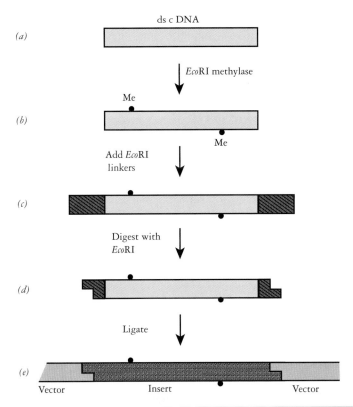

Fig. 6.7 Cloning cDNA in λ vectors using linkers. (*a*) The ds cDNA is treated with *Eco*RI methylase, which (*b*) methylates any internal *Eco*RI recognition sequences. (*c*) *Eco*RI linkers are then added to the ends of the methylated cDNA, and the linkers digested with *Eco*RI. (*d*) The methylation prevents digestion at internal sites, and the result is a cDNA with *Eco*RI cohesive ends. (*e*) This can be ligated into the *Eco*RI site of a λ vector such as λgt10.

insertion vector has usually a large enough cloning capacity. Vectors such as λgt10 and Charon 16A (Section 5.3.2) are suitable, with cloning capacities of some 7.6 and 9.0 kb, respectively. The cDNA may be size-fractionated prior to cloning to remove short cDNAs that may not be representative full-length copies of the mRNA. In the case of vectors such as λgt10, cDNA is usually ligated into the *Eco*RI site using linkers, as shown in Fig. 6.7. The recombinant DNA is packaged *in vitro* and plated on a suitable host for selection and screening.

6.3 | Cloning from genomic DNA

Although cDNA cloning is an extremely useful branch of gene manipulation technology, there are certain situations where cDNAs will not provide the answers to the questions that are being posed. If, for example, the overall structure of a particular **gene** is being investigated (as opposed to its RNA transcript), the investigator may wish to determine

If elements such as control sequences or introns are being investigated, or if genome sequencing is the goal, mRNA cannot be used (for cDNA cloning) and, thus, genomic DNA has to be isolated.

if there are introns present. He or she will probably also wish to examine the control sequences responsible for regulating gene expression, and these will not be present in the processed mRNA molecule that is represented by a cDNA clone. In such a situation clones generated from genomic DNA must be isolated. This presents a slightly different set of problems than those involved in cloning cDNA and, therefore, requires a different cloning strategy.

6.3.1 Genomic libraries

Cloning DNA, by whatever method, gives rise to a population of recombinant DNA molecules, often in plasmid or phage vectors, maintained either in bacterial cells or as phage particles. A collection of independent clones is termed a **clone bank** or **library**. The term **genomic library** is often used to describe a set of clones representing the entire genome of an organism, and the production of such a library is usually the first step in isolating a DNA sequence from an organism's genome.

What are the characteristics of a good genomic library? In theory, a genomic library should represent the entire genome of an organism as a set of overlapping cloned fragments, which will therefore enable the isolation of any sequence in the genome. The fragments for cloning should ideally be generated by a sequence-independent procedure, ensuring that the fragments are generated at random and thus there is no bias towards any particular sequence. Finally, the cloned fragments should be maintained in a stable form with no misrepresentation of sequences due to recombination or differential replication of the cloned DNAs during propagation of the recombinants. Whilst these criteria may seem rather demanding, the systems available for producing genomic libraries enable these requirements to be met more or less completely.

> A genomic library is a rich resource for the scientist, as it represents the entire genome of an organism and (at least in theory) should contain all the genes and their control sequences.

The first consideration in constructing a genomic library is the number of clones required. This depends on a variety of factors, the most obvious one being the size of the genome. Thus, a small genome such as that of *E. coli* will require fewer clones than a more complex one such as the human genome. The type of vector to be used also has to be considered, which will determine size of fragments that can be cloned. In practice, library size can be calculated quite simply on the basis of the probability of a particular sequence being represented in the library. There is a formula that takes account of all the factors and produces a 'number of clones' value. The formula is:

$$N = \ln(1 - P)/\ln(1 - a/b)$$

where N is the number of clones required, P is the desired probability of a particular sequence being represented (typically set at 0.95 or 0.99), a is the average size of the DNA fragments to be cloned, and b is the size of the genome (expressed in the same units as a).

Table 6.2. Genomic library sizes for various organisms

| Organism | Genome size (kb) | No. clones N, $P = 0.95$ | |
		20 kb inserts	45 kb inserts
Escherichia coli (bacterium)	4.0×10^3	6.0×10^2	2.7×10^2
Saccharomyces cerevisiae (yeast)	1.4×10^4	2.1×10^3	9.3×10^2
Arabidopsis thaliana (simple higher plant)	7.0×10^4	1.1×10^4	4.7×10^3
Drosophila melanogaster (fruit fly)	1.7×10^5	2.5×10^4	1.1×10^4
Stronglyocentrotus purpuratus (sea urchin)	8.6×10^5	1.3×10^5	5.7×10^4
Homo sapiens (human)	3.0×10^6	4.5×10^5	2.0×10^5
Triticum aestivum (hexaploid wheat)	1.7×10^7	2.5×10^6	1.1×10^6

Note: The number of clones (N) required for a probability (P) of 95% that a given sequence is represented in a genomic library is shown for a range of different organisms. Approximate genome sizes of the organisms are given (haploid genome size, if appropriate). Two values of N are shown, for 20 kb inserts (λ replacement vector size) and 45 kb inserts (cosmid vectors). The values should be considered as minimum estimates, as strictly speaking the calculation assumes (1) that the genome size is known accurately, (2) that the DNA is fragmented in a totally random manner for cloning, (3) that each recombinant DNA molecule will give rise to a single clone, (4) that the efficiency of cloning is the same for all fragments, and (5) that diploid organisms are homozygous for all loci. These assumptions are usually not all valid for a given experiment.

By using this formula, it is possible to determine the magnitude of the task ahead and to plan a cloning strategy accordingly. Some genome sizes and their associated library sizes are shown in Table 6.2. These library sizes should be considered as minimum values, as the generation of cloned fragments may not provide a completely random and representative set of clones in the library. Thus, for a human genomic library, we are talking of some 10^6 clones or more in order to be reasonably sure of isolating a particular single-copy gene sequence.

When dealing with this size of library, phage or cosmid vectors are usually essential, as the cloning capacity and efficiency of these vectors is much greater than that of plasmid vectors. Although cosmids, with the potential to clone fragments of up to 47 kb, would seem to be the better choice, λ replacement vectors are often used for library construction. This is because they are easier to use than cosmid vectors, and this outweighs the disadvantage of having only half the cloning capacity. In addition, the techniques for screening phage libraries are now routine and have been well characterised. This is an important consideration, particularly where workers new to the technology wish to use gene manipulation in their research. Alternatively, artificial chromosome vectors such as bacterial artificial

By linking genome size and the desired probability of a gene being isolated, the size of genomic library required can be calculated. However, this should only be used as a guide, as there is always the possibility of any given sequence not being present in the primary library.

A wide range of vectors may be used for cloning genomic DNA, although in practice the choice is determined by the requirements of the procedure and the size of the fragments to be cloned.

chromosomes (BACs) or yeast artificial chromosomes (YACs) may be used to clone large DNA fragments.

6.3.2 Preparation of DNA fragments for cloning

One of the most important aspects of library production is the generation of genomic DNA fragments for cloning. If a λ replacement vector such as EMBL4 is to be used, the maximum cloning capacity will be around 23 kb. Thus, fragments of this size must be available for the production of recombinants. In practice a range of fragment sizes is used, often between 17 and 23 kb for a vector such as EMBL4. At this stage it is important that smaller fragments are not used for ligation into the vector, as there is the possibility of multiple inserts being generated. These can arise by ligation of two (or more) small non-contiguous DNA sequences into the vector. This is obviously undesirable, as any such clones could be isolated and could give false information about the relative locations of particular sequences.

There are two main considerations when preparing DNA fragments for cloning: (1) the molecular weight of the DNA after isolation from the organism and (2) the method used to fragment the DNA. For a completely random library, the starting material should be very-high-molecular-weight DNA, and this should be fragmented by a totally random (*i.e.* sequence-independent) method. Isolation of DNA in excess of 100 kb in length is desirable, and this in itself can pose technical difficulties where the type of source tissue does not permit gentle disruption of cells. In addition, pipetting and mixing solutions of high-molecular-weight DNA can cause shearing of the molecules, and great care must be taken when handling the preparations. The problem of DNA size is even more important when cloning in BAC or YAC vectors, as inserts of around 300 kb may be required. Thus, DNA with an average size of more than 300 kb is needed if a representative set of fragments is to be generated. There is a slight paradox at this stage in that high-molecular-weight DNA is needed, but this is immediately fragmented into smaller pieces! The key point is that this method provides the best set of random fragments. It is of course possible to clone DNA of low average molecular weight, but there is then a greater chance of sequence misrepresentation unless the fragmentation has been entirely random (*e.g.* **mechanical shearing** during DNA isolation).

Assuming that sufficient DNA of 100 kb is available, **random fragmentation** can be carried out. This is usually followed by a size-selection procedure to isolate fragments in the desired range of sizes. Fragmentation can be achieved either by mechanical shearing or by partial digestion with a restriction enzyme. Although mechanical shearing (by forcing the DNA through a syringe needle, or by sonication) will generate random fragments, it will not produce DNA with cohesive termini. Thus, further manipulation such as trimming or filling in the ragged ends of the molecules will be required before the DNA can be joined to the vector, usually with linkers, adaptors, or homopolymer tails (see Fig. 6.1). In practice these additional

Genomic DNA for cloning should be isolated as intact as possible so that generation of fragments suitable for cloning can be controlled.

steps are often considered undesirable, and fragmentation by **partial restriction enzyme digestion** is used extensively in library construction. However, this is not a totally sequence-independent process, as the occurrence of restriction enzyme recognition sites is clearly sequence-dependent. Partial digestion is therefore something of a compromise, but careful design and implementation of the procedure can overcome most of the disadvantages. So how can this be achieved?

If a restriction enzyme is used to digest DNA to completion, the fragment pattern will obviously depend on the precise location of recognition sequences. Therefore, this approach has two drawbacks. First, a six-cutter such as *Eco*RI will have recognition sites on average about once every 4096 bp, which would produce fragments that are too short for λ replacement vectors. Second, any sequence bias, perhaps in the form of repetitive sequences, may skew the distribution of recognition sites for a particular enzyme. Thus, some areas of the genome may contain few sites, whilst others have an overabundance. This means that a complete digest will not be suitable for generating a representative library. If, however, a partial digest is carried out using an enzyme that cuts frequently (*e.g.* a four-cutter such as *Sau*3A, which cuts on average once every 256 bp), the effect is to produce a collection of fragments that are essentially random. This can be achieved by varying the enzyme concentration or the time of digestion, and a test run will produce a set of digests which contain different fragment size distribution profiles, as shown in Fig. 6.8.

> Although not strictly sequence-independent, partial digestion with a frequent-cutting restriction enzyme can generate fragments that are essentially randomly generated.

6.3.3 Ligation, packaging, and amplification of libraries

Having established the optimum conditions for partial digestion, a sample of DNA can be prepared for cloning. After digestion the sample may be fractionated, perhaps using a technique such as density gradient centrifugation or gel electrophoresis. Fragments in the range 17–23 kb can then be selected for ligation. If *Sau*3A (or *Mbo*I, which has the same recognition sequence) has been used as the digesting enzyme, the fragments can be inserted into the *Bam*HI site of a vector such as EMBL4, as the ends generated by these enzymes are complementary (Fig. 6.9). The insert DNA can be treated with phosphatase to reduce self-ligation or concatemer formation, and the vector can be digested with *Bam*HI and *Sal*I to generate the cohesive ends for cloning and to isolate the stuffer fragment and prevent it from re-annealing during ligation. The *Eco*RI site in the vector can be used to excise the insert after cloning. Ligation of DNA into EMBL4 is summarised in Fig. 6.10.

When ligation is carried out, concatemeric recombinant DNA molecules are produced, which are suitable substrates for packaging *in vitro*, as shown in Fig. 6.11. This produces what is known as a **primary library**, which consists of individual recombinant phage particles. Whilst this is theoretically the most useful type of library in terms of isolation of a specific sequence, it is a finite resource.

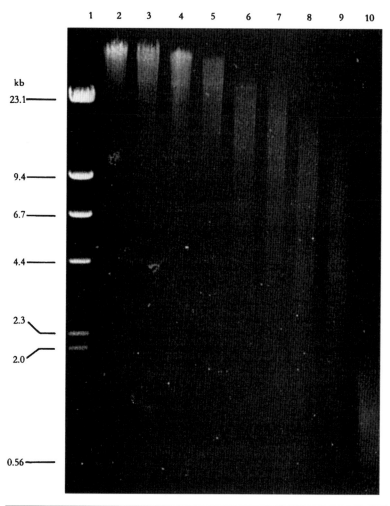

Fig. 6.8 Partial digestion and fractionation of genomic DNA. High-molecular-weight genomic DNA was digested with various concentrations of the restriction enzyme *Sau*3A. Samples from each digest were run on a 0.7% agarose gel and stained with ethidium bromide. Lane 1 shows λ *Hind*III markers, sizes as indicated. Lanes 2–10 show the effects of increasing concentrations of restriction enzyme in the digestions. As the concentration of enzyme is increased, the DNA fragments generated are smaller. From this information the optimum concentration of enzyme to produce fragments of a certain size distribution can be determined. These can then be run on a gel (as here) and isolated prior to cloning. Photograph courtesy of Dr N. Urwin.

Amplification of genomic libraries is often used to generate copies of the original primary clones. The library can then be used for many applications or distributed to colleagues in other laboratories.

Thus, a primary library is produced, screened, and then discarded. If the sequence of interest has not been isolated, more recombinant DNA will have to be produced and packaged. Whilst this may not be a problem, there are occasions where a genomic library may be screened for several different genes or may be sent to different laboratories who may share the resource (this is sometimes called 'cloning-by-phoning'!). In these cases it is therefore necessary to amplify the library. This is achieved by plating the packaged phage on a suitable

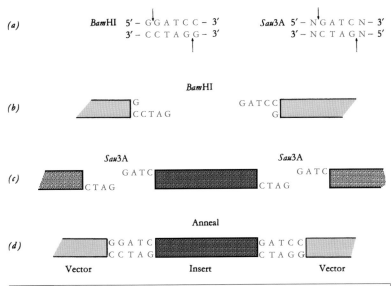

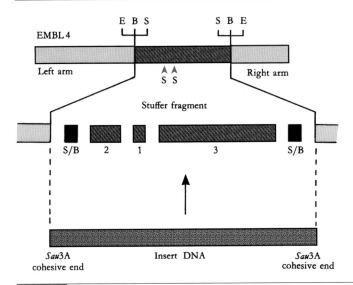

(a)

BamHI 5′ – G↓G A T C C – 3′   Sau3A 5′ – N↓G A T C N – 3′
        3′ – C C T A G G – 3′   3′ – N C T A G N – 5′
                    ↑   ↑

Fig. 6.9 Cloning *Sau*3A fragments into a *Bam*HI site. (*a*) The recognition sequences and cutting sites for *Bam*HI and *Sau*3A. In the *Sau*3A site, N is any base. (*b*) Vector DNA cut with *Bam*HI generates 5′ protruding termini with the four-base sequence 5′-GATC-3′. (*c*) Insert DNA cut with *Sau*3A also generates identical four-base overhangs. (*d*) Thus, DNA cut with *Sau*3A can be annealed to *Bam*HI cohesive ends to generate a recombinant DNA molecule.

Fig. 6.10 Ligation of *Sau*3A-cut DNA into the λ replacement vector EMBL4. Sites on the vector are *Eco*RI (E), *Bam*HI (B), and *Sal*I (S). The vector is cut with *Bam*HI and *Sal*I, which generates five fragments from the stuffer fragment (hatched in top panel). Removal of the very short *Sal*I/*Bam*HI fragments (filled boxes) prevents the stuffer fragment from re-annealing. In addition, the two internal *Sal*I sites cleave the stuffer fragment, producing three *Sal*I/*Sal*I fragments (1 to 3). If desired, the short fragments can be removed from the preparation by precipitation with isopropanol, which leaves the small fragments in the supernatant. On removal of the stuffer, *Sau*3A-digested insert DNA can be ligated into the *Bam*HI site of the vector (see Fig. 6.9).

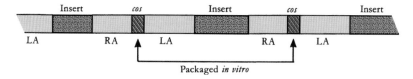

Fig. 6.11 Concatemeric recombinant DNA. On ligation of inserts into a vector such as EMBL4, a concatemer is formed. This consists of the left arm of the vector (LA), the insert DNA, and the right arm (RA). These components of the unit are repeated many times and are linked together at the *cos* sites by the cohesive ends on the vector arms. On packaging *in vitro*, the recombinant genomes are cut at the *cos* sites and packaged into phage heads.

host strain of *E. coli*, and then resuspending the plaques by gently washing the plates with a buffer solution. The resulting phage suspension can be stored almost indefinitely and will provide enough material for many screening and isolation procedures.

Although **amplification** is a useful step in producing stable libraries, it can lead to **skewing** of the library. Some recombinant phage may be lost, perhaps because of the presence of repetitive sequences in the insert, which can give rise to recombinational instability. This can be minimised by plating on a recombination-deficient host strain. Some phage may exhibit differential growth characteristics that may cause particular phage to be either over- or under-represented in the amplified library, and this may mean that a greater number of plaques have to be screened in order to isolate the desired sequence. This is not usually a major problem.

6.4 | Advanced cloning strategies

In Sections 6.2 and 6.3 we considered cDNA and genomic DNA cloning strategies using basic plasmid and phage vectors in *E. coli* hosts. These approaches have proved to be both reliable and widely applicable, and they still represent a major part of the technology of gene manipulation. However, advances made over the past few years have increased the scope (and often the complexity!) of cloning procedures. Such advances include more sophisticated vectors for *E. coli* and other hosts, increased use of expression vectors, and novel approaches to various technical problems, including the extensive use of PCR technology. Some examples of more advanced cloning strategies are discussed next.

6.4.1 Synthesis and cloning of cDNA

An elegant scheme for generating cDNA clones was developed by Hiroto Okayama and Paul Berg in 1982 and illustrates how careful design of the system can alleviate some of the problems that may occur with less sophisticated procedures. In their method the plasmid vector itself is used as the **priming** molecule, and the mRNA is annealed to this for cDNA synthesis. A second adaptor molecule is

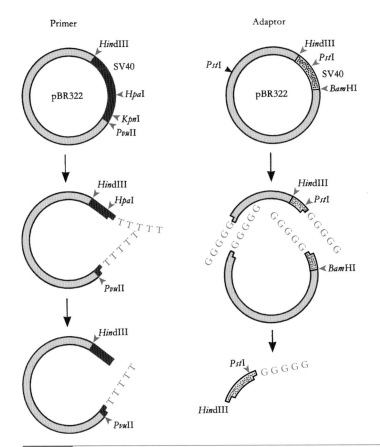

Primer

Adaptor

Fig. 6.12 Preparation of vector and adaptor molecules for the Okayama and Berg cDNA cloning procedure. The vector is made up from pBR322 plus parts of the SV40 genome (solid or shaded in the diagram). For the primer, the vector is cut with *KpnI* and tailed with dT residues. It is then digested with *HpaI* to create a vector in which one end is tailed. The adaptor molecule is generated by cutting the adaptor plasmid with *PstI*, which generates two fragments. These are tailed with dG residues and digested with *HindIII* to produce the adaptor molecule itself, which therefore has a *HindIII* cohesive end in addition to the dG tail. The fragment is purified for use in the protocol (see Fig. 6.13). From Old and Primrose (1989), *Principles of Gene Manipulation*, Blackwell. Reproduced with permission.

required to complete the process. Both adaptor and primer are based on pBR322, with additional sequences from the SV40 virus. Preparation of the vector and adaptor molecules involves restriction digestion, tailing with oligo(dT) and dG, and purification of the fragments to give the molecules shown in Fig. 6.12. The mRNA is then annealed to the plasmid and the first cDNA strand synthesised and tailed with dC. The terminal vector fragment (which is also tailed during this procedure) is removed and the adaptor added to circularise the vector prior to synthesis of the second strand of the cDNA. Second-strand synthesis involves the use of RNase H, DNA polymerase I, and DNA ligase in a strand-replacement reaction that converts the mRNA–cDNA hybrid into ds cDNA and completes the ligation of the ds cDNA into

As in many aspects of life in general, increased development activity often leads to more sophisticated products. Many current cloning protocols demonstrate this, as they are based on original techniques that have been modified and extended to achieve more specialist aims.

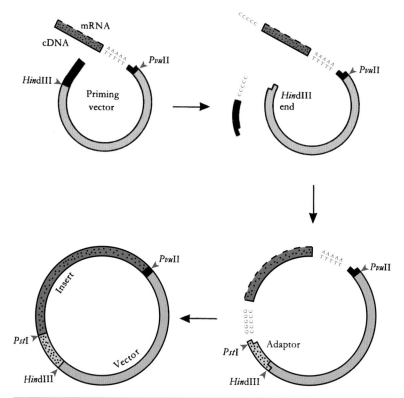

Fig. 6.13 The Okayama and Berg cDNA cloning protocol. The mRNA is annealed to the dT-tailed priming vector *via* the poly(A) tail, and the first cDNA strand synthesised using reverse transcriptase. The cDNA is then tailed with dC residues and the dC-tailed vector fragment removed by digestion with *Hind*III. The cDNA is annealed to the dG-tailed adaptor molecule, which is ligated into the vector using the cohesive *Hind*III ends on the vector and adaptor molecules. Finally the mRNA is displaced and the second cDNA strand synthesised using RNase H and DNA polymerase I to generate the complete vector/insert recombinant. From Old and Primrose (1989), *Principles of Gene Manipulation*, Blackwell. Reproduced with permission.

the vector. The end result is that recombinants are generated in which there is a high proportion of full-length cDNAs. Okayama and Berg's method is summarised in Fig. 6.13.

6.4.2 Expression of cloned DNA molecules

Many of the routine manipulations in gene cloning experiments do not require expression of the cloned DNA. However, there are certain situations in which some degree of genetic expression is needed. A transcript of the cloned sequence may be required for use as a probe, or a protein product (requiring transcription and translation) may be required as part of the screening process used to identify the cloned gene. Another common biotechnological application is where the recombinant DNA is used to produce a protein of commercial value. If eukaryotic DNA sequences are cloned, post-transcriptional and post-translational modifications may be required, and the type of host/vector system that is used is therefore very important in determining whether or not such sequences will be expressed effectively. The problem of RNA processing in prokaryotic host organisms may be obviated by cloning cDNA sequences; this is the most common approach where expression of eukaryotic sequences is desired. In this section we will consider some aspects of cloning cDNAs for expression, concentrating mainly on the characteristics of the vector/insert combinations that enable expression to be achieved. Further discussion of the topic is presented in Chapters 9 and 10.

Assuming that a functional cDNA sequence is available, a suitable host/vector combination must be chosen. The host cell type will usually have been selected by considering aspects such as ease of use, fermentation characteristics, or the ability to secrete proteins derived from cloned DNA. However, for a given host cell, there may be several types of expression vector, including both plasmid- and (for bacteria) phage-based examples. In addition to the normal requirements such as restriction site availability and genetic selection mechanisms, a key feature of expression vectors is the type of **promoter** that is used to direct expression of the cloned sequence. Often the aim will be to maximise the expression of the cloned sequence, so a vector with a highly efficient promoter is chosen. Such promoters are often termed **strong promoters**. However, if the product of the cloned gene is toxic to the cell, a **weak promoter** may be required to avoid cell death due to overexpression of the toxic product.

Promoters are regions with a specific base sequence, to which RNA polymerase will bind. By examining the base sequence lying on the 5′ (upstream) side of the coding regions of many different genes, the types of sequences that are important have been identified. Although there are variations, these sequences all have some similarities. The 'best fit' sequence for a region such as a promoter is known as the **consensus sequence**. In prokaryotes there are two main regions that are important. Some 10 base pairs upstream from the transcription start site (the -10 region, as the T_C start site is numbered $+1$) there is a region known as the **Pribnow box**, which has the consensus sequence 5′-TATAAT-3′. A second important region is located around position -35 and has the consensus sequence 5′-TTGACA-3′. These two regions form the basis of promoter structure in prokaryotic cells, where the precise sequences found in each region determine the strength of the promoter.

Sequences important for transcription initiation in eukaryotes have been identified in much the same way as for prokaryotes. Eukaryotic promoter structure is generally more complex than that found in prokaryotes, and control of transcription initiation can involve sequences (*e.g.* enhancers) that may be several hundreds or thousands of base pairs upstream from the T_C start site. However, there are important motifs closer to the start site. These are a region centred around position -25 with the consensus sequence 5′-TATAAAT-3′ (the

Cloned cDNA can be used to express genes by enabling the synthesis of the protein in a suitable host system. Many expression vectors are now available to facilitate this.

The use of a suitable promoter is a key element in achieving the expression of a gene by means of a cloned cDNA sequence.

Table 6.3. Some promotors that may be used in expression vectors.

Organism	Gene promoter	Induction by
E. coli	lac operon	IPTG
	trp operon	β-indolylacetic acid
	λP$_L$	λ cl protein
A. nidulans	Glucoamylase	Starch
S. cerevisiae	Acid phosphatase	Phosphate depletion
	Alcohol dehydrogenase	Glucose depletion
	Galactose utilisation	Galactose
	Metallothionein	Heavy metals
T. reesei	Cellobiohydrolase	Cellulose
Mouse	Metallothionein	Heavy metals
Human	Heat-shock protein	Temperature >40°C

Note: Some examples of various promoters that can be used in expression vectors are given, with the organism from which the gene promoter is taken. The conditions under which gene expression is induced from such promoters are also given.

Source: Collated from Brown (1990), *Gene Cloning*, Chapman and Hall; and Old and Primrose (1989), *Principles of Gene Manipulation*, Blackwell. Reproduced with permission.

TATA or **Hogness** box) and a sequence in the −75 region with the consensus 5′-GG(T/C)CAATCT-3′, known as the **CAAT box**.

In addition to the strength of the promoter, it may be desirable to regulate the expression of the cloned cDNA by using promoters from genes that are either **inducible** or **repressible**. Thus, some degree of control can be exerted over the transcriptional activity of the promoter; when the cDNA product is required, transcription can be 'switched on' by manipulating the system using an appropriate metabolite. Some examples of promoters used in the construction of expression vectors are given in Table 6.3.

In theory, constructing an expression vector is straightforward once a suitable promoter has been identified. In practice, as is often the case, the process is often highly complex, requiring many manipulations before a functional vector is obtained. The basic vector must carry an origin of replication that is functional in the target host cell, and there may be antibiotic resistance genes or other genetic selection mechanisms present. However, as far as expression of cloned sequences is concerned, it is the arrangement of restriction sites immediately downstream from the promoter that is critical. There must be a unique restriction site for cloning into, and this has to be located in a position where the inserted cDNA sequence can be expressed effectively. This aspect of vector structure is discussed further when the applications of recombinant DNA technology are considered in Part III.

Regulatory control of the expression of cloned genes can be achieved by selecting appropriate vector/host systems. Thus, an inducible system can be 'switched on' if the cells are grown under inducing conditions.

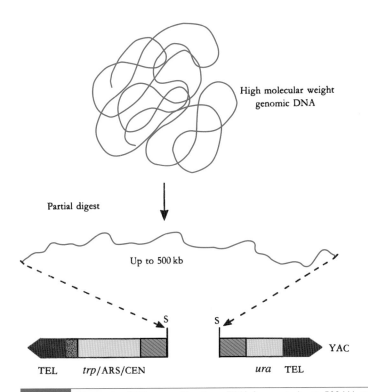

Fig. 6.14 Cloning in a YAC vector. Very large DNA fragments (up to 500 kb) are generated from high-molecular-weight DNA. The fragments are then ligated into a YAC vector (see Fig. 5.11) that has been cut with *Bam*HI and *Sma*I (S). The construct contains the cloned DNA and the essential requirements for a yeast chromosome: telomeres (TEL), an autonomous replication sequence (ARS), and a centromere region (CEN). The *trp* and *ura* genes can be used as dual selectable markers to ensure that only complete artificial chromosomes are maintained. From Kingsman and Kingsman (1988), *Genetic Engineering*, Blackwell. Reproduced with permission.

6.4.3 Cloning large DNA fragments in BAC and YAC vectors

BACs and YACs (see Fig. 5.11) can be used to clone very long pieces of DNA. The use of a BAC or YAC vector can reduce dramatically the number of clones needed to produce a representative genomic library for a particular organism, and this is a desirable outcome in itself. A consequence of cloning large pieces of DNA is that physical mapping of genomes is made simpler, as there are obviously fewer sequences to fit together in the correct order. For large-scale DNA sequencing projects BAC- or YAC-based cloning is useful if an ordered strategy (as opposed to a shotgun approach; see Fig. 3.7) is to be used.

A further advantage of cloning long stretches of DNA stems from the fact that many eukaryotic genes are much larger than the 47 kb or so that can be cloned using cosmid vectors in *E. coli*. Thus, with plasmid, phage, and cosmid vectors it may be impossible to isolate the entire gene. This makes it difficult to determine gene structure

Artificial chromosomes such as YACs or BACs can be used to clone large fragments of DNA (in excess of several hundred kb). This makes the isolation of large genes easier and is also important for large-scale DNA sequencing projects.

without using several different clones, which is not the ideal way to proceed. The use of BAC or YAC vectors can alleviate this problem and can enable the structure of large genes to be determined by providing a single DNA fragment to work from.

Let's consider using a YAC vector to clone DNA fragments. In practice, cloning in YAC vectors is similar to other protocols (Fig. 6.14). The vector is prepared by a double restriction digest, which releases the vector sequence between the telomeres and cleaves the vector at the cloning site. Thus, two arms are produced, as is the case with phage vectors. Insert DNA is prepared as very long fragments (a partial digest with a six-cutter may be used) and ligated into the cloning site to produce **artificial chromosomes**. Selectable markers on each of the two arms ensure that only correctly constructed chromosomes will be selected and propagated.

Concept map 6

Cloning strategies
— need methods for → Joining DNA fragments to vector
— require → Generation of DNA fragments

Joining DNA fragments to vector
— to give → Concatemeric phage/insert DNA
— or — Bimolecular recombinants
— by → Blunt-end ligation / Use of linkers / Use of adaptors / Homopolymer tailing / Ligation of cohesive termini

Blunt-end ligation / Use of linkers / Use of adaptors / Homopolymer tailing / Ligation of cohesive termini
— join to vector by → DNA fragments for cloning
— produces → cDNA library / Genomic library

cDNA library / Genomic library
— use to → Isolate genes

Generation of DNA fragments
— or by → Polymerase chain reaction
— from either → mRNA / Genomic DNA

Direct chemical synthesis — gives → DNA fragments for cloning

mRNA — by synthesis of → cDNA

Genomic DNA — by → Mechanical shearing / Restriction digests

cDNA / Mechanical shearing / Restriction digests — gives → DNA fragments for cloning

Chapter 7 summary

Aims

- To outline the principles of the PCR
- To describe the methodology used for the PCR process
- To illustrate the range of PCR variants
- To outline the applications of the PCR

Chapter summary/learning outcomes

When you have completed this chapter you will have knowledge of:

- The history of the PCR and its importance
- The basic principles of the PCR
- Designing PCR primers
- DNA polymerases for the PCR
- Principles of the standard PCR protocol for amplifying DNA
- PCR using mRNA templates
- Nested, inverse, RAPD, and other variants of the PCR
- Processing of PCR products

Key words

Polymerase chain reaction (PCR), oligonucleotide, DNA templates, DNA polymerase, DNA duplex, melting of the duplex, primer, thermostable, *Thermus aquaticus*, *Taq* polymerase, denature, anneal, thermal cycler, contaminants, quality control procedures, medical applications, forensic applications, primer design, sequence, code degeneracy, wobble position, inosine, primer length, melting temperature (T_m), unique sequence, thermolabile, recombinant *Taq* polymerase, processivity, fidelity, rate of synthesis, half-life, reverse transcriptase PCR (RT-PCR), competitor RT-PCR, nested PCR, external primers, internal (nested) primers, inverse PCR, random amplified polymorphic DNA PCR (RAPD-PCR), arbitrarily primed PCR (AP-PCR), amplicon.

Chapter 7

The polymerase chain reaction

Now and again a scientific discovery is made that changes the whole course of the development of a subject. In the field of molecular biology we can identify several major milestones – the emergence of bacterial genetics, the discovery of the mechanism of DNA replication, the double helix and the genetic code, restriction enzymes, and finally the techniques involved in the generation and analysis of recombinant DNA. Many of these areas of molecular biology have been recognised by the award of the Nobel prize in either Chemistry or in Medicine and Physiology. Some of the key discoveries recognised in this way are listed in Table 7.1.

The topic of this chapter is the **polymerase chain reaction** (**PCR**), which was discovered by Kary Mullis and for which he was awarded the Nobel prize in Chemistry in 1993. The PCR technique produces a similar result to DNA cloning – the selective amplification of a DNA sequence – and has become such an important part of the genetic engineer's toolkit that in many situations it has essentially replaced traditional cloning methodology. In this chapter we will look at some of the techniques and applications of PCR technology.

7.1 | History of the PCR

In addition to requiring someone to provide a spark of genius, major scientific breakthroughs depend on existing knowledge, the techniques available, and often a little luck. In genetics and molecular biology there have been many examples of scientists being in the right place, with the right 'mindset', investigating the right problem, and coming up with a seminal discovery. Gregor Mendel, James Watson, and Francis Crick are three names that stand out; Kary Mullis can legitimately be added to the list. In 1979 he joined the Cetus Corporation, based in Emeryville, California. By this time the essential prerequisites for the development of the PCR had been established. Mullis was working on **oligonucleotide** synthesis, which by the early 1980s had become an automated and somewhat tedious process. Thus, his mind was free to investigate other avenues. In his own words, he

Major scientific breakthroughs require a number of things to come together at the right time, and the discovery of the PCR is a good example of this.

Table 7.1.			**Some milestones in cell and molecular biology recognised by the award of the Nobel prize**

Year	Prize	Recipient(s)	Awarded for studies on
1958	C	Frederick Sanger	Primary structure of proteins
	P/M	Joshua Lederberg	Genetic recombination in bacteria
		George W. Beadle	Gene action
		Edward L. Tatum	
1959	P/M	Arthur Kornberg	Synthesis of DNA and RNA
		Severo Ochoa	
1962	C	John C. Kendrew	3D structure of globular proteins
		Max F. Perutz	
	P/M	Francis H. C. Crick	3D structure of DNA (the double helix)
		James D. Watson	
		Maurice H. F. Wilkins	
1965	P/M	Francois Jacob	Operon theory for bacterial gene expression
		Andre M. Lwoff	
		Jacques L. Monod	
1968	P/M	H. Gobind Khorana	Elucidation of the genetic code and its role
		Marshall W. Nirenberg	in protein synthesis
		Robert W. Holley	
1969	P/M	Max Delbrück	Structure and replication of viruses
		Alfred D. Hershey	
		Salvador E. Luria	
1975	P/M	David Baltimore	Reverse transcriptase and tumour viruses
		Renato Dulbecco	
		Howard M. Temin	
1978	P/M	Werner Arber	Restriction endonucleases
		Daniel Nathans	
		Hamilton O. Smith	
1980	C	Paul Berg	Recombinant DNA technology
		Walter Gilbert	DNA sequencing
		Frederick Sanger	
1982	C	Aaron Klug	Structure of nucleic acid/protein complexes
1984	P/M	George Kohler	Monoclonal antibodies
		César Milstein	
		Niels K. Jerne	Antibody formation
1989	C	Thomas R. Cech	Catalytic RNA
		Sidney Altman	
	P/M	J. Michael Bishop	Genes involved in malignancy
		Harold Varmus	
1993	C	Kary B. Mullis	Polymerase chain reaction
		Michael Smith	Site-directed mutagenesis
	P/M	Richard J. Roberts	Split genes and RNA processing
		Phillip A. Sharp	
1995	P/M	Edward B. Lewis	Genetic control of early embryonic development
		Christiane Nusslein-Volhard	
		Eric F. Wieschaus	

Year	Prize	Recipient(s)	Awarded for studies on
Table 7.1	(cont.)		
1997	C	Paul D. Boyer	Synthesis of ATP
		John E. Walker	
		Jens C. Skou	Sodium/potassium ATPase
2001	P/M	Leland H. Hartwell	Key regulators of the cell cycle
		R. Timothy Hunt	
		Sir Paul M. Nurse	
2002	P/M	Sydney Brenner	Genetic regulation of organ development and
		H. Robert Horvitz	programmed cell death
		John E. Sulston	
2003	C	Peter Agre	Channels in cell membranes
		Roderick MacKinnon	
2004	C	Aaron Ciechanover	Ubiquitin-mediated protein degradation
		Avram Hershko	
		Irwin Rose	
		Roger D. Kornberg	Molecular basis of eukaryotic transcription
2006	P/M	Andrew Z. Fire	RNA interference
		Criag C. Mello	

Note: 'C' and 'P/M' refer to Nobel prizes in Chemistry and Physiology or Medicine, respectively. Note also that Frederick Sanger is part of a very select group of Nobel laureates – he has been awarded *two* Nobel prizes for his work on proteins (1958) and DNA sequencing (1980). Information on past and current Nobel prizewinners can be found at www.nobelprize.org.

found himself 'puttering around with oligonucleotides', and the main thrust of his puttering was to try to develop a modified version of the dideoxy sequencing procedure. His thoughts were therefore occupied with oligonucleotides, **DNA templates**, and **DNA polymerase**.

Late one Friday night in April 1983, Mullis was driving to his cabin with a friend and was thinking about his modified sequencing experiments. He was in fact trying to establish if extension of oligonucleotide primers by DNA polymerase could be used to 'mop up' unwanted dNTPs in the solution, which would otherwise get in the way of his dideoxy experiment. Suddenly he realised that, if *two* primers were involved, and they served to enable extension of the DNA templates, the sequence would effectively be duplicated. Fortunately, he had also been writing computer programs that required reiterative loops – and realised that sequential repetition of his copying reaction (although not what was intended in his experimental system!) could provide many copies of the DNA sequence. Some hasty checking of the figures confirmed that the exponential increase achieved was indeed 2^n, where n is the number of cycles. The PCR had been discovered.

Subsequent work proved that the theory worked when applied to a variety of DNA templates. Mullis presented his work as a poster at the annual Cetus Scientific Meeting in the Spring of 1984. In his account of the discovery of PCR in *Scientific American* (April 1990), he recalls how Joshua Lederberg discussed his results and appeared to react in

The basic premise of the PCR is quite simple – two primers are used so that each strand of the DNA serves as a template; thus, the number of strands of DNA doubles on each cycle of the PCR.

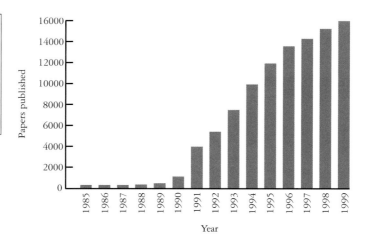

Fig. 7.1 Number of publications citing the polymerase chain reaction from 1985 to 1999. This shows the rapid spread of the use of the PCR in research. From McPherson and Moller (2000), *PCR*, Bios. Reproduced with permission.

a way that was to become familiar – the 'why didn't I think of that' acceptance of a discovery that is brilliant in its simplicity.

Over the past 20 years the PCR technique has been adopted by scientists in a pattern similar to that for recombinant DNA technology itself. The acceptance and use of a procedure can best be demonstrated by looking at the number of published scientific papers in which the technique is used. For the PCR, this is shown in Fig. 7.1. This impressive increase in use confirms the importance of PCR, which is now established as one of the major techniques for gene manipulation and analysis.

7.2 | The methodology of the PCR

As outlined earlier, the PCR is elegantly simple in theory. When a **DNA duplex** is heated, the strands separate or '**melt**'. If the single-stranded sequences can be copied by a DNA polymerase, the original DNA sequence is effectively duplicated. If the process is repeated many times, there is an exponential increase in the number of copies of the starting sequence. The length of the fragment is defined by the 5′ ends of the primers, which helps to ensure that a homogeneous population of DNA molecules is produced. Thus, after relatively few cycles, the target sequence becomes greatly amplified, which generates enough of the sequence for identification and further processing.

7.2.1 The essential features of the PCR
In addition to a DNA sequence for amplification, there are two requirements for PCR. First, a suitable **primer** is required. In practice, two primers are necessary, one for each strand of the duplex. The primers should flank the target sequence, so some sequence information is required if selective amplification is to be achieved. The primers are synthesised as oligonucleotides and are added to the reaction in excess so that each of the primers is always available following the

denaturation step. A second requirement is a suitable form of DNA polymerase. Whilst it is possible to use a standard DNA polymerase (as was done in the early days of the PCR), this is inactivated by the heat denaturation step of the process and, thus, fresh enzyme has to be added after each cycle of operation. The availability of a **thermostable** form of DNA polymerase makes life much easier for the operator. This is purified from the thermophilic bacterium *Thermus aquaticus*, which inhabits hot springs. The use of *Taq* **polymerase** means that the PCR procedure can be automated, as there is no need to add fresh polymerase after each denaturation step. In addition to the DNA, primers, and polymerase, the usual mix of the correct buffer composition and the availability of the four dNTPs is needed to ensure that copying of the DNA strands is not stalled because of inactivation of the enzyme or lack of monomers.

> The key requirements for the PCR are a DNA template, a pair of suitable oligonucleotide primers, and a DNA polymerase.

In operation, the PCR is straightforward. The target DNA and reaction components are usually mixed together at the start of the process, and the tube heated to around 90°C to **denature** the DNA. As the temperature drops, primers will **anneal** to their target sequences on the single-stranded DNA, and *Taq* polymerase will begin to copy the template strands. The cycle is completed (and re-started) by a further denaturation step. The operational sequence is shown in Fig. 7.2.

Automation of the PCR cycle of operations is achieved by using a programmable heating system known as a **thermal cycler**. This takes small microcentrifuge tubes (96-well plates or glass capillaries can also be used) in which the reactants are placed. Thin-walled tubes permit more rapid temperature changes than standard tubes or plates. Various thermal cycling patterns can be set according to the particular reaction conditions required for a given experiment, but in general the cycle of events shown in Fig. 7.2 forms the basis of the amplification stage of the PCR process. Although thermal cyclers are simple devices, they have to provide accurate control of temperature and similar rates of heating and cooling for tubes in different parts of the heating block. More sophisticated devices provide a greater range of control patterns than the simpler versions, such as variable rates of heating and cooling and heated lids to enclose the tubes in a sealed environment. If a heated lid is not used, there is the possibility of evaporation of liquid during the PCR. A layer of mineral or silicone oil on top of the reactants can avoid this, although sometimes contamination of the tube contents with oil can be a problem.

> The PCR can be automated because it is in essence a simple series of defined steps that is repeated many times.

A final practical consideration in setting up a PCR protocol is general housekeeping and manipulation of samples. As the technique is designed to amplify small amounts of DNA, even trace **contaminants** (perhaps from a tube that has been lying open in a lab, or from an ungloved finger) can sometimes ruin an experiment. Thus, the operator needs to be fastidious (or even a little paranoid!) about cleanliness when carrying out PCR. Also, even the aerosols created by pipetting reagents can lead to cross-contamination, so good technique is essential. It is best if a sterile hood or flow cabinet can be set aside for setting up the PCR reactions, with a separate area used for

> Because the PCR is so efficient at making lots of copies of a DNA sequence, great care must be taken to avoid contamination with minute amounts of unwanted nucleic acid, which would be amplified along with the target sequence.

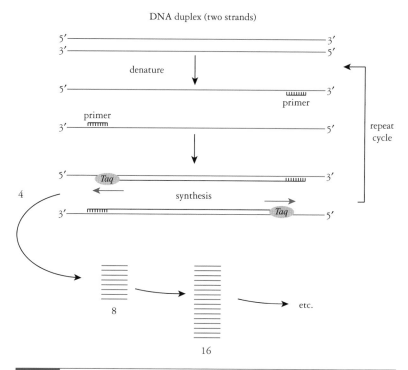

Fig. 7.2 The basic polymerase chain reaction (PCR). Duplex DNA is heat denatured to give single strands, and two oligonucleotide primers are annealed to their complementary sequences on the target DNA. *Taq* polymerase (thermostable) is used to synthesise complementary strands from the template strands by primer extension. The cycle is then repeated by denaturation of the DNA, and the denature/prime/copy programme is repeated many times. A typical temperature profile would be denaturation at 95°C, primer annealing at 55°C, and extension by *Taq* polymerase at 72°C. The numbers refer to the number of DNA strands in the reaction; there are 4 at the end of the first cycle, 8 at the end of the second, 16 at the end of the third, and so on. In theory, after 30 cycles, 5.4×10^9 strands are produced. Thus, DNA sequences can be amplified very quickly.

post-reaction processing. Accurate labelling of tubes (primers, target DNAs, nucleotide mixes, *etc.*) is also needed, and **quality control procedures** are important, particularly where analysis is being carried out in **medical** or **forensic applications**.

7.2.2 The design of primers for PCR

Oligonucleotide primers are available from many commercial sources and can be synthesised to order in a few days. In **primer design**, there are several aspects that have to be considered. Perhaps most obvious is the **sequence** of the primer – more specifically, where does the sequence information come from? It may be derived from amino acid sequence data, in which case the genetic **code degeneracy** has to be considered, as shown in Fig. 7.3. In synthesising the primer, two approaches can be taken. By incorporating a mixture of bases at the **wobble position**, a mixed primer can be made, with the 'correct' sequence represented as a small proportion of the mixture.

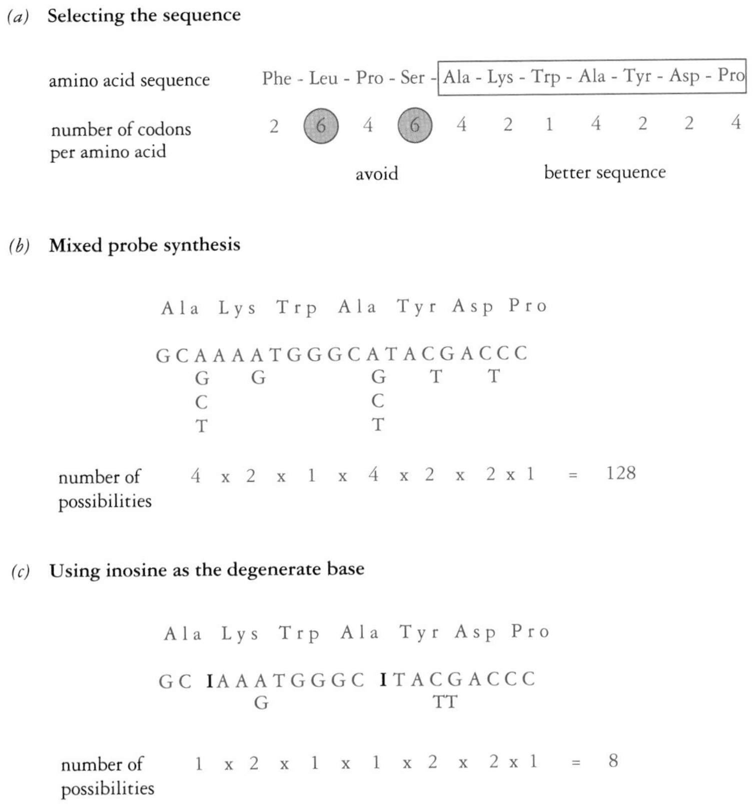

(*a*) Selecting the sequence

amino acid sequence	Phe - Leu - Pro - Ser -	Ala - Lys - Trp - Ala - Tyr - Asp - Pro

number of codons   2   ⑥   4   ⑥   4   2   1   4   2   2   4
per amino acid

              avoid   better sequence

(*b*) Mixed probe synthesis

          Ala Lys Trp Ala Tyr Asp Pro

      G C A A A A T G G G C A T A C G A C C C
          G   G   G   T   T
          C   C
          T   T

number of   4 x 2 x 1 x 4 x 2 x 2 x 1   =   128
possibilities

(*c*) Using inosine as the degenerate base

          Ala Lys Trp Ala Tyr Asp Pro

      G C I A A A T G G G C I T A C G A C C C
          G   T T

number of   1 x 2 x 1 x 1 x 2 x 2 x 1   =   8
possibilities

Fig. 7.3 Primer design. In (*a*) the amino acid sequence and number of codons per amino acid are shown. Those amino acids with 6 codons (circled) are best avoided. The boxed sequence is therefore selected. In (*b*) a mixed probe is synthesised by including the appropriate mixture of dNTPs for each degenerate position. Note that in this example the final degenerate position for proline is not included, giving an oligonucleotide with 20 bases (a 20-mer). There are 128 possible permutations of sequence in the mixture. In (*c*) inosine (I) is used to replace the four-fold degenerate bases, giving eight possible sequences.

Alternatively, the base **inosine** (which pairs equally well with any of the other bases) can be incorporated as the third base in degenerate codons.

If the primer sequence is taken from an already-determined DNA sequence, it may be from the same gene from a different organism or may be from a cloned DNA that has been sequenced during previous experimental work.

Regardless of the source of the sequence information for the primers, there are some general considerations that should be addressed. The **primer length** is important; it should be long enough to ensure stable hybridisation to the target sequence at the required temperature. Although calculation of the **melting temperature (T_m)** can be used to provide information about annealing temperatures, this is often best determined empirically. The primer must also be long enough to ensure that it is a **unique sequence** in the genome

When designing primers for the PCR there are many aspects to be considered. The stringency of primer binding can have a great effect on the number of PCR products in any one amplification.

from which the target DNA is taken. Primer lengths of around 20–30 nucleotides are usually sufficient for most applications. With regard to the base composition and sequence of primers, repetitive sequences should be avoided, and also regions of single-base sequence. Primers should obviously not contain regions of internal complementary sequence or regions of sequence overlap with other primers.

Because extension of PCR products occurs from the 3′ termini of the primer, it is this region that is critical with respect to fidelity and stability of pairing with the target sequence. Some 'looseness' of primer design can be accommodated at the 5′ end, and this can sometimes be used to incorporate design features such as restriction sites at the 5′ end of the primer.

7.2.3 DNA polymerases for PCR

In theory any DNA polymerase would be suitable for use in a PCR procedure. Originally the Klenow fragment of DNA polymerase was used, but this is **thermolabile** and requires addition of fresh enzyme for each extension phase of the cycle. This was inefficient in that the operator had to be present at the machine for the duration of the process, and a lot of enzyme was needed. Also, because extension was carried out at 37°C, primers could bind to non-target regions, generating a high background of non-specific amplified products. The availability of *Taq* polymerase solved these problems. Today, a wide variety of thermostable polymerases is available for PCR, sold under licence from Hoffman LaRoche (who holds the patent rights for the use of *Taq* polymerase in PCR). These include several versions of **recombinant *Taq* polymerase**, as well as enzymes from *Thermus flavus* (*Tfl* polymerase) and *Thermus thermophilus* (*Tth* polymerase). In addtion to *Thermus*-derived polymerases, a thermostable DNA polymerase from *Pyrococcus furiosus* is available from Stratagene. This organism is classed as a hyperthermophile, with an optimal growth temperature of 100°C, and is found in deep-sea vents.

> Thermostable DNA polymerases such as *Taq* polymerase have enabled automation of the PCR process without the need for adding fresh polymerase after the denaturation stage of each cycle.

The key features required for a DNA polymerase include **processivity** (affinity for the template, which determines the number of bases incorporated before dissociation), **fidelity** of incorporation, **rate of synthesis**, and the **half-life** of the enzyme at different temperatures. In theory the variations in these aspects shown by different enzymes should make choice of a polymerase a difficult one; in reality, a particular source is chosen and conditions adjusted empirically to optimise the activity of the enzyme.

Of the features just mentioned, fidelity of incorporation of nucleotides is perhaps the most critical. Obviously an error-prone enzyme will generate mutated versions of the target sequence out of proportion to the basal rate of misincorporation, given the repetitive cycling nature of the reaction. In theory, an error rate of 1 in 10^4 in a million-fold amplification would produce mutant sequences in around one-third of the products. Thus, steps often need to be taken to identify and avoid such mutated sequences in cases where high-fidelity copying is essential.

> Any errors introduced during polymerase copying of DNA strands will become amplified to significant levels over a typical number of PCR cycles.

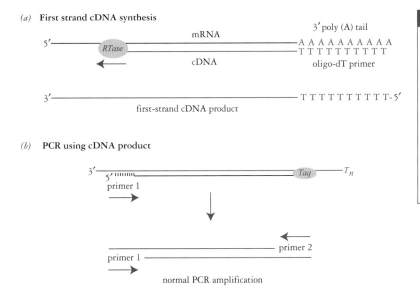

(a) **First strand cDNA synthesis**

mRNA

3′ poly (A) tail

5′————————————————————————A A A A A A A A A A
RTase
←——————— cDNA ————————T T T T T T T T T T
oligo-dT primer

3′————————————————————————T T T T T T T T T T-5′
first-strand cDNA product

(b) **PCR using cDNA product**

3′————————————————————————*Taq*——*T*$_n$
5′ ‖‖‖‖‖
primer 1
——→

primer 2
←——
———————————————————————————
primer 1 ———————————————————————————
——→
normal PCR amplification

Fig. 7.4 RT-PCR. (*a*) Reverse transcriptase is used to synthesise a cDNA copy of the mRNA. In this example oligo(dT)-primed synthesis is shown. (*b*) The cDNA product is amplified using gene-specific primers. The initial PCR synthesis will copy the cDNA to give a duplex molecule, which is then amplified in the usual way. In many kits available for RT-PCR the entire procedure can be carried out in a single tube.

7.3 | More exotic PCR techniques

As the PCR technique became established, variations of the basic procedure were developed. This is still an area of active development, and new techniques and applications for PCR appear regularly. As we have already seen for other aspects of recombinant DNA technology, there are many kit-based PCR protocols for both standard and more sophisticated PCR applications. In this section we will look at some of the variations on the basic PCR process.

7.3.1 PCR using mRNA templates

Using mRNA as the template is a common variant of the basic PCR, known as **reverse transcriptase PCR** (**RT-PCR**). It can be useful in determining low levels of gene expression by analysing the PCR product of cDNA prepared from the mRNA transcript. In theory a single mRNA molecule can be amplified, but this is unlikely to be achieved (or required) in practice. The process involves copying the mRNA using reverse transcriptase, as in a standard cDNA synthesis. Oligo(dT)-primed synthesis is often used to generate the first-strand cDNA. PCR primers are then used as normal, although the first few cycles may be biased in favour of copying the cDNA single-stranded product until enough copies of the second strand have been generated to allow exponential amplification from both primers. This has no effect on the final outcome of the process. An overview of RT-PCR is shown in Fig. 7.4.

One use of RT-PCR is in determining the amount of mRNA in a sample (**competitor RT-PCR**). A differing but known amount of competitor RNA is added to a series of reactions, and the target and competitor amplified using the same primer pair. If the target and competitor products are of different sizes, they can be separated on a gel

There are many variants of the PCR protocol; one of the most important is RT-PCR for amplification of cDNA derived from mRNA samples.

(a) Spike RNA samples and convert to cDNA

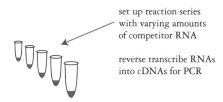

set up reaction series
with varying amounts
of competitor RNA

reverse transcribe RNAs
into cDNAs for PCR

(b) Perform PCR using same primer pair

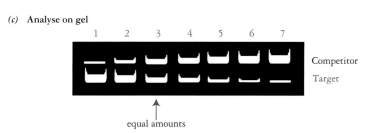

(c) Analyse on gel

equal amounts

Fig. 7.5 Competitor RT-PCR. (*a*) A series of samples is spiked with increasing amounts of a competitor RNA that will produce a PCR product that is different in size to the prospective target fragment. (*b*) The competitor RNA 'competes' with the target RNA for the primers and other resources in the reaction during PCR. (*c*) When analysed on an electrophoresis gel, the products can be distinguished on the basis of size. In this case the equal band intensities in lane 3 (boxed) enable the amount of target mRNA in the original sample to be determined, as this is the point where the competitor and target were present at equal concentrations at the start of the PCR process.

and the amount of target estimated by comparing with the amount of competitor product. When the two bands are of equal intensity, the amount of target sequence in the original sample is the same as the amount of competitor added. This approach is shown in Fig. 7.5.

7.3.2 Nested PCR

Nested PCR is a useful way of overcoming some of the problems associated with a large number of PCR cycles, which can lead to error-prone synthesis. The technique essentially increases both the sensitivity and the fidelity of the basic PCR protocol. It involves using two sets of primers. The first **external** set generates a normal PCR product. Primers that lie inside the first set are then used for a second PCR reaction. These **internal** or **nested** primers generate a shorter product, as shown in Fig. 7.6.

(a) First PCR using external primers

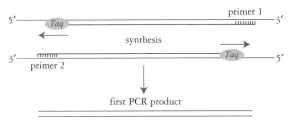

(b) Second PCR using internal (nested) primers

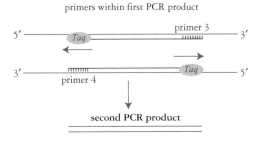

Fig. 7.6 Nested PCR. (*a*) Standard PCR amplification of a fragment using a primer pair. (*b*) In the second PCR, a set of primers that lie inside the first pair is used to prime synthesis of a shorter fragment. Nested PCR can be used to increase the specificity and fidelity of the PCR procedure.

7.3.3 Inverse PCR

Often a stretch of DNA sequence is known, but the desired target sequence lies outside this region. This causes problems with primer design, as there may be no way of determining a suitable primer sequence for the unknown region. **Inverse PCR (IPCR)** involves isolating a restriction fragment that contains the known sequence plus flanking sequences. By circularising the fragment, and then cutting inside the known sequence, the fragment is effectively inverted. Primers can then be synthesised using the known sequence data and used to amplify the fragment, which will contain the flanking regions. Primers that face away from each other (with respect to direction of product synthesis) in the original known sequence are required so that, on circularisation, they are in the correct orientation. The technique can also be used with sets of primers for nested PCR. Deciphering the result usually requires DNA sequencing to determine the areas if interest. An illustration of IPCR is shown in Fig. 7.7.

7.3.4 RAPD and several other acronyms

Normally the aim of PCR is to generate defined fragments from highly specific primers. However, there are some techniques based on low-stringency annealing of primers. The most widely used of these is **RAPD-PCR**. This stands for **random amplified polymorphic DNA PCR**. The technique is also known as **arbitrarily primed PCR (AP-PCR)**. It is a useful method for genomic profiling and involves using a single primer at low stringency. The primer binds to many sites in the genome, and fragments are amplified from these. The stringency of primer binding can be increased after a few cycles, which effectively

Although high-fidelity copying is usually the desired outcome of a PCR experiment, there are some procedures such as RAPD where low stringency of primer binding enables a DNA profile (made up of a number of amplified fragments) to be generated.

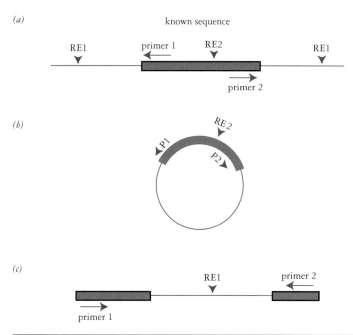

Fig. 7.7 Inverse PCR. (a) A region of DNA in which part of the sequence is known (shaded). If the areas of interest lie outside the known region, inverse PCR can be used to amplify these flanking regions. Primers are selected to bind within the known sequence in the opposite orientation to that normally required. Restriction sites are noted as RE1 and RE2. If RE1 is used to cut the DNA, the resulting fragment can be circularised to give the construct shown in (b). If this is cut with RE2, the linear fragment that results has the unknown sequence in the middle of two known sequence fragments. Note that the unknown region is non-contiguous; it is made up of two flanking regions joined at RE1. The PCR product will therefore contain these two regions plus the regions of known sequence.

means that only the 'best mismatches' are fully amplified. By careful design the protocol can yield genome-specific band patterns that can be used for comparative analysis. However, there can be problems associated with the reproducibility of the technique, as it is difficult to obtain similar levels of primer binding in different experiments. This is largely due to the mismatched binding of primers at low stringency that is the basis of the technique.

There are many other variants of the PCR protocol, which have given rise (as we have already seen) to many new PCR acronyms. Some of these are listed in Table 7.2. These variants will not be discussed further in this book; the reader is directed to the texts in the Suggestions for Further Reading for further information on the increasingly novel and complex procedures and applications associated with PCR.

7.4 | Processing of PCR products

Once the PCR process has been completed, the DNA fragments that have been amplified (which are often called **amplicons**) can be

Acronym	Technique name	Application
Table 7.2.	Some variants of the basic PCR process	
RT-PCR	Reverse transcriptase PCR	PCR from mRNA templates
IPCR	Inverse PCR	PCR of sequences lying outside primer binding sites
RAPD-PCR	Random amplified polymorphic DNA PCR	Genomic fingerprinting under low-stringency conditions
AP-PCR	Arbitrarily primed PCR	As RAPD-PCR
TAIL-PCR	Thermal asymmetric interlaced PCR	PCR using alternating high/low primer binding stringency with arbitrary and sequence-specific primers
mrPCR	Multiplex restriction site PCR	PCR using primers with restriction site recognition sequences at their 3′ ends
AFLP	Amplified fragment length polymorphism	Genome analysis by PCR of restriction digests of genomic DNA
CAPS	Cleaved amplified polymorphic sequence analysis	Genome analysis for single nucleotide polymorphisms (SNPs) by PCR and restriction enzyme digestion
GAWTS	Gene amplification with transcript sequencing	PCR coupled with transcript synthesis and sequencing using reverse transcriptase
RAWTS	RNA amplification with transcript sequencing	Variant of GAWTS using mRNA templates
RACE	Rapid amplification of cDNA ends	Isolation of cDNAs from low-abundance mRNAs

analysed. Gel electrophoresis of PCR products is usually the first step in many post-reaction processes. This can verify the fragment size and can also give some idea of its purity and homogeneity. A typical PCR result is shown in Fig. 7.8. The fate of PCR products depends on the experiment – often a simple visualisation of the product on an electrophoresis gel, with confirmation of the length of the sequence, will be sufficient. This may be coupled with blotting and hybridisation techniques to identify specific regions of the sequence. Perhaps the PCR product is to be cloned into an expression vector, in which case specific vectors can be used. In one type of vector the addition of a single base onto the ends of the amplified sequence is required to enable the product to be inserted into the vector. Other cloning vectors for PCR products utilise blunt-end ligation to facilitate cloning. In this case the ends of the amplicons are usually filled in or 'polished'

If the products of a PCR amplification are to be cloned, specific vectors and processes are available to ensure that high cloning efficiency is achieved.

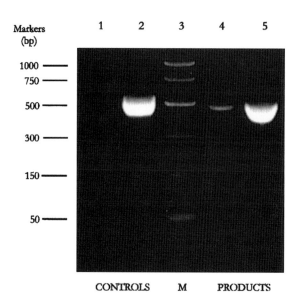

Fig. 7.8 Visualisation of PCR products of ornithine decarboxylase on an agarose gel. Lane 1, negative control (no DNA); lane 2, positive control (cloned ornithine decarboxylase fragment, 460 bp); lane 3, PCR size markers; lanes 4 and 5, PCR product using rat liver genomic DNA and the ornithine decarboxylase primers used in lane 2. Lane 4 shows product after 15 cycles, lane 5 after 30 cycles of PCR. Photograph courtesy of Drs F. McKenzie and E. Kelso.

to ensure maximum efficiency for the process. Often DNA sequencing is carried out on the PCR product to ensure that the 'correct' sequence has been cloned, which is particularly important when the potential error rate is considered (see Section 7.2.3).

7.5 | Applications of the PCR

This chapter started with a rather bold pronouncement – 'Now and again a scientific discovery is made that changes the whole course of the development of a subject.' In the case of the PCR and its impact, this statement is undoubtedly true. The applications of PCR technology are many and diverse, and new procedures are developed on a regular basis. Sometimes these involve the development of a more sophisticated form of PCR-based technology that has wide application potential. At other times it may be that an existing technique is used in a novel way, or on an organism that has not been studied in detail.

The PCR can be used to clone specific sequences, although in many cases it is in fact not necessary to do this, as enough material for subsequent manipulations may be produced by the PCR process itself. It can be used to clone genes from one organism by using priming sequences from another if some sequence data are available for the gene in question. Another use of the PCR process is in forensic and diagnostic procedures such as the examination of body fluid stains, or in antenatal screening for genetic disorders. These areas are particularly important in the wider context of the applications of gene technology and will be discussed further in Part III.

There are many diverse applications of the PCR technique, which has had a major impact on the whole area of molecular biology in general and recombinant DNA technology in particular.

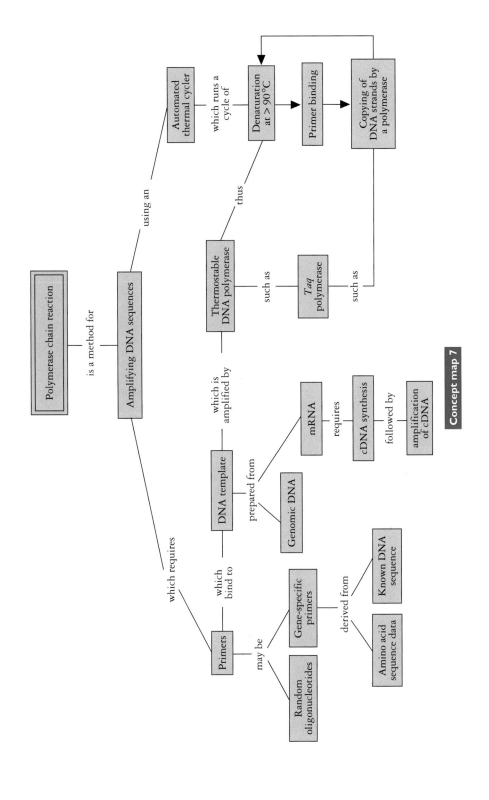

Concept map 7

Chapter 8 summary

Aims

- To define the terms selection and screening with reference to clone anaysis
- To describe a range of genetic selection and screening methods
- To illustrate the use of nucleic acid probes in screening clone banks
- To outline the techniques used for analysis of cloned DNA fragments

Chapter summary/learning outcomes

When you have completed this chapter you will have knowledge of:

- The definition of the terms selection and screening
- Techniques for selection of clones carrying DNA inserts
- The range of methods available for screening clone banks
- The use of chromogenic substrates in screening techniques
- Screening using nucleic acid hybridisation
- The use of PCR in screening procedures
- Immunological screening for expressed genes
- Analysis of cloned DNA fragments using mRNA hybridization methods
- Blotting techniques – Southern, Northern, and Western blots
- DNA sequencing for clone analysis

Key words

Selection, screening, automation, microtitre plate, antibiotic, pBR322, ampicillin, tetracycline, chromogenic substrate, X-gal, β-galactosidase, IPTG, *lacZ* gene, α-peptide, insertional inactivation, plaque morphology, λ *cI* gene, cI repressor, complementation, mutant, auxotrophic, hybridisation, probe, homology, homologous, heterologous, plus/minus screening, chromosome walking, chromosome jumping, *in silico*, stringency, polymerase chain reaction, combinatorial screening, immunological screening, antigen, polyclonal antibodies, monoclonal antibodies, hybrid-arrest translation, hybrid-release translation, hybrid-select translation, SDS-polyacrylamide gel electrophoresis, restriction map, Southern blotting, Northern blotting, dot blotting, slot blotting, Western blotting, DNA sequencing.

Chapter 8

Selection, screening, and analysis of recombinants

Success in any cloning experiment depends on being able to identify the desired gene sequence among the many different recombinants that may be produced. Given that a large genomic library may
contain a million or more cloned sequences, which are not readily
distinguishable from each other by simple analytical methods, it is
clear that identification of the target gene is potentially the most
difficult part of the cloning process. Fortunately, there are several
selection/identification methods that can be used to overcome most
of the problems that arise.

In this chapter, the various techniques that can be used to identify cloned genes will be described. We will look at some examples
that illustrate the principles of gene identification and characterisation. As has been the case in previous chapters, we will include some
examples that have been an important part of the development of
the technology, although they are perhaps not so widely used today.

There are two terms that require definition before we proceed:
selection and **screening**. Selection is where some sort of pressure
(*e.g.* the presence of an antibiotic) is applied during the growth of
host cells containing recombinant DNA. The cells with the desired
characteristics are therefore selected by their ability to survive. This
approach ranges in sophistication, from simple selection for the presence of a vector up to direct selection of cloned genes by complementation of defined mutations. Screening is a procedure by which
a population of viable cells is subjected to some sort of analysis that
enables the desired sequences to be identified. Because only a small
proportion of the large number of bacterial colonies or bacteriophage
plaques being screened will contain the DNA sequence(s) of interest,
screening requires methods that are highly sensitive and specific. In
practice, both selection and screening methods may be required in
any single experiment and may even be used at the same time if the
procedure is designed carefully.

In recent years the **automation** of various procedures in gene
manipulation has become more widespread, particularly in large research groups and biotechnology companies. In smaller labs, with perhaps only a few research projects at different stages of development,

Identifying a clone can involve
using some sort of selective
mechanism (e.g. antibiotic or
β-galactosidase status) and/or
screening a population of clones
with a specific probe (e.g. a
radioactive cDNA or
oligonucleotide).

there may not be the same requirement for high-efficiency through-put as there might be in a commercially based company. Robotic systems can be used to carry out many routine tasks, often in 96-well **microtitre plates**. In many cases the technological advances in sample preparation, processing, and handling mean that large numbers of clones can be handled much more easily than was the case just a few years ago, and many protocols are now developed specifically for use with automated procedures.

8.1 Genetic selection and screening methods

Genetic selection and screening methods rely on the expression (or non-expression) of certain traits. Usually these traits are encoded by the vector, or perhaps by the desired cloned sequence if a direct selection method is available.

One of the simplest genetic selection methods involves the use of **antibiotics** to select for the presence of vector molecules. For example, the plasmid **pBR322** contains genes for **ampicillin** resistance (Ap^r) and **tetracycline** resistance (Tc^r). Thus, the presence of the plasmid in cells can be detected by plating potential transformants on an agar medium that contains either (or both) of these antibiotics. Only cells that have taken up the plasmid will be resistant, and these cells will therefore grow in the presence of the antibiotic. The technique can also be used to identify mammalian cells containing vectors with selectable markers.

Genetic selection methods can be simple (as just described) or complex, depending on the characteristics of the vector/insert combination and on the type of host strain used. Such methods are extremely powerful, and there is a wide variety of genetic selection and screening techniques available for many diverse applications. Some of these are described in the following subsections.

Antibiotic resistance marker genes provide a simple and reliable way to select for the presence of vectors in cells.

8.1.1 The use of chromogenic substrates

The use of **chromogenic substrates** in genetic screening methods has been an important aspect of the development of the technology. The most popular system uses the compound **X-gal** (5-bromo-4-chloro-3-indolyl-β-D-galactopyranoside), which is a colourless substrate for **β-galactosidase**. The enzyme is normally synthesised by *E. coli* cells when lactose becomes available. However, induction can also occur if a lactose analogue such as **IPTG** (*iso*-propyl-thiogalactoside) is used. This has the advantage of being an inducer without being a substrate for β-galactosidase. On cleavage of X-gal a blue-coloured product is formed (Fig. 8.1); thus, the expression of the *lacZ* (β-galactosidase) **gene** can be detected easily. This can be used either as a screening method for cells or plaques or as a system for the detection of tissue-specific gene expression in transgenics.

The X-gal detection system can be used where a functional β-galactosidase gene is present in the host/vector system. This can occur

The β-galactosidase/X-gal system has become a popular method of indentifying cloned DNA fragments on the basis of an easy-to-read blue/white colour difference.

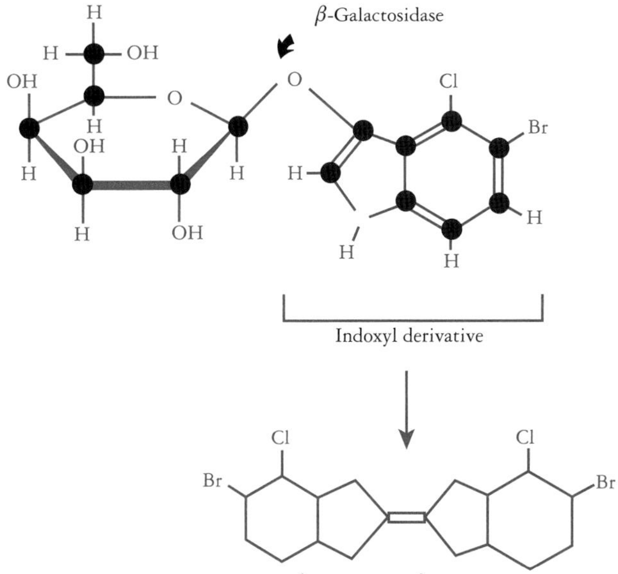

Fig. 8.1 Structure of X-gal and cleavage by β-galactosidase. The colourless compound X-gal (5-bromo-4-chloro-3-indolyl-β-D-galactopyranoside) is cleaved by β-galactosidase to give galactose and an indoxyl derivative. This derivative is in turn oxidised in air to generate the dibromo–dichloro derivative, which is blue.

in two ways. First, an intact β-galactosidase gene (*lacZ*) may be present in the vector, as is the case for the λ insertion vector Charon 16A (see Fig. 5.8). Host cells that are Lac⁻ are used for propagation of the phage, so that the Lac⁺ phenotype will only arise when the vector is present. A second approach is to employ the α-complementation system, in which only part of the *lacZ* gene (encoding a peptide called the **α-peptide**) is carried by the vector. The remaining part of the gene sequence is carried by the host cell. The region coding for the smaller, vector-encoded α-peptide is designated *lacZ′*. Host cells are therefore designated *lacZ′⁻*. Blue colonies or plaques will only be produced when the host and vector fragments complement each other to produce functional β-galactosidase.

The α-peptide of β-galactosidase can be used in a vector to enable a more sophisticated form of X-gal blue/white identification to be achieved by complementation.

8.1.2 Insertional inactivation

The presence of cloned DNA fragments can be detected if the insert interrupts the coding sequence of a gene. This approach is known as **insertional inactivation** and can be used with any suitable genetic system. Three systems will be described to illustrate the use of the technique.

Antibiotic resistance can be used as an insertional inactivation system if DNA fragments are cloned into a restriction site within an anitbiotic-resistance gene. For example, cloning DNA into the *Pst*I site of pBR322 (which lies within the Apʳ gene) interrupts the coding sequence of the gene and renders it non-functional. Thus, cells that harbour a recombinant plasmid will be ApˢTcʳ. This can be used to identify recombinants as follows: if transformants are plated first onto a tetracycline-containing medium, all cells that contain the

Antibiotic resistance status can be used to identify a vector carrying a cloned insert. If the insert is cloned into the coding sequence of an antibiotic resistance gene, the gene will be inactivated and the cell will be sensitive to the antibiotic. This is a very powerful method of selecting cells for further analysis.

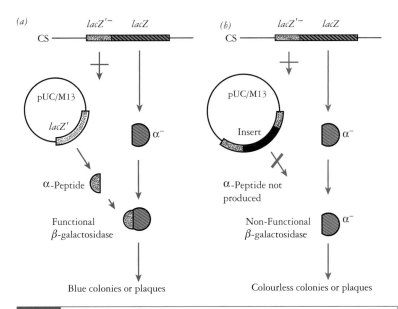

Fig. 8.2 Insertional inactivation in the α-complementation system. (a) The chromosome (CS) has a defective *lacZ* gene that does not encode the N-terminal α-peptide of β-galactosidase (specified by the *lacZ'⁻* gene fragment). Thus, the product of the chromosomal *lacZ* region is an enzyme lacking the α-peptide (α⁻, hatched). If a non-recombinant pUC plasmid or M13 phage is present in the cell, the *lacZ'* gene fragment encodes the α-peptide, which enables functional β-galactosidase to be produced. In the presence of X-gal, blue colonies or plaques will appear. If a DNA fragment is cloned into the vector, as shown in (b), the *lacZ'* gene is inactivated and no complementation occurs; thus, colonies or plaques will not appear blue.

plasmid will survive and form colonies. If a replica of the plate is then taken and grown on ampicillin-containing medium, the recombinants (Ap^s Tc^r) will not grow, but any non-recombinant transformants (Ap^r Tc^r) will. Thus, recombinants are identified by their absence from the replica plate and can be picked from the original plate and used for further analysis.

The X-gal system can also be used as a screen for cloned sequences. If a DNA fragment is cloned into a functional β-galactosidase gene (*e.g.* into the *Eco*RI site of Charon 16A), any recombinants will be genotypically *lacZ⁻* and will therefore not produce β-galactosidase in the presence of IPTG and X-gal. Plaques containing such phage will therefore remain colourless. Non-recombinant phage will retain a functional *lacZ* gene and, therefore, give rise to blue plaques. This approach can also be used with the α-complementation system; in this case the insert DNA inactivates the *lacZ'* region in vectors such as the M13 phage and pUC plasmid series. Thus, complementation will not occur in recombinants, which will be phenotypically Lac⁻ and will therefore give rise to colourless plaques or colonies (Fig. 8.2).

Plaque morphology can also be used as a screening method for certain λ vectors such as λgt10, which contain the *cI* **gene**. This gene encodes the **cI repressor**, which is responsible for the formation of lysogens. Plaques derived from cI^+ vectors will be slightly turbid because of the survival of some cells that have become lysogens. If the cI gene is inactivated by cloning a fragment into a restriction site within the gene, the plaques are clear and can be distinguished from the turbid non-recombinants. This system can also be used as a selection method (see Section 8.1.4).

8.1.3 Complementation of defined mutations

Direct selection of cloned sequences is possible in some cases. An example is where antibiotic resistance genes are cloned, as the presence of cloned sequences can be detected by plating cells on a medium that contains the antibiotic in question (assuming that the host strain is normally sensitive to the antibiotic). The method is also useful where specific mutant cells are available, as the technique of **complementation** can be employed, where the cloned DNA provides the function that is absent from the mutant. There are three requirements for this approach to be successful. First, a **mutant** strain that is deficient in the particular gene that is being sought must be available. Second, a suitable selection medium is required on which the specific recombinants will grow. The final requirement, which is often the limiting step as far as this method is concerned, is that the gene sequence must be expressed in the host cell to give a functional product that will complement the mutation. This is not a problem if, for example, an E. coli strain is used to select cloned E. coli genes, as the cloned sequences will obviously function in the host cells. This approach has been used most often to select genes that specify nutritional requirements, such as enzymes of the various biosynthetic pathways. Thus, genes of the tryptophan operon can be selected by plating recombinants on mutant cells that lack specific functions in this pathway (these are known as **auxotrophic** mutants or just auxotrophs). In some cases, complementation in E. coli can be used to select genes from other organisms, such as yeast, if the enzymes are similar in terms of their function and they are expressed in the host cell. Complementation can also be used if mutants are available for other host cells, as is the case for yeast and other fungi.

> Direct complementation of a defined mutation may be feasible if the gene system is well known and an appropriate mutant is available. In this case the defective gene function would be 'supplied' by the cloned sequence.

> There are certain constraints on direct selection methods; some are more difficult to overcome than others.

Selection processes can also be used with higher eukaryotic cells. The gene for mouse dihydrofolate reductase (DHFR) has been cloned by selection in E. coli using the drug trimethoprim in the selection medium. Cells containing the mouse DHFR gene were resistant to the drug and were, therefore, selected on this basis.

8.1.4 Other genetic selection methods

Although the methods just outlined represent some of the ways by which genetic selection and screening can be used to detect the presence of recombinants, there are many other examples of such techniques. These are often dependent on the use of a particular vector/host combination, which enables exploitation of the genetic characteristics of the system. Two examples will be used to illustrate this approach; many others can be found in some of the texts noted in Suggestions for Further Reading.

The use of the cI repressor system of λgt10 can be extended to provide a powerful selection system if the vector is plated on a mutant strain of E. coli that produces lysogens at a high frequency. Such strains are designated hfl (high frequency of lysogeny), and any phage that encodes a functional cI repressor will form lysogens on these hosts. These lysogens will be immune to further infection by phage. DNA fragments are inserted into the λgt10 vector at a restriction site in the cI gene. This inactivates the gene and, thus, only recombinants (genotypically cI^-) will form plaques.

A second example of genetic selection based on phage/host characteristics is the Spi selection system that can be used with vectors such as EMBL4. Wild-type λ will not grow on cells that already carry a phage, such as phage P2, in the lysogenic state. Thus, the λ phage is said to be Spi^+ (sensitive to P2 inhibition). The Spi^+ phenotype is dependent on the red and gam genes of λ, and these are arranged so that they are present on the stuffer fragment of EMBL4. Thus, recombinants, which lack the stuffer fragment, will be red^- gam^- and will therefore be phenotypically Spi^-. Such recombinants will form plaques on P2 lysogens, whereas non-recombinant phage that are red^+ gam^+ will retain the Spi^+ phenotype and will not form plaques.

Some selection/screening methods are quite sophisticated and make use of key aspects of the host/vector interaction.

8.2 | Screening using nucleic acid hybridisation

General aspects of nucleic acid **hybridisation** are described in Section 3.5. It is a very powerful method of screening clone banks and is one of the key techniques in gene manipulation. The production of a cDNA or genomic DNA library is often termed the 'shotgun' approach, as a large number of essentially random recombinants is generated. By using a defined nucleic acid **probe**, such libraries can be screened and the clone(s) of interest identified. The conditions for hybridisation are now well established, and the only limitation to the method is the availability of a suitable probe.

The complementary nature of DNA strands means that a sequence-specific probe is a very powerful screening tool, as it can be used to identify the target sequence clones among thousands or hundreds of thousands of other recombinants.

8.2.1   Nucleic acid probes

The power of nucleic acid hybridisation lies in the fact that complementary sequences will bind to each other with a very high degree of fidelity (see Fig. 2.9). In practice this depends on the degree of **homology** between the hybridising sequences, and usually the aim is to use a probe that has been derived from the same source as the target DNA. However, under certain conditions, sequences that are not 100% **homologous** can be used to screen for a particular gene, as may be the case if a probe from one organism is used to detect clones prepared using DNA from a second organism. Such **heterologous** probes have been extremely useful in identifying many genes from different sources.

Homologous or heterologous probes may be used in screening protocols, if genes have sequences similar enough to enable hybridisation to be stable under relatively high stringency conditions.

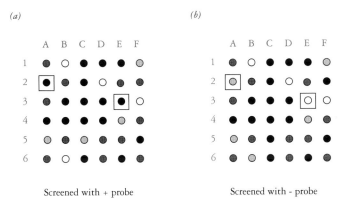

(a) *(b)*

Screened with + probe   Screened with - probe

Fig. 8.3 Screening by the 'plus/minus' method. In this example clones have been spotted on duplicate filters in a 6 × 6 array. (*a*) One copy of the filter has been hybridised with a radioactive cDNA probe prepared from cells expressing the gene of interest (the + probe). (*b*) The other copy of the filter has been hybridised with a probe prepared from cells that are not expressing the gene (the − probe). Some clones hybridise weakly with both probes (e.g. B1), some strongly with both (e.g. C1). Some show a strong signal with the − probe and a weaker signal with the + probe (e.g. C5). Clones A2 and E3 (boxed) show a strong signal with the + probe and a weaker signal with the − probe. These would be selected for further analysis.

There are three main types of DNA probe: (1) cDNA, (2) genomic DNA, and (3) oligonucleotides. Alternatively, RNA probes can be used if these are suitable. The availability of a particular probe will depend on what is known about the target gene sequence. If a cDNA clone has already been obtained and identified, the cDNA can be used to screen a genomic library and isolate the gene sequence itself. Alternatively, cDNA may be made from mRNA populations and used without cloning the cDNAs. This is often used in what is known as **'plus/minus' screening**. If the clone of interest contains a sequence that is expressed only under certain conditions, probes may be made from mRNA populations from cells that are expressing the gene (the plus probe) and from cells that are not expressing the gene (the minus probe). By carrying out duplicate hybridisations, the clones can be identified by their different patterns of hybridisation with the plus and minus probes. Although this method cannot usually provide a definitive identification of a particular sequence, it can be useful in narrowing down the range of candidate clones. The principle of the plus/minus method is shown in Fig. 8.3.

Genomic DNA probes are usually fragments of cloned sequences that are used either as heterologous probes or to identify other clones that contain additional parts of the gene in question. This is an important part of the techniques known as **chromosome walking** and **chromosome jumping** and can enable the identification of overlapping sequences which, when pieced together, enable long stretches of DNA to be characterised. We will look at chromosome walking and jumping in more detail in Chapter 12 when we look at medical applications of gene manipulation.

Oligonucleotides can be used as probes if designed carefully, taking the degeneracy of the genetic code into account.

The use of oligonucleotide probes is possible where some amino acid sequence data are available for the protein encoded by the target gene. Using the genetic code, the likely gene sequence can be derived and an oligonucleotide synthesised. The degenerate nature of the genetic code means that it is not possible to predict the sequence with complete accuracy, but this is not usually a major problem if the least degenerate sequence is used. As we saw when looking at primer design for PCR (Section 7.2.2), a mixed probe can be synthesised that covers all the possible sequences by varying the base combinations at degenerate wobble positions. Alternatively, inosine can be used in highly degenerate parts of the sequence. The great advantage of oligonucleotide probes is that only a short stretch of sequence is required for the probe to be useful and, thus, genes for which clones are not already available can be identified by sequencing peptide fragments and constructing probes accordingly.

The availability of genome sequences means that computer-based searches can be used to find particular genes.

A major development in clone identification procedures has appeared hand-in-hand with genome sequencing projects. Assuming that the genome sequence for your target organism is available, a computer can be used to search for any particular sequence in the genome. This type of 'experiment' has become established as a very powerful method, and is sometimes referred to as doing experiments '*in silico*' (as in a silicon chip in the computer, rather than *in vivo* or *in vitro*). In some cases this might even remove the need for cloning at all, as the information might be used to prepare PCR primers so that the target fragment could then be amplified specifially. For a more conventional approach, the sequence data could be used to prepare an oligonucleotide probe to isolate the cloned fragment from a cDNA or genomic DNA library.

When a suitable probe has been obtained, it can be labelled with a radioactive isotope such as ^{32}P (as described in Section 3.4). This produces a radiolabelled fragment of high specific activity that can be used as an extremely sensitive screen for the gene of interest. Alternatively, non-radioactive labelling methods such as fluorescent tags may be used if desired.

8.2.2 Screening clone banks

Colonies or plaques are not suitable for direct screening, so a replica is made on either nitrocellulose or nylon filters. This can be done either by growing cells directly on the filter on an agar plate (colonies) or by 'lifting' a replica from a plate (colonies or plaques). To do this, the recombinants are grown and a filter is placed on the surface of the agar plate. Some of the cells/plaques will stick to the filter, which therefore becomes a mirror image of the pattern of recombinants on the plate (Fig. 8.4). Reference marks are made so that the filters can be orientated correctly after hybridisation. The filters are then processed to denature the DNA in the samples, bind this to the filter, and remove most of the cell debris.

Clone banks can be screened by producing a copy of the colonies (or plaques) on a nitrocellulose or nylon filter. This is then used to identify cloned sequences of interest using a specific nucleic acid probe.

The probe is denatured (usually by heating), the membrane filter is placed in a sealed plastic bag (or a plastic tube with capped ends),

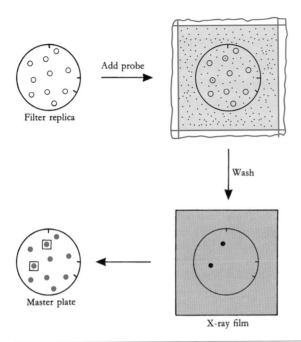

Add probe

Wash

Master plate

X-ray film

Fig. 8.4 Screening clone banks by nucleic acid hybridisation. A nitrocellulose or nylon filter replica of the master petri dish containing colonies or plaques is made. Reference marks are made on the filter and the plate to assist with correct orientation. The filter is incubated with a labelled probe, which hybridises to the target sequences. Excess or non-specifically bound probe is washed off and the filter exposed to X-ray film to produce an autoradiograph. Positive colonies (boxed) are identified and can be picked from the master plate.

and the probe is added and incubated at a suitable temperature to allow hybrids to form. The **stringency** of hybridisation is important and depends on conditions such as salt concentration and temperature. For homologous probes under standard conditions incubation is usually around 65–68°C; the incubation time may be up to 48 h in some cases, depending on the predicted kinetics of hybridisation. After hybridisation, the filters are washed (again the stringency of washing is important) and allowed to dry. They are then exposed to X-ray film to produce an autoradiograph, which can be compared with the original plates to enable identification of the desired recombinant.

An important factor in screening genomic libraries by nucleic acid hybridisation is the number of plaques that can be screened on each filter. Often an initial high-density screen is performed, and the plaques picked from the plate. Because of the high plaque density, it is often not possible to avoid contamination by surrounding plaques. Thus, the mixture is re-screened at a much lower plaque density, which enables isolation of a single recombinant (Fig. 8.5). This approach can be important if a large number of plaques has to be screened, as it cuts down the number of filters (and hence the amount of radioactive probe) required.

The conditions for nucleic acid hybridisation have to be controlled carefully if optimum results are to be obtained. Key aspects include salt concentration and temperature.

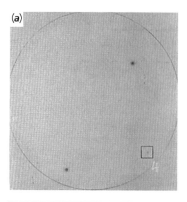
(a)

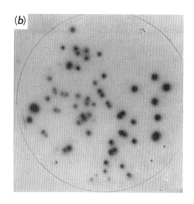
(b)

Fig. 8.5 Screening plaques at high and low densities. A radiolabelled probe was used to screen a genomic library in the λ vector EMBL3. (a) Initial screening was at a high density of plaques, which identified two positive plaques on this plate. The boxed area shows a false positive. (b) The plaques were picked from the positive areas and re-screened at a lower density to enable isolation of individual plaques. Many more positives are obtained because of the high proportion of 'target' plaques in the re-screened sample. Photograph courtesy of Dr M. Stronach.

8.3 | Use of the PCR in screening protocols

PCR can be used for screening clone libraries. Often clones are pooled from the rows and columns of microtitre plates and a PCR carried out on the pooled samples.

As we have already seen in Chapter 7, the **polymerase chain reaction** (PCR) has had a profound effect on gene manipulation and molecular biology. The PCR can be used as a method for screening clone banks, although there are many cases in which the more traditional approach to library screening may be more efficient. Although PCRs are easy to set up, it is still a major task if several hundreds (or thousands!) of clones have to be screened. There are, however, ways around this problem, and because the PCR is so specific (assuming that a suitable primer pair is available), it can be a useful way to identify clones.

One way of reducing the number of individual PCRs that have to be performed is to carry out PCRs on pooled samples of clones. This is known as **combinatorial screening**, and at first the idea can seem a little strange – why pool clones if the aim is to isolate only one? The basis of the technique is shown in Fig. 8.6, in which we consider clones that have been grown up in a single 96-well microtitre tray. As a library is likely to be much larger than 96 clones, the method can be used with many trays, with a vertical dimension added to the horizontal pooling of rows and columns in a single tray. Clones in the rows and columns of the plate are pooled and a PCR performed on the mixed samples. If a clone is present in a 'row' pool, it must be in one of the 12 wells in that row. Similarly, if in a 'column' pool, the clone must be in that column. If a single positive is obtained, the clone is identified at the intersection of the row and column PCRs. If more than one positive is obtained, additional PCRs will need to be carried out to enable unequivocal identification of the clone.

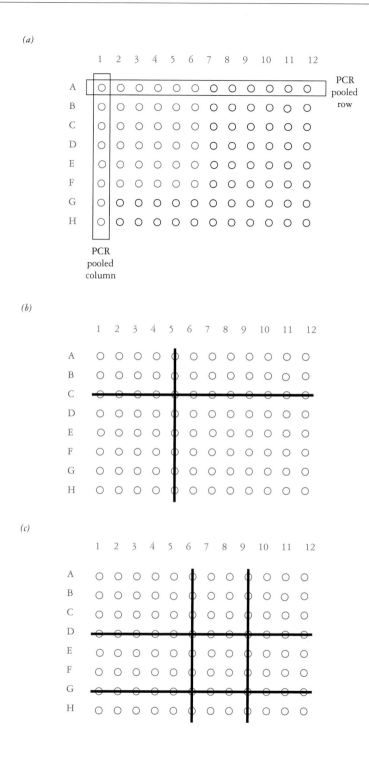

Fig. 8.6 The principle of combinatorial screening using PCR in a single 96-well microtitre plate. (a) Pooling of row and column clones. Row A clones (A1–A12) are pooled and a single PCR is performed. This is also done for column 1 clones (A1–H1). A single PCR is carried out for each pooled row and column; thus, 20 PCRs are needed for a 96-well plate (8 rows and 12 columns). (b) A result is shown where column 5 and row C PCRs show a positive result; thus, a positive clone must be in well C5. (c) A different result, with positives in rows D and G and columns 6 and 9. In this case four clones (D5, D9, G5, and G9) would be re-tested to determine which were positives. The technique can be extended 'vertically' by adding more plates in a stacked arrangement and pooling in all three dimensions (row, column, and vertical). The overall aim is to reduce the number of PCRs that need to be performed to achieve a specific identification.

8.4 | Immunological screening for expressed genes

An alternative to screening with nucleic acid probes is to identify the protein product of a cloned gene by **immunological screening**. The technique requires that the protein is expressed in recombinants, and is most often used for screening cDNA expression libraries that have been constructed in vectors such as λgt11. Instead of a nucleic acid probe, a specific antibody is used.

Antibodies are produced by animals in response to challenge with an **antigen**, which is normally a purified protein. There are two main types of antibody preparation that can be used. The most common are **polyclonal antibodies**, which are usually raised in rabbits by injecting the antigen and removing a blood sample after the immune response has occurred. The immunoglobulin fraction of the serum is purified and used as the antibody preparation (antiserum). Polyclonal antisera contain antibodies that recognise all the antigenic determinants of the antigen. A more specific antibody can be obtained by preparing **monoclonal antibodies**, which recognise a single antigenic determinant. However, this can be a disadvantage in some cases. In addition, monoclonal antibody production is a complicated technique in its own right, and good-quality polyclonal antisera are often sufficient for screening purposes.

There are various methods available for immunological screening, but the technique is most often used in a similar way to 'plaque lift' screening with nucleic acid probes (Fig. 8.7). Recombinant λgt11 cDNA clones will express cloned sequences as β-galactosidase fusion proteins, assuming that the sequence is present in the correct orientation and reading frame. The proteins can be picked up onto nitrocellulose filters and probed with the antibody. Detection can be carried out by a variety of methods, most of which use a non-specific second binding molecule such as protein A from bacteria, or a second antibody, which attaches to the specifically bound primary antibody. Detection may be by radioactive label (^{125}I-labelled protein A or second antibody) or by non-radioactive methods, which produce a coloured product.

> Antibody screening enables the specific identification of proteins that are synthesised from expression libraries. This technique can be used when a specific antibody is available.

8.5 | Analysis of cloned genes

Once clones have been identified by techniques such as hybridisation or immunological screening, more detailed characterisation of the DNA can begin. There are many ways of tackling this, and the choice of approach will depend on what is already known about the gene in question, and on the ultimate aims of the experiment.

8.5.1 Characterisation based on mRNA translation *in vitro*

In some cases the identity of a particular clone may require confirmation. This is particularly true when the plus/minus method of

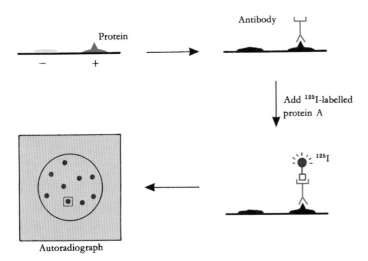

Fig. 8.7 Immunological screening for expressed genes. A filter is taken from a Petri dish containing the recombinants (usually cDNA/λ constructs). Protein and cell debris adhere to the filter. Plaques expressing the target protein (+) are indistinguishable from the others (−) at this stage. The filter is incubated with a primary antibody that is specific for the target protein. This is then complexed with radiolabelled protein A, and an autoradiograph is prepared. As in nucleic acid screening, positive plaques can be identified and picked from the master plate. Chromogenic/enzymatic detection methods may also be used with this type of system.

screening has been used, as the results of such a process are usually somewhat ambiguous and cannot give a definitive identification. If the desired sequence codes for a protein, and the protein has been characterised, it is possible to identify the protein product by two methods based on translation of mRNA *in vitro*. The two methods are known as **hybrid-arrest translation** (HART) and **hybrid-release translation** (HRT, sometimes called **hybrid-select translation**). Although these techniques are no longer widely used in gene analysis, they do illustrate how a particular problem can be approached by two different variations of a similar theme – a central part of good scientific method. A comparison of HART and HRT is shown in Fig. 8.8.

Both HART and HRT rely on hybridising cloned DNA fragments to mRNA prepared from the cell or tissue type from which the clones have been derived. In hybrid arrest, the cloned sequence blocks the mRNA and prevents its translation when placed in a system containing all the components of the translational machinery. In hybrid release, the cloned sequence is immobilised and used to select the clone-specific mRNA from the total mRNA preparation. This is then released from the hybrid and translated *in vitro*. If a radioactive amino acid (usually [35S]methionine) is incorporated into the translation mixture, the proteins synthesised from the mRNA(s) will be labelled and can be detected by autoradiography or fluorography after **SDS-polyacrylamide gel electrophoresis**. In hybrid arrest, one protein band should be absent, whilst in hybrid release there should be a

Matching the cloned sequence with the protein that it encodes can sometimes be useful in clone identification. Two classic methods are 'hybrid-arrest' and 'hybrid-release' translation.

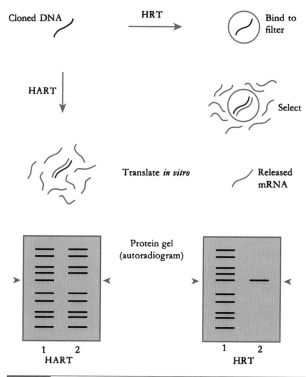

Fig. 8.8 Hybrid-arrest and hybrid-release translation to identify the protein product of a cloned fragment. In hybrid arrest (HART) the cloned fragment is mixed with a preparation of total mRNA. The hybrid formed effectively prevents translation of the mRNA to which the cloned DNA is complementary. After translation *in vitro*, the protein products of the translation are separated on a polyacrylamide gel. The patterns of the control (lane 1, HART gel) and test (lane 2, HART gel) translations differ by one band because of the absence of the protein encoded by the mRNA that has hybridised with the DNA. In hybrid release (HRT) the cloned DNA is bound to a filter and used to select its complementary mRNA from total mRNA. After washing to remove unbound mRNAs and releasing the specifically bound mRNA from the filter, translation *in vitro* generates a single band (lane 2, HRT gel) as opposed to the multiple bands of the control (lane 1, HRT gel). In both cases the identity of the protein (and hence the gene) can be determined by examination of the protein gels. The protein band of interest is arrowed.

single band. Thus, hybrid release gives a cleaner result than hybrid arrest and is the preferred method.

8.5.2 Restriction mapping

Obtaining a **restriction map** for cloned fragments is usually essential before additional manipulations can be carried out. This is particularly important where phage or cosmid vectors have been used to clone relatively large pieces of DNA. If a restriction map is available, smaller fragments can be isolated and used for various procedures, including subcloning into other vectors, the preparation of probes for chromosome walking, and DNA sequencing. If a cloning protocol has involved the use of bacterial or yeast artificial chromosome vectors, with inserts in the size range of 300–500 kb, then so-called long-range restriction mapping may be needed initially.

The basic principle of restriction mapping is outlined in Section 4.1.3. In practice, the cloned DNA is usually cut with a variety of restriction enzymes to determine the number of fragments produced by each enzyme. If an enzyme cuts the fragment at frequent intervals it will be difficult to decipher the restriction map, so enzymes with multiple cutting sites are best avoided. Enzymes that cut the DNA into two to four pieces are usually chosen for initial experiments. By performing a series of single and multiple digests with a range of enzymes, the complete restriction map can be pieced together. This provides the essential information required for more detailed characterisation of the cloned fragment.

> Generating a restriction map of a cloned DNA fragment is usually one of the first tasks in analysing the clone.

8.5.3 Blotting techniques

Although a clone may have been identified and its restriction map determined, this information in itself does not provide much of an insight into the fine structure of the cloned fragment and the gene that it contains. Ultimately the aim may be to obtain the gene sequence (see Section 8.5.4), but it may be that it is not appropriate to begin sequencing straight away. If, for example, a 20 kb fragment of genomic DNA has been cloned in a λ replacement vector, and the area of interest is only 2 kb in length, then much effort would be wasted by sequencing the entire clone. In many experiments it is therefore essential to determine which parts of the original clone contain the regions of interest. This can be done by using a variety of methods based on blotting nucleic acid molecules onto membranes and hybridising with specific probes. Such an approach is in some ways an extension of clone identification by colony or plaque hybridisation, with the refinement that information about the structure of the clone is obtained.

> Blotting techniques, when used with restriction enzyme digestion and gel electrophoresis, can be used to identify particular regions of a gene in a cloned fragment of DNA.

The first blotting technique was developed by Ed Southern, and is eponymously known as **Southern blotting**. In this method fragments of DNA, generated by restriction digestion, are subjected to agarose gel electrophoresis. The separated fragments are then transferred to a nitrocellulose or nylon membrane by a 'blotting' technique. The original method used capillary blotting, as shown in Fig. 8.9. Although other methods such as vacuum blotting and electroblotting have been devised, the original method is still used extensively because it is simple and inexpensive. Blots are often set up with whatever is at hand, and precarious-looking versions of the blotting apparatus are a common sight in many laboratories.

When the fragments have been transferred from the gel and bound to the filter, it becomes a replica of the gel. The filter can then be hybridised with a radioactive probe in a similar way to colony or plaque filters. As with all hybridisation, the key is the availability of a suitable probe. After hybridisation and washing, the filter is exposed to X-ray film and an autoradiograph prepared, which provides

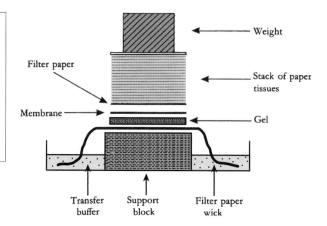

Fig. 8.9 Blotting apparatus. The gel is placed on a filter paper wick and a nitrocellulose or nylon filter placed on top. Further sheets of filter paper and paper tissues complete the setup. Transfer buffer is drawn through the gel by capillary action, and the nucleic acid fragments are transferred out of the gel and onto the membrane.

information on the structure of the clone. An example of the use of Southern blotting in clone characterisation is shown in Fig. 8.10.

Although Southern blotting is a very simple technique, it has many applications and has been an invaluable method in gene analysis. The same technique can also be used with RNA, as opposed to DNA, and in this case it is known as **Northern blotting**. It is most useful in determining hybridisation patterns in mRNA samples and can be used to determine which regions of a cloned DNA fragment

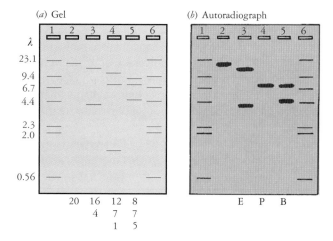

Fig. 8.10 Southern blotting. A hypothetical 20 kb fragment from a genomic clone is under investigation. A cDNA copy of the mRNA is available for use as a probe. (a) Gel pattern of fragments produced by digestion with various restriction enzymes; (b) autoradiograph resulting from the hybridisation. Lanes 1 and 6 contain λ HindIII markers, sizes as indicated. These have been marked on the autoradiograph for reference. The intact fragment (lane 2) runs as a single band to which the probe hybridises. Lanes 3, 4, and 5 were digested with EcoRI (E), PstI (P), and BamHI (B). Fragment sizes are indicated under each lane in (a). The results of the autoradiography show that the probe hybridises to two bands in the EcoRI and BamHI digests; therefore, the clone must have internal sites for these enzymes. The PstI digest shows hybridisation to the 7 kb fragment only. This might, therefore, be a good candidate for subcloning, as the gene may be located entirely on this fragment.

will hybridise to a particular mRNA. However, it is more often used as a method of measuring transcript levels during expression of a particular gene.

There are further variations on the blotting theme. If nucleic acid samples are not subjected to electrophoresis but are spotted onto the filters, hybridisation can be carried out as for Northern and Southern blots. This technique is known as **dot blotting** and is particularly useful in obtaining quantitative data in the study of gene expression. In some variants of the apparatus the nucleic acid is applied in a slot rather than as a dot. Not surprisingly, this is called **slot blotting**! The final technique is known as **Western blotting**, which involves the transfer of electrophoretically separated protein molecules to membranes. Often used with SDS-PAGE (polyacrylamide gel electrophoresis under denaturing conditions), Western blotting can identify proteins specifically if the appropriate antibody is available. The membrane with the proteins fixed in position is probed with the antibody to detect the target protein in a similar way to immunological screening of plaque lifts from expression libraries. Sometimes Western blots can be useful to measure the amount of a particular protein in the cell at any given time. By comparing with other data (such as amount of mRNA, and/or enzyme activity) it is possible to build up a picture of how the expression and metabolic control of the protein is regulated.

Blotting without separation of sequences can be a useful technique for determining the amount of a specific sequence in a sample – often this is used to measure transcript levels in gene expression studies.

8.5.4 DNA sequencing

The development of rapid methods for **DNA sequencing**, as outlined in Section 3.7, has meant that this task has now become routine practice in most laboratories where cloning is carried out. Sequencing a gene provides much useful information about coding sequences, control regions, and other features such as intervening sequences. Thus, full characterisation of a gene will inevitably involve sequencing, and a suitable strategy must be devised to enable this to be achieved most efficiently. The complexity of a sequencing strategy depends on a number of factors, the main one being the length of the fragment that is to be sequenced. Most manual sequencing methods enable about 300–400 bases to be read from a sequencing gel. If the DNA is only a few hundred base pairs long, it can probably be sequenced in a single step. However, it is more likely that the sequence will be several kilobase pairs in length; thus, sequencing is more complex.

There are basically two ways of tackling large sequencing projects, as outlined in Fig. 3.7. Either a random or 'shotgun' approach is used, or an ordered strategy is devised in which the location of each fragment is known prior to sequencing. In the shotgun method, random fragments are produced and sequenced. Assembly of the complete sequence relies on there being sufficient overlap between the sequenced fragments to enable computer matching of sequences from the raw data.

Determining the sequence of a cloned fragment is usually essential if gene structure is being studied. Sequencing can range from a small-scale project involving a single gene up to sequencing entire genomes.

An ordered sequencing strategy is usually more efficient than a random fragment approach. There are several possible ways of generating defined fragments for sequencing. Examples include (1) isolation and subcloning of defined restriction fragments and (2) generation of a series of subclones in which the target sequence has been progressively deleted by nucleases. If defined restriction fragments are used, the first requirement is for a detailed restriction map of the original clone. Using this, suitably sized fragments can be identified and subcloned into a sequencing vector such as M13 or pBluescript. Each subclone is then sequenced, usually by the dideoxy method (see Section 3.7.4). Both strands of the DNA should be sequenced independently, so that any anomalies can be spotted and re-sequenced if necessary. The complete sequence is then assembled by using a suitable computer software package. This is made easier if overlapping fragments have been isolated for subcloning, as the regions of overlap enable adjoining sequences to be identified easily.

By devising a suitable strategy and paying careful attention to detail, it is possible to derive accurate sequence data from most cloned fragments. The task of sequencing a long stretch of DNA is not trivial, but it is now such an integral part of gene manipulation technology that most gene cloning projects involve sequence determination at some point. The technology has improved greatly to the point where entire genomes are being sequenced, which has ushered in a whole new era of molecular genetics. This aspect of sequencing will be discussed further in Chapter 10.

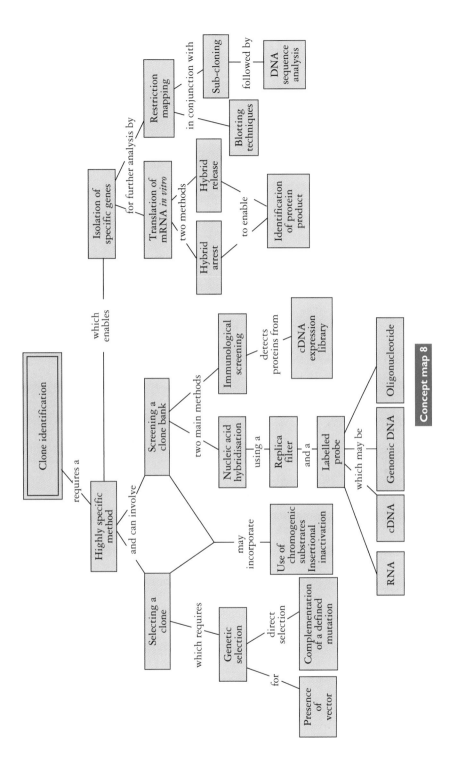

Concept map 8

Chapter 9 summary

Aims

- To define and outline the scope of bioinformatics
- To describe the use of computers in bioinformatics
- To illustrate the generation and use of biological data sets
- To outline the impact of bioinformatics on the biosciences

Chapter summary/learning outcomes

When you have completed this chapter you will have knowledge of:

- The definition of the term bioinformatics
- The range of disciplines in bioinformatics and the scope of the subject
- The use of computers in analysing biological data
- The generation and use of biological data sets
- Nucleic acid databases
- Protein databases
- Using bioinformatics as a tool for research

Key words

Bioinformatics, computer science, molecular biology, World Wide Web, uniform resource locator (URL), information technology, biological data sets, *in vivo*, *in vitro*, *in silico*, sequence data for nucleic acids and proteins, data warehouse, data mining, text mining, computer hardware, software, co-ordinated database management, organise, collate, annotate, primary database, secondary database, bacteriophage phi-X174, bacteriophage lambda, European Molecular Biology Laboratory (EMBL), GenBank, DNA Data Bank of Japan (DDBJ), International Nucleotide Sequence Database Collaboration (INSDC), EMBL-Bank, European Bioinformatics Institute (EBI), National Center for Biotechnology Information (NCBI), National Institute of Genetics, megabase, gigabase, Edman degradation, mass spectrometry, UniProt, Protein Information Resource (PIR), SwissProt, UniParc, Uniprot KB, GIGO effect, single-pass sequence data, finished sequence data, File Transfer Protocol (FTP), Ensembl, Vega, Entrez Mapview.

Chapter 9

Bioinformatics

In this final chapter of Part II, we will look at the topic of **bioinformatics**, which is emerging as a major discipline in its own right, representing as it does the interface between **computer science** and **molecular biology**. In writing an introduction to bioinformatics, there were two difficulties that appeared. The first (as is often the case) was to decide how much detail is needed to gain an insight into the topic. The second was slightly different, and perhaps gives a clue to how the field is developing. Because of the nature of the discipline, much of the information on bioinformatics is to be found on the **World Wide Web** (WWW). This applies not just to supporting information but also to the actual core data on which the subject is based. So how can we best deal with this without simply listing a series of website addresses (known as **uniform resource locators** or **URLs**)?

In trying to answer this I have tried to stress some of the most important central concepts and procedures rather than deal with a range of topics in great detail. There are many excellent texts and websites that provide detailed information about bioinformatics, and these can be very useful once an initial knowledge of the subject has been gained. Inevitably there are lists of URLs, but I have tried to keep these to some of the key sites that enable the interested reader to get started (see Table 9.1 for examples). The best way to learn more about the subject is to dive in and see the extent to which the use of data presentation and management tools has transformed biological databases in terms of level of complexity and ease of use (relatively speaking!). In this chapter we will look at the scope of the subject, and in Part III we will see how bioinformatics is having a major impact on many areas of bioscience when we look at some of the applications of gene manipulation technology.

Bioinformatics involves the use of computers to store, organise, and interpret biological information, usually in the form of sequence data. The World Wide Web is used extensively to share information with the global scientific community.

9.1 | What is bioinformatics?

Bioinformatics can be described in simple terms as the use of **information technology** (**IT**) for the analysis of **biological data sets**; it links the areas of bioscience and computer science, although it can be

Table 9.1. | Useful 'gateway' websites for bioinformatics

Site/page	URL
European Bioinformatics Institute (EBI)	http://www.ebi.ac.uk/
EBI education site	http://www.ebi.ac.uk/2can/
The Wellcome Trust Sanger Institute	http://www.sanger.ac.uk/
National Center for Biotechnology Information (NCBI)	http://www.ncbi.nlm.nih.gov/
NCBI education site	http://www.ncbi.nlm.nih.gov/Education/
Swiss Institute of Bioinformatics (SIB)	http://www.isb-sib.ch/
The proteomics server (Expert Protein Analysis System) of SIB	http://www.expasy.org/
The Bioinformatics Organisation	http://www.bioinformatics.org/
Bios Scientific Publishers Ltd.	http://www.bios.co.uk/inbioinformatics/

Note: For those new to the topic, the education sites of EBI and NCBI are good places to start. The Bioinformatics Organisation is a user-group community site and offers some useful introductory information. The Bios website includes a useful list of links to a wide range of resources. All websites provide extensive links to other relevant sites.

difficult to define the limits of the subject. This is fairly typical of emerging interdisciplinary subjects, as they by definition have no long-established history and generally push the boundaries of research in the topic. When two rapidly developing subjects such as bioscience and computer science are involved, there can be excitement and confusion in equal measure at times!

Although it can be difficult to define what bioinformatics *is*, it is relatively easy to define what bioinformatics is *not* – it is not simply using computers to look at sequence data. The IT requirements are of course central to and critical for the development of the discipline, but in fact bioinformatics has very quickly become established as a major branch of modern bioscience, with a range of sophisticated tools for analysing genes and proteins *in vivo*, *in vitro*, and *in silico*. The field of bioinformatics has broadened considerably over the past few years, with novel interactive and predictive applications emerging to supplement the original function of data storage and analysis. I have attempted to summarise the key elements of bioinformatics in Fig. 9.1.

> The use of information technology to store and organise sequence data is co-ordinated by an international network of centres of excellence in bioinformatics, each of which has a specific role to play in the maintenance and development of the information.

9.2 | The role of the computer

Computers are ideal for the task of analysing complex data sets, such as **sequence data for nucleic acids and proteins**. This often requires a series of computing operations, each of which may be relatively simple in isolation. However, the computation may need to be repeated thousands or millions of times; therefore, it is important that this is achieved quickly and accurately. Even a short sequence is tedious

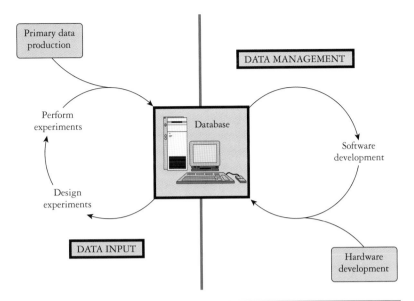

Fig. 9.1 The core of bioinformatics is the database. The two key supporting branches are data input (here shown to the left of the vertical line) and data management (shown to the right of the vertical line). Data management relies on hardware and software developments, with data input requiring the generation of data using experimental techniques such as DNA sequencing.

to analyse without the help of a computer, and there is always the possibility of error due to misreading of the sequence or to a loss in concentration. Computers don't get tired, they generally don't make mistakes if programmed correctly, and they can store large amounts of information in digital form. The emergence of genome sequencing as a realistic proposition required the concomitant development of IT systems to deal with the output of large-scale sequencing projects and, thus, bioinformatics became fully established as a discipline in its own right. New terms such as **data warehouse** (a store of information) and **data mining** (interrogation of databases of various types) have been coined to describe aspects of bioinformatics, with **text mining** (interrogation of bibliographic databases) providing an important supporting role.

Over the past 10–15 years or so, the staggering developments in desktop and server-based **computer hardware** have meant that bioinformatics is not the preserve of those with access to major computing facilities. Even a relatively low-cost desktop machine is powerful enough to access the various databases that hold the information, and the requirements of a host server are no longer prohibitive in terms of purchase and maintenance costs. Thus, there is essentially no constraint on accessing and using the information from a hardware point of view. The **software** has, as might be expected, been developed in tandem with the computing power available and the requirements imposed by the nature of the data sets that are generated. Much of

Computer hardware has developed to the point where desktop machines are powerful enough to act as the interface between the scientist and the various sequence databases and interrogation programmes. Software and sequence data are usually available free of charge to the scientific community.

(a)   AGGTCGCGATGCTATTCGCCTACGTAGCAGGATGGCGAGGCTAGTAGCGC
        TCCAGCGCTACGATAAGCGGATGCATCGTCCTACCGCTCCGATCATCGCG

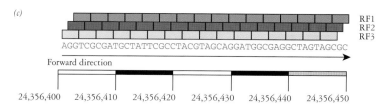

```
                    10        20        30        40        50
(b)  i   AGGTCGCGATGCTATTCGCCTACGTAGCAGGATGGCGAGGCTAGTAGCGC

     ii  AGGTCGCGAT GCTATTCGCC TACGTAGCAG GATGGCGAGG CTAGTAGCGC

     iii aggtcgcgat gctattcgcc tacgtagcag gatggcgagg ctagtagcgc
```

(c)

```
                                                              RF1
                                                              RF2
                                                              RF3
        AGGTCGCGATGCTATTCGCCTACGTAGCAGGATGGCGAGGCTAGTAGCGC

        Forward direction


    24,356,400   24,356,410   24,356,420   24,356,430   24,356,440   24,356,450
```

Fig. 9.2 (a) A short DNA sequence is shown in double-stranded format. (b) Three different ways of writing the sequence are shown: i, ii, and iii. By convention only one strand is listed. In (b)i uppercase type is used with numbering above every 10 bases. In (b)ii spaces are used to separate groups of 10 bases, and in (b)iii lowercase type is used with separation. Lowercase type avoids any confusion between G and C, as g and c are more easily distinguished. (c) An annotated version of the sequence, with the three reading frames identified as RF1, RF2, and RF3. The numbering shows bases from the start of the chromosome. The region shown here is therefore some 24 million base pairs from the end of the chromosome.

the software that is needed for the manipulation and interrogation of information is now made available free of charge by the various host organisations who run the bioinformatics network, although some packages are available from various companies on a commercial basis.

9.3 | Biological data sets

Database management requires a high level of co-operation to establish common procedures for the storage, organisation, and annotation of the information. The current state of development of the bioinformatics network represents an astonishing achievement that rivals the Human Genome Project in terms of significance.

It could be argued that bioinformatics really began to take off with the advent of rapid DNA sequencing methods in the late 1970s (see Section 3.7). Up until this point, the rate of acquisition of biological sequence data was limited; thus, there was relatively little pressure to develop common storage and analysis methods to cope with the information. As sequencing techniques became established and used more extensively, the rate of data generation obviously increased and, thus, the need for **co-ordinated database management** became greater. This need can be illustrated by looking at a simple example using some DNA sequence data (Fig. 9.2). This figure shows a short sequence of 50 bases, listed in various formats. Even with this length of sequence, it is clear that there are several requirements if sense is to be made of the information. These are:

- the need for accurate generation and capture of information,
- a clear and logical presentation of the sequence in visual format,
- annotation of the sequence to enable orientation and identification of features, and
- consistent use of annotation by different users/providers.

As sequence databases now deal with billions of bases rather than hundreds, it is clearly a major task to **organise**, **collate**, and **annotate** the information – the sequence itself is actually one of the simplest components in terms of organisational complexity.

There are obviously many different ways to collate and store sequence information and, thus, many different databases exist. It can be useful to think of databases as falling into two main categories. Repositories for experimentally derived sequence information are often known as **primary databases**. The main primary databases today are nucleic acid sequence databases, although it is interesting to note that protein sequences were collated in the early 1960s and thus protein databases actually pre-date large-scale nucleic acid sequence databases! **Secondary databases** are variants that have been derived from primary databases in some way. They may represent collections of sequence data arranged by organism or phylogenetic group, or may be based on genome projects. Alternatively, derived protein sequences, gene expression data, or data restricted to a particular group of nucleic acids (*e.g.* ribosomal RNA sequences) can be used as the basis for establishing a secondary database.

Biological databases can be classified as either primary or secondary databases, although the distinction between these categories is becoming less important as database technology improves and annotation becomes more sophisticated.

9.3.1 Generation and organisation of information

Data generation is a central part of any scientific procedure, and forms the core of bioinformatics. As we have seen, DNA sequencing was essentially the first procedure to begin to generate data in sufficiently large amounts to require serious co-ordination and organisation of biological databases. Significant milestones occurred with the sequencing of the genomes of the **bacteriophages phi-X174** (5 386 base pairs, 1977) and **lambda** (48 502 base pairs, 1982). During this period those involved in sequence determination realised that there was a need for a co-ordinated approach to database management if full benefit was to be obtained from the sequences that were being determined. We can illustrate the development of sequence databases by looking briefly at the two major types, those dealing with nucleic acid and protein sequence data.

The need for coordinated database management was evident soon after techniques for rapid DNA sequencing became available and began to produce significant amounts of sequence data.

9.3.2 Nucleic acid databases

The first nucleic acid database to be established was set up by the **European Molecular Biology Laboratory** (EMBL) in 1980. Soon afterwards **GenBank** was established in the USA, and in 1986 the **DNA Data Bank of Japan** (DDBJ) began collating information. These three database providers now make up an international collaboration that

Table 9.2.	Nucleic acid sequence database websites
Site/page	URL
Nucleic acid sequence database sites	
International Nucleotide Sequence Database Collaboration (INSDC)	http://www.insdc.org/
EMBL Nucleotide Sequence Database (EMBL-Bank)	http://www.ebi.ac.uk/embl/
GenBank	http://www.ncbi.nlm.nih.gov/Genbank/
DNA Data Bank of Japan	http://www.ddbj.nig.ac.jp/
Genome sequencing and other database sites	
The Wellcome Trust Sanger Institute	http://www.sanger.ac.uk/
Databases at the European Bioinformatics Institute (EBI)	http://www.ebi.ac.uk/Databases/
Databases at the National Center for Biotechnology Information (NCBI)	http://www.ncbi.nlm.nih.gov/Databases/

The International Nucleotide Sequence Database Collaboration (INSDC) helps to oversee the activities of the three main primary sequence databases.

is the major source of primary annotated nucleic acid sequence data, with the **International Nucleotide Sequence Database Collaboration (INSDC)** playing a key role in overseeing the collaboration. The EMBL database is known as the **EMBL Nucleotide Sequence Database** (or simply **EMBL-Bank**) and is hosted and maintained by the **European Bioinformatics Institute (EBI)**. In the USA the **National Center for Biotechnology Information (NCBI)** runs GenBank, and the DDBJ is hosted by the **National Institute of Genetics** in Japan. More information about these databases can be found in the websites listed in Table 9.2.

One of the most striking aspects of nucleic acid databases is the increase in size since the first sequences were entered manually in the early 1980s. The early growth of the data deposited in the global nucleic acid sequence database is shown in Fig. 9.3. An impressive rate of growth is shown, with the total entries reaching some 250 Mb (**megabases**) by the mid 1990s. At this point DNA sequencing moved from relatively small-scale projects to much larger genome projects and, thus, the rate of data acquisition increased even more. Recent growth in sequence data (*i.e.* over the past 10 years or so) is shown in Figs. 9.4 and 9.5. Fig. 9.4 shows data as recorded by GenBank. In addition to the sequence data, the figure also shows how the increase in use of these data has developed. This is presented in the form of 'database searches per day', and (as might be expected) this measure of the use of the database closely mirrors the growth in the database itself.

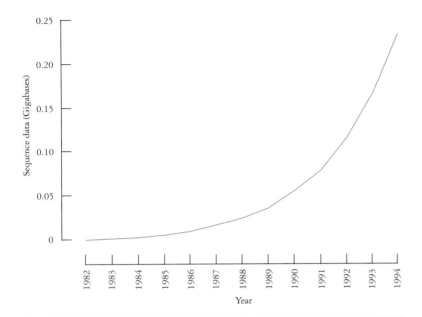

Fig. 9.3 Early growth of global nucleic acid sequence database entries. Entries are presented as gigabases of sequence data accumulating over the period 1982–1994.

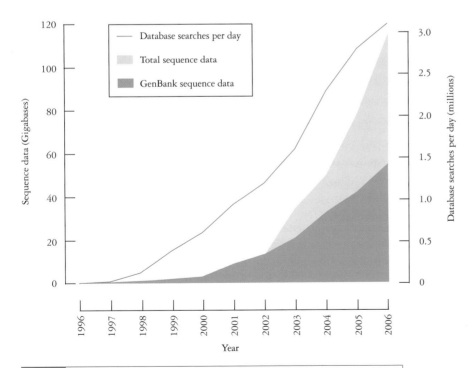

Fig. 9.4 Growth in the Genbank database from 1995. Total sequence data and GenBank-derived sequence data are shown. A measure of the use of the information is shown by the number of database searches requested per day. Modified from data supplied by NCBI (www.ncbi.nlm.nih.gov/Genbank), with permission.

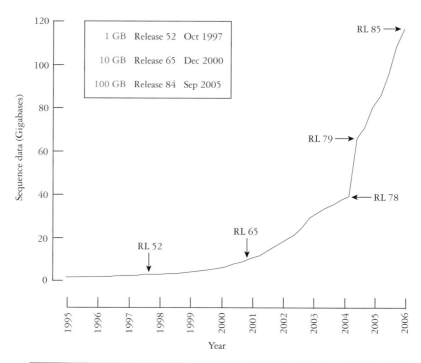

Fig. 9.5 Growth of the EMBL nucleic acid sequence database (EMBL-Bank) since 1995. The release version of the database is indicated by RL, with key release versions 52, 65, 78, 79, and 85 shown. These represent the 1, 10, and 100 gb milestones (releases 52, 65, and 85) and also a period of exceptional growth of the database (between releases 78 and 79). The figure was produced using data provided by the EMBL-European Bioinformatics Institute, Hinxton, UK (www.ebi.ac.uk), with permission.

> Growth of the collated nucleic acid sequence database has been quite staggering, reaching the 100 Gb milestone in September 2005.

The final figure showing nucleic acid database growth is Fig. 9.5, in which data are shown as recorded by EMBL. This shows the total sequence entries as released in different versions of the database, which is constructed by sequence deposition to one of the three main database hosts (GenBank, EMBL-Bank, and DDBJ). Daily exchange of data between these three organisations means that a continually updated version of the entire database is available at all times. The database passed the 100 Gb (**gigabase**) milestone in September 2005 with release 84 – a staggering achievement involving widespread international collaboration, developed over a 25-year period. By October 2006, the database held over 148 billion nucleotides in some 81 million entries from around 273 000 organisms – you might like to check the current statistics at the time you are reading this text!

9.3.3 Protein databases

In looking at protein sequence databases, it is important to distinguish between database entries that have been determined by direct methods, and protein sequences derived from nucleic acid database information by translation of the predicted mRNA sequence. The direct method is similar to nucleic acid sequencing in that primary data are generated from the source protein, and additional biochemical and physical analysis is often possible. Thus, a reasonably complete characterisation of the protein, perhaps with biological activity fairly well established, is the standard to aim for. With derived sequence data, modifications to the amino acids may not be picked up, and there is a risk that the derived sequence may not in fact be a functional protein in the cell. One of the key challenges for database developers is to ensure that databases can 'talk' to each other, so that complex data conversion is avoided where possible.

Direct sequencing of proteins by the classical **Edman degradation** method was first established in 1950. However, the rate of growth of primary protein databases has been somewhat less spectacular than for nucleic acids. This is largely due to the difficulties in determining the amino acid sequence of a protein – amino acid sequences can't be cloned like genes, and proteins are complex three-dimensional structures with biological activity dependent on structural aspects beyond the primary sequence. Although recent developments in protein sequencing techniques (*e.g.* automation of the Edman procedure, use of **mass spectrometry**) have improved the situation greatly, sequencing cloned DNA is always going to provide data more quickly than sequencing amino acid residues in a peptide fragment.

Despite the constraints on protein sequence acquisition, the databases currently available represent an invaluable resource for the biological community. The key resource is the Universal Protein Resource or **UniProt**, which represents the product of close collaboration between the major protein sequence repositories in Europe and the USA. UniProt was set up in 2002, with three major partners involved. The **Protein Information Resource (PIR)**, hosted by Georgetown University in the USA, joined forces with the **SwissProt** database, developed and maintained by the Swiss Institute for Bioinformatics and the EBI. The central core of UniProt is the **UniParc** database and **UniProt Knowledgebase (UniProt KB)**, which maintains a high level of consistency, accuracy, and annotation of the sequences deposited. Although falling a little short of nucleic acid databases in terms of entries, the scale of currently available protein sequence data is still very impressive. At the time of writing, the latest release of UniParc had some 7.7 million entries, and UniProt KB some 3.5 million entries. Further details about UniProt and its associated components can be found in the websites listed in Table 9.3.

> Although protein sequencing has not produced primary data at the same rate as DNA sequencing, a wealth of information about protein form and function is available in the various protein databases.

9.4 | Using bioinformatics as a tool

Having illustrated the scope of bioinformatics by looking at the basic organisation of nucleic acid and protein databases, the obvious question is 'what use can be made of such information?' At this point things become a little difficult! Given that this is only part of a single chapter, it is almost impossible to present a reasoned analysis of the range of activities that use bioinformatics-derived information at

> New applications of bioinformatics are constantly being developed and made available to the scientific community on an open-access basis.

Table 9.3.	Protein sequence database websites
Site/page	URL
Database sites	
UniProt – the key gateway site for protein database information	http://www.expasy.uniprot.org/
SwissProt (the SIB protein resource and part of UniProt)	http://www.expasy.org/sprot/
The Protein Information Resource (PIR, hosted by Georgetown University, and part of UniProt)	http://pir.georgetown.edu/pirwww/
Analysis and proteomics sites	
The proteomics server (Expert Protein Analysis System) of SIB	http://www.expasy.org/
The bioinformatics toolbox of the EBI	http://www.ebi.ac.uk/Tools

some stage. In addition, as the field itself develops, new applications and developments emerge on an almost weekly basis. The interested reader can quite easily log onto the same sites and access the same range of facilities as those carrying out advanced research in the field, even if it can be a little confusing for the newcomer. This open access is one of the great strengths of the bioinformatics network and is a major contributing factor to the overall success of the discipline in terms of new discoveries and insights.

9.4.1 The impact of the Internet and the World Wide Web

In some ways the history of computing parallels that of the technologies for analysing and sequencing DNA. As the structure of DNA was being worked out in the early 1950s, transistors were being developed. The first Internet connections were put in place in 1969, when the tools for constructing recombinant DNA were being established. In the late 1970s the first personal computers appeared, and rapid DNA sequencing techniques were devised. In the early 1990s the WWW became a reality, as did genome sequencing. Thus, the developing field of molecular biology benefited from concurrent developments in computer science, and today the computational aspects of bioinformatics are an integral part of the subject. In fact, computational developments are now often the drivers of progress in bioinformatics rather than the biological or technical aspects of the subject.

In addition to the developments in the actual computer hardware, use of the Internet and WWW has been essential for the rapid expansion of bioinformatics. I recall receiving an early release of the GenBank database on 44 large floppy disks, each of which had to be loaded to access the data. Today the WWW provides real-time access to the major databases, which are continuously updated. This remote storage in host servers means that maintenance of large databases is not

The Internet and World Wide Web are essential parts of the bioinformatics network, as the major databases are maintained on remote servers and made available for consultation by anyone with an Internet connection.

required at a local level. The availability of a search/compare function, often as a free service or through downloadable free software, means that scientists can have instant access to the information and the tools needed to interrogate it. It is this free-access ethos that is one of the great strengths of the Internet-based bioinformatics network.

9.4.2 Avoiding the 'GIGO' effect – real experiments

'Garbage in–garbage out' (GIGO) is a well-known phrase in computing, which was coined to stress the importance of valid and reliable data entry when programming or entering information. GIGO can apply to any situation where there is the possibility of data generation and/or entry errors and is particularly appropriate when considering bioinformatics. Although many aspects of bioinformatics could give rise to data corruption, there are essentially three classes of problem, covering data generation, entry, and processing.

The generation of primary data is obviously dependent on the availability of a reliable and reproducible experimental method. For example, DNA sequencing techniques are now well established and generally demonstrate a very high degree of validity and reliability. Automation has reduced the incidence of human error, and with multiple 'reads' of both strands of DNA, the accuracy of the resulting data is close to 100%. Some human intervention to resolve anomalies may be required for tricky stretches of sequence, but this adds to the fidelity of the process rather than detracting from it.

Having generated the sequence, the transfer of data from the sequencing software to the database is the next step where data migration could present problems. Although the days of keyboard entry for the bulk of primary sequence data are now over, there can be issues due to the interface between different computer network systems. However, these are often of a technical nature and if resolved should not affect the stage of data collection.

Data processing is an area where there is a further risk of introducing errors. There is the possibility of incorrect information, derived experimentally, persisting through the data generation and entry processes. Thus, any computation carried out can often compound the problem by 'cloning' the incorrect data and establishing it in the database. There is a real issue with the 'typewritten therefore must be correct' tendency, in which many people tend to believe any data that appear to be part of a formal data set. Thus, the sequence could be 'wrong', but the incorrect data could be perfectly plausible and thus very difficult to identify. Any such incorrect data could be used for the production of derived sequence data sets or other purposes, making the problem worse. The best quality control for sequencing involves careful design of techniques and procedures, followed by careful performance of the various manipulations and rigorous cross-checking of sequence data. A distinction is usually made between **'single-pass' sequence data**, which may be inaccurate in places, and **'finished' sequence data**, where the inconsistencies have been removed by several sequencing runs and rigorous validation of the sequence.

Although the core information for bioinformatics is based on data sets of various sorts, it is vital that the information in these data sets is determined by careful experimentation to avoid propagation of errors. Thus, there is a high value placed on 'finished' sequence data that has been rigorously checked and confirmed.

9.4.3 Avoiding the test tube – computational experimentation

One of the great recent advances in bioscience has been the establishment of a range of computer-based techniques that enable work to be carried out that was not possible, say, 20 years ago. Assuming that the GIGO effect can be reduced to an acceptable level, the interrogation of databases can provide some novel insights. This form of *in silico* experimentation does not replace 'real' practical work, but it can offer a very quick route to establishing what experiments might prove most incisive and informative. This cycle of experimental data generation and analysis, as outlined in Fig. 9.1, is a central pillar of bioinformatics. One obvious example of this approach is the translation of a DNA sequence into the predicted protein sequence. However, identification of a protein-sized open reading frame by computer analysis does not necessarily mean that the protein will be produced in the cell. To establish the presence of such a protein, identification and analysis will be required using a range of experimental techniques to characterise biological activity and function of the protein.

The use of computer analysis and prediction to carry out 'experiments' *in silico* enables researchers to search, identify, and modify gene and/or protein sequences easily. Results can then be tested by experiments *in vitro* or *in vivo* to determine biological significance.

9.4.4 Presentation of database information

Our final consideration in this chapter is about data presentation, without which any database would be of very limited use. In particular, the availability of presentation interfaces that do not require any computer programming skills has been essential in ensuring widespread access to the large nucleic acid and protein databases. Sequence data can be stored simply as data files, which can be transferred easily using the **File Transfer Protocol (FTP)** to download the file. An accession number and some basic annotation often provide sufficient information to enable extraction and manipulation of the data. However, a more active annotation and presentation is often helpful and is usually necessary where complex data are presented (*e.g.* genome sequence information). Also, the generation of secondary databases, or databases for specific purposes, requires careful planning at the setup stage if clear rules for annotation and presentation are to be established and adhered to by contributors and database managers. The overall aim will be to enable users to navigate through the data by means of a clear user interface that is easy to use. As might be expected, there are many different variants of such interfaces. Some may have specific functions built in, such as search tools for restriction enzyme recognition sites or protocols for translating nucleic acid sequence data into proteins. To illustrate the type of presentation that is possible using the Web, we will look at a few examples that are used to present data from nucleic acid sequence databases.

Although a basic data file may be useful, complex data sets are much more accessible with clear user interfaces, annotation, and presentation. This aspect of database management is a major part of bioinformatics and is perhaps the critical step that has transformed the subject into a universally accessible resource.

The human genome is perhaps the most interesting sequence to consider (largely because of what it represents, rather than for any inherent bioinformatics reason). As we are considering Web-based presentation tools, the best way to read this part of the book is with the

Table 9.4.	Presentation interfaces for sequence databases	
Site/page	URL
The National Center for Biotechnology Information (NCBI) site that describes the Entrez database presentation system and lists a series of databases using this method	http://www.ncbi.nlm.nih.gov/
The NCBI Mapview site with direct entry to graphical presentation of database information	http://www.ncbi.nlm.nih.gov/mapview/
The Wellcome Trust Sanger Institute automated annotation tool (Ensembl) for genome database presentation	http://www.ensembl.org/
The Sanger Institute site for Vega, a manually annotated genome database presentation tool	http://vega.sanger.ac.uk/
The European Bioinformatics Institute gateway site for TREMBL (translated EMBL)	http://www.ebi.ac.uk/trembl/

book propped up beside your computer, so that you can access the websites and see for yourself how the various features work. Some suitable websites are listed in Table 9.4. These provide some useful background information and also access to the sequence data. Three important presentation tools are **Ensembl**, **Vega**, and **Entrez Mapview**. Ensembl has been developed by EMBL and the Wellcome Trust Sanger Institute and is an automated annotation system that aims to provide continually updated accurate annotation of a range of genome databases. Vega is a similar tool, differing from Ensembl in two main ways – only a few genomes are covered, and manual rather than automated curation is used. The NCBI version of genome annotation is the Mapview tool.

Let's consider presentation of the human genome using Ensembl. When you log on to the Ensembl site (see Table 9.4), a list of the available mammalian genomes is presented. By selecting the *Homo sapiens* button, a graphic of the chromosomes (the karyotype) appears, along with some additional information. Clicking on a chromosome opens up a more detailed graphic of the chromosome, and clicking on an area of the chromosome takes you into the detailed presentation of that region. There are several levels of detail, down to the actual sequence itself. Now, all this mouse-clicking and instant access to information seems straightforward – but we need to think a little about what is actually being presented to get some sort of perspective. Let's consider chromosome 1. Opening up the page for chromosome 1 from the karyotype view gives us some information about this chromosome. It has around 250 million base pairs, with about 2000 genes and 800 000 single nucleotide polymorphisms (SNPs, which are positions where single nucleotide differences exist between non-related members of the same species and are useful in genetic mapping studies). If we navigate to the base-pair view, and run this at maximum 'magnification', 25 base pairs are displayed across 150 mm of computer screen – that's 6 mm per base pair. We can now carry out a couple of interesting calculations. If we think of the computer screen as a window on one part of chromosome 1, we would need a 'page' some 1500 *kilo*metres in width to present all the base pairs contiguously! If we were to print the sequence onto A4 paper at 25 base pairs per line and 20 lines per page, we would need half a million pages to accommodate the chromosome 1 sequence – and there are 22 other chromosomes to add.

I think that these calculations demonstrate that it would be impossible to present genome sequence information in any meaningful way if we did not have computers, databases, and presentation tools that can make some sort of sense of the data. As the information is so readily accessible, and the navigation straightforward, it is easy to become a little blasé when looking at presentation tools such as Ensembl. If this happens, think of the 1500 km of sequence presentation for chromosome 1, and reflect on the advances in sequencing and computing technology that have made such a rich resource available to all.

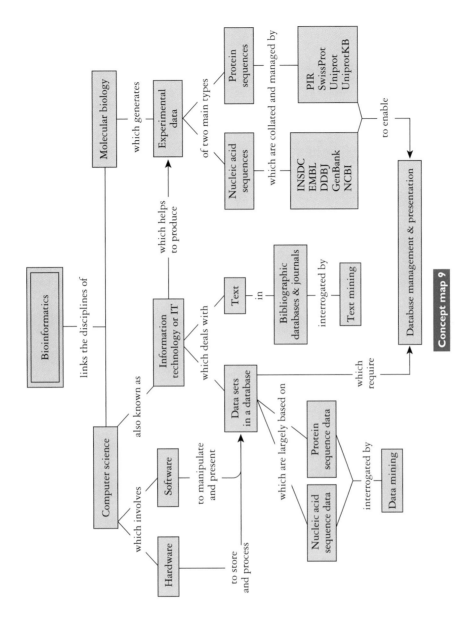

Concept map 9

Part III

Genetic engineering in action

Chapter 10 summary

Aims

- To describe the methods available for the analysis of gene structure and function
- To outline the shift in emphasis from genes to genomes
- To illustrate genome sequencing projects and the human genome sequence
- To describe the 'otheromes' that are now described and studied
- To indicate some of the likely future developments in genome research

Chapter summary/learning outcomes

When you have completed this chapter you will have knowledge of:

- The methods used to investigate gene structure and function
- The development and use of DNA microarrays
- Mapping and sequencing of whole genomes
- The Human Genome Project
- The use of genome browsers such as Ensembl
- The transcriptome, proteome, metabolome, and interactome

Key words

Genome, gene structure, gene function, DNA/protein binding, gel retardation, DNA footprinting, DNase protection, ChIP assay, primer extension, S_1 mapping, deletion analysis, Northern blot, dot blot, genomics, DNA microarray, DNA chips, renaturation kinetics, hyperchromic effect, abundance classes, repetitive sequence DNA, genetic mapping, physical mapping, recombination frequency, linkage mapping, genetic marker, physical marker, sequence-tagged site (STS), fluorescent labelling, integrated strategy, information technology, *Escherichia coli*, *Saccharomyces cerevisiae*, *Caenorhabditis elegans*, *Drosophila melanogaster*, *Arabidopsis thaliana*, *Mus musculus*, comparative genome analysis, *Homo sapiens*, Human Genome Project (HGP), draft sequence, finished sequence, polymorphism, Ensembl, Centre d'Etude Polymorphism Humain (CEPH), monogenic, Online Mendelian Inheritance in Man, pedigree analysis, neutral molecular polymorphism, restriction fragment length polymorphism (RFLP), minisatellite, microsatellite, variable number tandem repeat (VNTR), genetic fingerprinting, DNA profiling, sequence-ready clone, shotgun sequencing, directed shotgun method, clone contig method, whole genome shotgun, p53, tumour suppressor, biome, proteome, transcriptome, expressome, molecular machine, metabolome, interactome, European Bioinformatics Institute, Cold Spring Harbour Laboratory, Gene Ontology Consortium, Reactome, single nucleotide polymorphism (SNP), copy number variations (CNVs), ethical, legal, ELSI.

Chapter 10

Understanding genes, genomes, and 'otheromes'

In Parts I and II of this book we examined some of the basic techniques and methods of gene manipulation. These techniques, and many more sophisticated variations of them, give the scientist the tools that enable genes to be isolated and characterised. In this final section of the book we will consider some of the applications of gene manipulation. Of necessity this will be a highly selective treatment; the aim is to give some idea of the immense scope of the subject whilst trying to include some detail in certain key areas. We will also look more broadly at some of the ethical problems that gene manipulation poses, and at the topic of organismal cloning.

In many ways genetic engineering has undergone a shift in emphasis over the past few years, away from the technical problems that had to be solved before the technology became 'user friendly' enough for widespread use. Gene manipulation is now used as a tool to address many diverse biological problems that were previously intractable, and the applications of the subject appear at times to be limited only by the imagination of the scientists who use the technology in basic research, medicine, biotechnology, and other related disciplines. In this chapter we will look at some of the methods that can be used to investigate gene structure and function. We will also examine how modern gene manipulation technology has opened up the world of the **genome** in a way that was not thought possible a few years ago and has led to the development of bioinformatics-based techniques for analysing genomes and gene expression.

> Gene manipulation has now matured and has developed into a sophisticated enabling technology that has revolutionised bioscience and its applications.

10.1 | Analysis of gene structure and function

In terms of 'pure' science, the major impact of gene manipulation has been in the study of **gene structure** and **gene function**. The organisation of genes within genomes is a fast-developing area that is essentially an extension of the early work on gene structure. Although the contribution of classical genetic analysis should not be underestimated, much of the fine detail regarding gene structure and expression remained a mystery until the techniques of gene cloning enabled the isolation of individual genes.

As discussed in Section 8.5, many of the techniques used to characterise cloned DNA sequences provide information about gene structure, with one of the aims of most experiments being the determination of the gene sequence. However, even when the sequence is available, there is still much work to be done to interpret the various structural features of the sequence in the context of its function *in vivo*. Whilst the advent of the powerful techniques of bioinformatics has moved gene analysis to a different level, there is still a need for initial characterisation of cloned genes and for accurate sequence determination.

10.1.1   A closer look at sequences

When a gene sequence has been determined, a number of things can be done with the information. Searches can be made for regions of interest, such as promoters, enhancers, and so on, and for sequences that code for proteins. Coding regions can be translated to give the amino acid sequence of the protein. Restriction maps can be generated easily and printed in a variety of formats. The sequence can be compared with others from different organisms and the degree of homology between them may be determined, which can assist in studying the phylogenetic relationships between groups of organisms.

Although computer analysis of a sequence is a very useful tool, it usually needs to be backed up with experimental evidence of structure or function. For example, if a previously unknown gene is being characterised, it will be necessary to carry out experiments to determine where the important regions of the gene are. Often the data produced will confirm the function inferred from the sequence analysis, although experiments sometimes generate novel findings that may not have been predicted. Thus, it is important that the computational and experimental sides of sequence analysis are used in concert.

It is important that any computer-based analysis of sequence data is linked with experimentally derived information where possible, particularly where functional aspects of genes and proteins are being investigated.

10.1.2   Finding important regions of genes

One of the key aspects in the control of gene expression concerns **DNA/protein binding** interactions. Thus, it is important to find the regions of a sequence to which the various types of regulatory proteins will bind. A relatively simple way to do this is to prepare a restriction map of the cloned DNA and generate a set of restriction fragments. The protein under investigation (perhaps RNA polymerase, a repressor protein, or some other regulatory molecule) is added and allowed to bind to its site. If the fragments are then subjected to electrophoresis, the DNA/protein hybrid will run more slowly than a control fragment without protein and can be detected by its reduced mobility. This technique is known as **gel retardation**, and it provides information about the location of particular binding sites on DNA molecules.

Techniques based on gel electrophoresis can be useful in determining which regions of genes are involved in binding regulatory proteins.

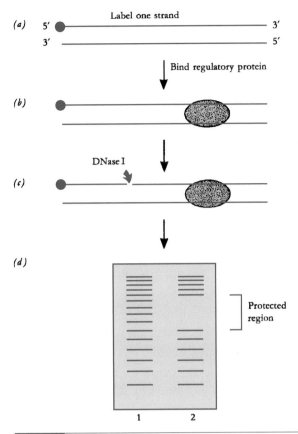

Fig. 10.1 DNA footprinting. (a) A DNA molecule is labelled at one end with ^{32}P. (b) The suspected regulatory protein is added and allowed to bind to its site. A control reaction without protein is also set up. (c) DNase I is used to cleave the DNA strand. Conditions are chosen so that on average only one nick will be introduced per molecule. The region protected by the bound protein will not be digested. Given the large number of molecules involved, a set of nested fragments will be produced. (d) The reactions are then run on a sequencing gel. When compared to the control reaction (lane 1), the test reaction (lane 2) indicates the position of the protein on the DNA by its 'footprint'.

Although gel retardation is a useful technique, its range of application is somewhat limited. This is largely due to the precision of the restriction map and the sizes of fragments that are generated. A more accurate way of identifying regions of protein binding is the technique of **DNA footprinting** (sometimes called the **DNase protection** method). The technique is elegantly simple and relies on the fact that a region of DNA that is complexed with a protein will not be susceptible to attack by DNase I (Fig. 10.1). The DNA fragment under investigation is radiolabelled and mixed with a suspected regulatory protein. DNase I is then added so that limited digestion occurs; on average, one DNase cut per molecule is achieved. Thus, a set of nested fragments will be generated, and these can be run on a sequencing gel. The region that is protected from DNase digestion gives a 'footprint' of the binding site within the molecule.

A more recent technique known as the **chromatin immunopre-cipitation (ChIP) assay** uses an alternative approach to examining protein binding sequences. In this technique, the protein and DNA to which it is bound are cross-linked using formaldehyde and the DNA sheared into fragments. Specific proteins can be precipitated using antibodies, and the DNA sequence determined.

It is often necessary to locate the start site of transcription for a particular gene, and this may not be apparent from the gene sequence data. Two methods can be used to locate the T_C start site: **primer extension** and **S_1 mapping**. In primer extension a cDNA is synthesised from a primer that hybridises near the 5′ end of the mRNA. By sizing the fragment that is produced, the 5′ terminus of the mRNA can be identified. If a parallel sequencing reaction is run using the genomic clone and the same primer, the T_C start site can be located on the gene sequence. In S_1 mapping the genomic fragment that includes the T_C start site is labelled and used as a probe. The fragment is hybridised to the mRNA and the hybrid then digested with single-strand-specific S_1 nuclease. The length of the protected fragment will indicate the location of the T_C start site relative to the end of the genomic restriction fragment.

> Transcription start sites can be identified by primer extension and S_1 mapping techniques.

10.1.3 Investigating gene expression

Recombinant DNA technology can be used to study gene expression in two main ways. First, genes that have been isolated and charac-terised can be modified and the effects of the modification studied. Second, probes that have been obtained from cloned sequences can be used to determine the level of mRNA for a particular protein under various conditions. These two approaches, and extensions of them, have provided much useful information about how gene expression is regulated in a wide variety of cell types.

One method of modifying genes to determine which regions are important in controlling gene expression – **deletion analysis** – involves deleting sequences lying upstream from the T_C start site. If this is done progressively using a nuclease such as exonuclease III or Bal 31 (see Section 4.2.1), a series of deletions is generated (Fig. 10.2). The effects of the various deletions can be studied by monitoring the level of expression of the gene itself or of a 'reporter' gene such as the *lacZ* gene. In this way, regions that increase or decrease transcription can be located, although the complete picture may be difficult to decipher if multiple control sequences are involved in the regulation of transcription.

> Deletion analysis can be used to determine which parts of upstream control regions are important in regulating gene expression.

Measurement of mRNA levels is an important aspect of studying gene expression and is often done using cDNA probes that have been cloned and characterised. The mRNA samples for probing may be from different tissue types or from cells under different physiological conditions, or may represent a time course if induction of a particular protein is being examined. If the samples have been subjected to elec-trophoresis a **Northern blot** can be prepared, which gives informa-tion about the size of transcripts as well as their relative abundance.

> The amount of a particular mRNA can be measured using cDNA probes and Northern- or dot-blot techniques.

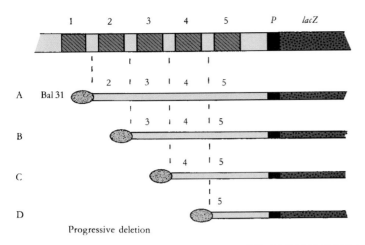

Fig. 10.2 Deletion analysis in the study of gene expression. In this hypothetical example a gene has five suspected upstream controlling regions (1 to 5, hatched). The gene promoter is labelled P. Often the *lacZ* gene is used as a reporter gene for detection of gene expression using the X-gal system. Deletions are created using an enzyme such as Bal 31 nuclease. In this example four deletion constructs have been made (labelled A to D). In A, region 1 has been deleted, with progressively more upstream sequence removed in each construct so that in D regions 1, 2, 3, and 4 have been deleted and only region 5 has been retained. The effects of these deletions can be monitored by the detection of β-galactosidase activity and, thus, the positions of upstream controlling elements can be determined. As an alternative to using Bal 31, restriction fragments can be removed from the controlling region.

Alternatively, a **dot blot** can be prepared and used to provide quantitative information about transcript levels by determining the amount of radioactivity in each 'dot'. This provides an estimate of the amount of specific mRNA in each sample (Fig. 10.3).

Northern- and dot-blotting techniques can provide a lot of useful data about transcript levels in cells under various conditions. When considered along with information about protein levels or activities, derived from Western blots or enzyme assays, a complete picture of gene expression can be built.

10.2 | From genes to genomes

Although much emphasis is still placed on the analysis of individual genes, advances in gene manipulation technology have opened up the study of genomes to the point where this is emerging as a discipline in its own right. Often called simply **genomics**, the emphasis here is on a holistic approach to how genomes function. Therefore, we are now much more likely to assess the function of a gene within the context of its role in the genome, as opposed to considering gene structure and expression in isolation. The development and use of **DNA microarrays** (sometimes referred to as **DNA chips**) for investigating gene structure and expression is a nice example of how a new

In recent years the emphasis has shifted towards analysing genomes rather than genes, using the powerful techniques available for looking at gene expression in a whole-genome context.

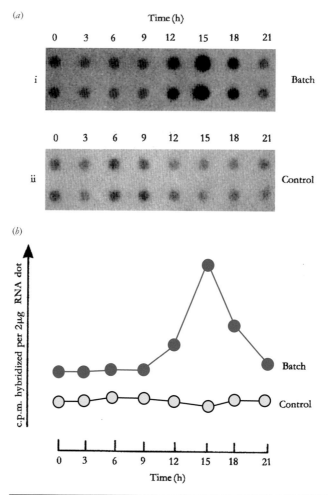

Fig. 10.3 Dot-blot analysis of mRNA levels. Samples of total RNA from synchronous cell cultures of *Chlamydomonas reinhardtii* grown under batch culture and turbidostat (control) culture conditions were spotted onto a membrane filter. The filter was probed with a radiolabelled cDNA specific for an mRNA that is expressed under conditions of flagellar regeneration. (a) An autoradiograph was prepared after hybridisation. Batch conditions (i) show a periodic increase in transcript levels with a peak at 15 h. Control samples (ii) show constant levels. Data shown in (b) were obtained by counting the amount of radioactivity in each dot. This information can be used to determine the effect of culture conditions on the expression of the flagellar protein. Photograph courtesy of Dr J. Schloss. From Nicholl et al. (1988), *Journal of Cell Science* **89**, 397–403. Copyright (1988) The Company of Biologists Limited. Reproduced with permission.

technology can move an area of science forward dramatically. Let's look at this aspect to illustrate the point.

10.2.1 Gene expression in a genome context

Early studies on gene expression tended to concentrate on one or two genes rather than several or many genes in a particular cell or

Fig. 10.4 A DNA microarray on a glass slide. A spot on the slide with a DNA sequence is called a field. Each of the 48 small squares contains a 45 × 45 array of DNA sequences, giving a total of 97 200 fields. Photograph courtesy of the Wellcome Trust Medical Photographic Library. Reproduced with permission.

tissue type – perhaps a cDNA probe was available to facilitate the measurement of transcript levels, or an inducible system enabled differential gene expression to be examined. In such applications the protein and gene have usually been characterised, at least to some extent; thus, the investigator can be sure of what is being measured. However, only a few genes can be investigated in this way in any reasonable scale of experiment. Thus, extending gene expression studies to try to analyse gene expression at the *genome* level (rather than at the single-gene level) required something of a technological breakthrough, which came with the development of the DNA microarray.

The story behind this is really one of technological development to exploit the base-pairing features of complementary nucleic acid sequences. In essence the technique requires immobilisation of a large number of different sequences on a support medium, such as a glass microscope slide (Fig. 10.4). This can be achieved in different ways, such as spotting DNA or cDNA sequences directly onto the slide using a robotic device to enable precise location of the microscopic spots of nucleic acid. An alternative is to synthesise oligonucleotide sequences directly on the slide, using technology based on computer chip manufacturing methods. When the array is ready, the pattern of sequences can be used to hybridise with a sample (often cDNA labelled with fluorescent tags). Binding of complementary sequences can be analysed and information about the pattern of gene expression can be

The development of DNA microarrays and gene chips has revolutionised the way that gene structure and expression can be analysed.

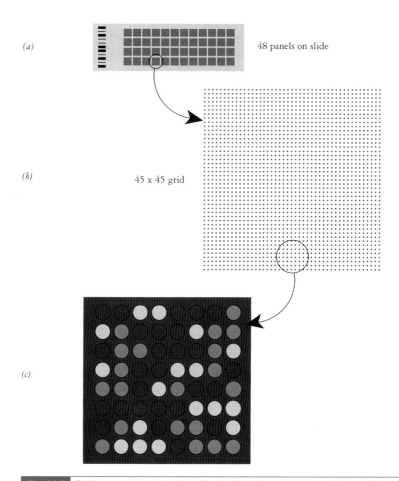

(a)

48 panels on slide

(b)

45 x 45 grid

(c)

Fig. 10.5 DNA microarray technology. The 97 200 field array (*a*) has **48** blocks of 45 × 45 fields (*b*). Samples to be analysed (often cDNA prepared from mRNA and labelled with fluorescent dyes) are pumped onto the array and allowed to hybridise. Excess is washed off and the array is 'read' in a laser-activated scanning device. The pattern of hybridisation is presented on a computer display, part of which is indicated in (*c*). The different levels of signal in each field indicate the level of gene expression and are correlated with the genes on the microarray using computer analysis.

deduced by the level of signal detected using laser scanning of the fluorescent tags and capture by computer (Fig. 10.5).

There are many different ways that DNA microarrays can be used. Two of the most important are in the analysis of gene expression and in screening for mutations or looking for polymorphisms. We will look at how microarrays can be used to analyse the transcripts produced by cells in Section 10.5.1.

10.2.2　Analysing genomes

The analysis of a genome is obviously much more complex than the analysis of a gene sequence and involves a variety of techniques. Structural features such as repetitive sequence elements and intervening sequences can make the task much more difficult than might be

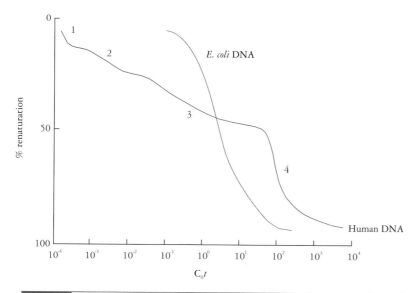

Fig. 10.6 Abundance classes found in different genomic DNA samples. Following denaturation, samples of DNA are allowed to cool and the percent re-naturation is measured. This is plotted against log C_0t (this is functionally a measure of time, with the added context of DNA concentration). The dotted line shows *E. coli* DNA, with a simple one-step pattern. Human DNA (solid line) shows a complex pattern with four classes recognised: (1) foldback DNA, (2) highly repetitive sequences, (3) moderately repetitive sequences, and (4) unique or single-copy sequences.

supposed when simply considering the number of base pairs, even though this is a major factor in determining the strategy for studying a particular genome. Some comparative genome sizes were presented in Table 2.3; these data show that eukaryotic genomes may be several orders of magnitude larger than bacterial genomes.

Some of the earliest indications of genome complexity were obtained by using the technique of **renaturation kinetics**. In this technique, a sample of DNA is heated to denature the double-stranded DNA. The strands are then allowed to re-associate as the mixture cools, and the ultraviolet absorbance (A_{260}) is monitored. Single-stranded DNA has a higher absorbance than double-stranded DNA (this is called the **hyperchromic effect**), so the degree of re-naturation can be assessed easily. Using this type of analysis, eukaryotic DNA was shown to be composed of several different **abundance classes**, as shown in Fig. 10.6. In the case of human DNA, about 40% of the total is either highly or moderately **repetitive sequence DNA**, which can often cause problems in the cloning and analysis of genes. Of the remaining 60%, which represents unique sequence and low-copy-number sequence elements, only around 3% constitutes the actual coding sequence. This immediately poses a problem in the analysis of the human genome, in that around 97% of the DNA could perhaps be 'avoided' if the genes themselves could be identified for further study.

Although some useful knowledge about genomes can be obtained by using kinetic analysis and identifying individual genes, the amount

Renaturation kinetics provided some early indication of the complexity of eukaryotic genomes and the presence of repetitive sequences.

of detailed information provided by these methods is relatively small. The ultimate aim of genome analysis is determination of the DNA sequence, a task that represents a major undertaking even for a small genome. However, in recent years great advances have been made in the technology of DNA sequencing, and large-scale genome sequencing is now well established for many organisms. We will discuss the human genome in more detail in Section 10.3, after we have examined some of the methods for mapping and sequencing genomes.

10.2.3   Mapping genomes

As we saw in Fig. 3.7, large-scale DNA sequencing can be done using a 'shotgun' method in which random fragments are sequenced. The sequences are then pieced together by matching overlaps. Despite its simplicity and occasional limitations, this approach has been successful with large-scale sequencing projects, including the human genome. An alternative approach to sequencing large genomes is to employ an ordered strategy involving **genetic mapping** and **physical mapping** before sequence analysis is attempted. This approach requires access to large clone banks (yeast artificial chromosome (YAC) or bacterial artificial chromosome (BAC) vectors are often used) and a well-developed system for recording the information that is generated. An analogy may help to put the whole process in context.

Genome mapping is a bit like using a road map. Say you wish to travel by car from one city to another – perhaps Glasgow to London, or New York to Denver. You could (theoretically) get hold of all the street maps for towns that exist along a line between your start and finish points, and follow these. However, there would be gaps, and you might go in the wrong direction from time to time. A much more sensible strategy would be to look at a map that showed the whole of the journey, and pick out a route with major landmarks – perhaps other cities or major towns along the way. At this initial planning stage of your journey you would not be too concerned about accurate distances – but you would like to know which roads to take. You would then split the journey into stages where local detail becomes very important – you need to know how far it is to the next petrol station, for example. For a stopover, the address of a hotel would guide you to the precise location. In genome analysis, the techniques of genetic and physical mapping provide the equivalent of large- and small-scale maps to enable progress to be made. Determination of the sequence itself is the level of detail that enables precise location of genome 'landmarks' such as genes and control regions.

Genetic mapping has provided a lot of information about the relative positions of genes on the chromosomes of organisms that can be used to set up experimental genetic crosses. The technique is based on the analysis of **recombination frequency** during meiosis and is often called **linkage mapping**. This approach relies on having **genetic markers** that are detectable. The marker alleles must be heterozygous so that meiotic recombination can be detected. If two genes are on different chromosomes, they are unlinked and will sort independently during meiosis. However, genes on the same chromosome

> Genetic mapping is similar to geographical mapping in many respects, with scale and level of detail being two important considerations. For a map to be useful the level of detail has to be appropriate for the map to function.

> The classic method of mapping genes using recombination frequency analysis is still useful as a low-resolution method of gene localisation.

are physically linked together, and a crossover between them during prophase I of meiosis can generate non-parental genotypes. The chance of this happening depends on how far apart they are – if very close together, it is unlikely that a crossover will occur between them. If far apart, they may behave as though they are essentially unlinked. By working out the recombination frequency it is therefore possible to produce a map of the relative locations of the marker genes.

Physical mapping of genomes builds on genetic mapping and adds a further level of detail. As with genetic maps, construction of a physical map requires markers that can be mapped to a specific location on the DNA sequence. The aim of a physical map is to cover the genome with these identifiable **physical markers** that are spaced appropriately. If the markers are too far apart, the map will not provide sufficient additional information to be useful. If markers are not spread across the genome, there may be sections that have too few markers, whilst others have more than might be required for that particular stage of the investigation.

Physical mapping of genomes is not a trivial task, and until the early 1980s it was thought that a physical map of the human genome was unlikely to be achieved. However, this proved to be incorrect, and techniques for physical mapping of genomes were developed relatively quickly. Physical maps of the genome can be constructed in a number of ways, all of which have the aim of generating a map in which the distances between markers are known with reasonable accuracy. The various methods that can be used for physical mapping are shown in Table 10.1. **Sequence-tagged site (STS)** mapping has become the most useful method, as it can be applied to any area of sequence that is unique in the genome and for which DNA sequence information is available. One advantage of STS mapping is that it can tie together physical map information with that generated by other methods. The technique essentially uses the sequence itself as the marker, identifying it either by hybridisation techniques or, more easily, by amplifying the sequence using the polymerase chain reaction (PCR). With regard to the human genome, STS mapping (with other methods) enabled the construction of a useful genetic and physical map of the human genome by the late 1990s.

The final level of detail in genome mapping is usually provided by some sort of restriction map (see Section 4.1.3 for an illustration of restriction mapping). Long-range restriction mapping using enzymes that cut infrequently is a useful physical mapping technique (Table 10.1), but more detailed restriction maps are required when cloned fragments are being analysed prior to sequencing. The overall link among genetic, physical, and clone maps is shown in Fig. 10.7.

> The development of reliable and efficient physical mapping techniques was an important part of the ordered strategy for sequencing the human genome.

> Restriction mapping provides further detail at the physical level. Genetic, physical, and restriction maps are used in concert to provide a detailed picture of where genes are located.

10.3 | Genome sequencing

The final stage of any large-scale sequencing project (such as a genome project) is to determine and assemble the actual DNA sequence itself.

Table 10.1.	Some methods for physical mapping of genomes
Technique name	Application
Clone mapping	Define the order of cloned DNA fragments by matching up overlapping areas in different clones. This generates a set of contiguous clones known as contigs. Clone mapping may be used with large or small cloned fragments as appropriate.
Radiation hybrid mapping	Fragment the genome into large pieces and locate markers on the same piece of DNA. The technique requires rodent cell lines to construct hybrid genomes as part of the process.
Fluorescent in situ hybridisation (FISH)	Locate DNA fragments by hybridisation, plotting the chromosomal position of the sequence by analysing the fluorescent markers used.
Long-range restriction mapping	Method is similar to any other restriction mapping procedure, but enzymes that cut infrequently in the DNA are used to enable long-range maps to be constructed.
Sequence-tagged site (STS) mapping	The most powerful technique, which can complement the other techniques of genetic and physical mapping. Can be applied to any part of the DNA sequence, as long as some sequence information is available. An STS is simply a unique identifiable region in the genome.
Expressed sequence tag (EST) mapping	A variant of STS mapping in which expressed sequences (*i.e.* gene sequences) are mapped. More limited than STS mapping, but useful in that genes are located by this method.

There are several critical requirements for this part. The DNA sequencing technology has to be accurate and fast enough to do the job in the proposed timescale, and there must be a library of cloned fragments available for sequencing. These two requirements will differ in scale for different projects.

DNA sequencing technology had to improve by orders of magnitude to enable the scale of genome sequencing to be accommodated in a reasonable time frame.

10.3.1 Sequencing technology

When considering sequence determination itself, it was apparent from the start of large-scale projects that the traditional laboratory methods could not deliver the rate of progress that was required to

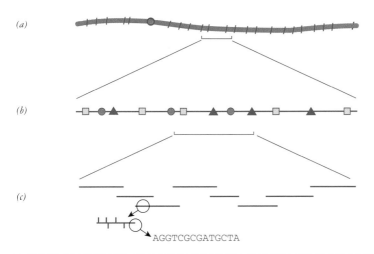

Fig. 10.7 Genome mapping. (*a*) The genetic map is shown, with genetic markers assigned to positions on the chromosome. (*b*) A section of the chromosome is shown, with the physical map of this region. Different types of physical marker are represented as different shapes. Various methods may be used to assign physical markers to their chromosomal locations (STS, FISH, *etc.*; see Table 10.1). (*c*) The clone map of a section of the physical map is shown, with large overlapping DNA fragments. From this a more detailed restriction map and the DNA sequence itself can be determined. From Nicholl (2000), *Cell & Molecular Biology*, Advanced Higher Monograph Series, Learning and Teaching Scotland. Reproduced with permission.

complete the task in a sensible timescale. Thus, automated sequencing methods were developed in which the standard chain-termination method of sequencing was adapted by using **fluorescent labelling** instead of radioactive methods. In one method, the ddNTPs can be tagged with different labels, and one reaction can be carried out where all four ddNTPs are used together. The products are separated by gel electrophoresis, and the fluorescent labels are detected as they come off the bottom of the gel. This gives a direct readout of the sequence. The process is much faster than conventional sequencing and can be run continuously to provide large amounts of data in a short time. It is summarised in Fig. 10.8.

For large sequencing projects such as the human genome, an **integrated strategy** is required, with many centres being involved, each with specific jobs to do. In large centres there may be banks of hundreds of automated sequencing machines, all running essentially non-stop. To cope with this level of data output, an appropriate **information technology** strategy is required to handle the bioinformatics side of sequencing projects (as we discussed in Chapter 9). Sequence information can be published directly to the World Wide Web, where it is immediately available to researchers around the world.

The Internet and World Wide Web have been essential in that they have provided the means for publishing and sharing digital information from genome projects. Data are made available to the scientific community almost immediately following submission to the appropriate database.

10.3.2   Genome projects

The technology for large-scale DNA sequencing has enabled scientists to undertake genome sequencing in a realistic timescale. The roots of

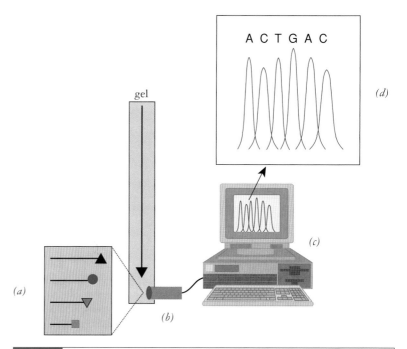

Fig. 10.8 Automated DNA sequencing using fluorescent marker dyes. Each ddNTP is tagged with a different dye, which generates a set of differently labelled nested fragments as shown in (*a*). On separation in a single lane of a sequencing gel, the DNA fragments pass through a detector (*b*) and the fluorescent labels are monitored. (*c*) A computer captures the data and displays the sequence as a series of peaks, from which the sequence is read as shown in (*d*). From Nicholl (2000), *Cell & Molecular Biology*, Advanced Higher Monograph Series, Learning and Teaching Scotland. Reproduced with permission.

genome sequencing go back to the early 1980s, with the publication of the sequence of bacteriophage lambda in 1982. This was the first 'large' sequence to be completed – 48 502 base pairs. Since then there has been a lot of activity in the mapping and sequencing of genomes. In the next section we will look at the Human Genome Project, but there are many other projects that have been either partially or fully completed. In fact, the number of different projects, and the enormous amount of effort that is put into these, illustrates how the emphasis in sequence analysis is changing from genes to genomes.

How are different organisms selected for sequencing? In most cases there is already a well-established history of research, and many 'model organisms' have a worldwide community of scientists working on many diverse aspects of their molecular biology, biochemistry, physiology, and ecology. The range covers many different groups – bacteria, slime moulds, yeasts, nematodes, fruit flies, plants, and mammals. Some notable examples include the bacterium *Escherichia coli*, the yeast *Saccharomyces cerevisiae*, the nematode worm *Caenorhabditis elegans*, the fruit fly *Drosophila melanogaster*, the plant (some call it a weed!) *Arabidopsis thaliana*, and the mouse *Mus musculus*. Naturally, molecular biologists working with these organisms want to sequence their genomes, to provide detailed information

The first genome sequencing projects were based on so-called 'model organisms' that were already well characterised; genome sequencing was the next logical stage in the investigation of these organisms.

Table 10.2. Selected genome sequencing projects

Organism (type)	Sequence completed	Information	Relevant website(s)
Escherichia coli (bacterium)	1997	Genome size 4.64 Mb, which encodes 4405 genes.	http://www.genome.wisc.edu/
Bacillus subtilis (bacterium)	1997	First gram-positive bacterial genome to be sequenced.	http://genolist.pasteur.fr/SubtiList/
Saccharomyces cerevisiae (yeast)	1996	First eukaryotic genome sequenced. Genome size 12 Mb with around 6 680 genes.	http://www.ensembl.org/
Caenorhabditis elegans (nematode worm)	1998	First multicellular organism's genome to be sequenced. Genome size 100 Mb.	http://www.wormbase.org/ http://www.ensembl.org/
Drosophila melanogaster (fruit fly)	2000	Shotgun approach used for this complex genome. Some 14 700 genes located to the genome in about 133 Mb.	http://flybase.bio.indiana.edu/ http://www.ensembl.org/
Arabidopsis thaliana (plant)	2000	Genome size 117 Mb, the first plant genome sequence to be determined.	http://www.arabidopsis.org/
Mus musculus (mouse)	2002	2.7 Gb with some 22 000 genes	http://www.informatics.jax.org/ http://www.ensembl.org/
Homo sapiens	2003	3.1 Gb with some 22 000 genes	http://www.ensembl.org/

Note: Website URLs are correct as of January 2008. There are usually multiple websites for each genome project; these can be accessed from link pages in the sites listed. Alternatively a search using [organism's name] and [genome sequence] (or some such combination of terms) will often get to relevant sites.

about genes, their organisation in the genome, and how they are expressed. Some non-human genome sequencing projects are listed in Table 10.2.

As work progresses, different levels of 'completeness' can be recognised in any genome project. It is perhaps a little unrealistic to say that any project is 'finished' – in fact, the complete sequence is in many ways just the beginning, and understanding how the genome functions is a task that will occupy scientists for many years to come. Major milestones in genome projects include the establishment of the genetic and physical maps, the production of unfinished or first draft sequences, and confirmation of the final version that is accepted as being the most accurate sequence possible. This usually involves sequencing each part of the genome several times and cross-referencing the sequences to ensure that any gaps are filled in and sequence anomalies are cleared up.

In addition to deciphering the genetics of any particular organism, genome sequencing also opens up the field of **comparative genome analysis**, which can help to understand how genomes evolve and how many genes are similar in all organisms. Thus, humans have genome sequences that are about 0.1% different between individuals, 2% different from the chimpanzee, and which contain many of the same genes as the bacterial genome. Given such a wide range of interests, the future for bioinformatics, genomics, and proteomics looks to be secure, and there will undoubtedly be many surprises in store as we seek to understand how cells and organisms function by examining their genetic information in detail.

10.4 | The Human Genome Project

The human genome has now been sequenced – it took around 13 years from inception to completion. The sequence is a resource that will provide the answers to many key biological questions, some as yet unknown.

Worms, weeds, and fruitflies are all very well, but public attention is of course usually focused on any developments that involve *Homo sapiens*. Over the past 15 years or so, the major problems of human genome mapping and sequencing have been addressed, and the **Human Genome Project (HGP)** has now effectively been completed as far as the generation and collation of 'one genome's worth' of sequence information is concerned. However, as noted earlier, obtaining a complete sequence is only one aspect of genome sequencing, and data production is still an inportant part of the HGP today.

The date of Monday, 26 June 2000, will be noted as perhaps one of the most important dates in our history, as this was the date that the 'first draft' of the human genome sequence was announced. Although this is of course a largely symbolic date (for reasons already discussed), it does rank alongside the discovery of the double helix in 1953 as one of the most exciting events in recent history. These two dates, with DNA as the focus, serve to illustrate how technology has changed science. Watson and Crick were part of a relatively small community of people who knew about DNA in the 1950s, and they constructed models using clamps and stands and bases made in their workshop at Cambridge. In contrast, sequencing the genome has required a major international collaborative effort, with around 20 major sequencing centres worldwide, and many thousands of molecular biologists, technicians, and computer scientists involved either directly or indirectly. The project also brought some level of controversy, with some of the key players having differing views of how the sequencing could or should be tackled. This gave rise to a pronounced split between the public consortium and a private company set up to sequence the genome and release the data on a commercial basis.

Although the first **draft sequence** was announced in 2000, and obviously made a great news story, the date of 14 April 2003 is arguably more important. This is when the 'gold standard' sequence was announced, thus completing the conversion of draft sequence into

Table 10.3.	Some interesting facts about our genome

The information would fill two hundred 500-page telephone directories.

Between humans, our DNA differs by only 0.2%, or 1 in 500 bases (letters). (This takes into account that human cells have two copies of the genome.)

If we recited the genome at one letter per second for 24 hours a day it would take a century to recite the book of life.

If two different people started reciting their individual books at a rate of one letter per second, it would take nearly eight and a half minutes (500 seconds) before they reached a difference.

A typist typing at 60 words per minute (around 360 letters) for 8 hours a day would take around 50 years to type the book of life.

Our DNA is 98% identical to that of chimpanzees.

The vast majority of DNA in the human genome – 97% – has no known function.

The first chromosome to be completely decoded was chromosome 22 at the Sanger Centre in Cambridgeshire, in December 1999.

There are 6 feet of DNA in each of our cells packed into a structure only 0.0004 inches across (it would easily fit on the head of a pin).

There are 3 billion (3000000000) letters in the DNA code in every cell in your body.

If all the DNA in the human body was put end to end it would reach to the sun and back over 600 times (100 trillion × 6 feet divided by 93 million miles = 1200).

Source: Information taken from the Sanger Centre website [http://www.sanger.ac.uk]. Reproduced with permission. The Sanger Centre is supported by the Wellcome Trust.

finished sequence. The date sits rather nicely alongside that of another major breakthrough made in Cambridge 50 years earlier, and the journey from double helix to genome sequence represents one of the most astonishing achievements of our time.

The idea that the human genome could be sequenced gained credibility in the mid 1980s, and by the end of the decade the project had acquired sufficient momentum to ensure that it would be supported. The initial impetus had come from the USA, and the formation of the Human Genome Organisation (HUGO) in 1988 marked the birth of the project on an international scale; the role of HUGO was to co-ordinate the efforts of the many countries involved. The HGP was officially launched in October 1990 and presented a task of almost unimaginable complexity and scale. Molecular biology has traditionally involved small groups of workers in individual laboratories, and most of the key discoveries have been made in this way. Sequencing the 3×10^9 base pairs of the human genome was in another league altogether. The analogies shown in Table 10.3 may help put the scale of the project into some sort of context.

The Human Genome Project provides some nice time frames: 100 years from Mendelian genetics to draft sequence (1900–2000) and 50 years from double helix to finished sequence (1953–2003). This 50-year period is likely to prove one of the most significant in human history.

10.4.1 Whose genome, and how many genes does it contain?

A question that is often asked is, 'whose genome was used for the project, and how does this relate to the other 6 billion or so variations that exist?' As might be expected, there is no simple answer to this. The positive aspect of this question is that it can largely be ignored – it does not really matter *whose* genome is sequenced, as the phenotypic differences between individuals are generated from very little overall variation in the sequence itself (around 0.1% – this still represents around 3 million base-pair differences between individuals!). Such differences or **polymorphisms** are in fact one area that is of great interest when examining how the genome functions; thus, particular loci may be sequenced many times for different reasons. The reality is that many individual genomes contributed DNA for the mapping and sequencing studies that led eventually to the generation of the entire sequence.

The other question that has been the subject of much debate concerns the number of genes in the genome. The figure of around 100 000 was often quoted as a first estimate in the early days of genome research. Estimates made in the early 1990s suggested around 50 000 genes, but this was revised upwards as more gene sequences were determined. Some people even put the number as high as 150 000. In short, when scientists began the genome project they did not really know with any certainty how many genes our genome would disclose when we finally were able to read the information fully. As genome sequencing began to generate large amounts of data, the estimate of the number of human genes began to creep downwards, through 40 000, then 30 000–35 000, and then to between 20 000 and 25 000. At one point a website of one of the major database presenting centres was taking bets on the actual number! The November 2006 release of the **Ensembl** annotated database (release 41) listed 22 205 known genes; thus, we appear to be a lot simpler than we first thought.

Estimates of the number of genes in the human genome have been reduced significantly with each phase of development of the genome project, dropping from around 100 000 to the 22 000 that have so far been identified from the finished sequence.

10.4.2 Genetic and physical maps of the human genome

In many ways the critical phase of the genome project was not the actual sequencing, but the genetic and physical mapping required to enable the sequence to be compiled against a reference map. Genetic mapping has been hampered by the fact that experimental crosses cannot be set up in humans, and therefore mapping must be a retrospective activity. In many cases tracing the inheritance pattern of genetic markers associated with disease can provide much useful information. The **Centre d'Etude Polymorphism Humain** (**CEPH**) in Paris maintains a set of reference cell lines extending over three generations (families of four grandparents, two parents, and at least six children). These have been extremely useful in mapping studies.

Many thousands of diseases of genetic origin have been identified in which the defect is traceable as a **monogenic** (single-gene) disorder. Currently over 17 000 entries (of various types) are listed in **Online Mendelian Inheritance in Man**, which is the central datastore

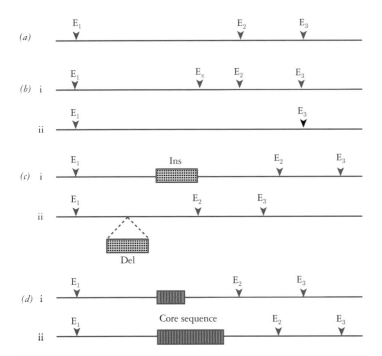

Fig. 10.9 Possible ways of generating restriction fragment length polymorphisms (RFLPs). (*a*) Consider a DNA fragment with three *Eco*RI sites (E$_1$, E$_2$, and E$_3$). On digestion with *Eco*RI, one of the fragments produced is E$_1$ → E$_2$. RFLPs can be generated if the relative positions of these two sites are altered in any way. (*b*) The effect of point mutations. If a point mutation creates a new *Eco*RI site, ((*b*) i, marked E$_X$), fragment E$_1$ → E$_2$ is replaced with two shorter fragments, E$_1$ → E$_X$ and E$_X$ → E$_2$. If a point mutation removes an *Eco*RI site ((*b*) ii, site E$_2$ removed), the fragment becomes E$_1$ → E$_3$, which is longer than the original fragment. (*c*) The effect of insertions or deletions. If additional DNA is inserted between E$_1$ and E$_2$ (Ins in (*c*) i), fragment E$_1$ → E$_2$ becomes larger. Insertions might also carry additional *Eco*RI sites, which would affect fragment lengths. If DNA is deleted (Del in (*c*) ii), the fragment is shortened. (*d*) The effect of variable numbers of repetitive core sequence motifs. This variation can be considered as a type of RFLP, which can be used as the basis of genetic profiling. (*d*) i has 10 copies of the repeated core sequence element; thus, fragment E$_1$ → E$_2$ is smaller than that shown in (*d*) ii, which has 24 copies of the core sequence. Differences in the lengths of fragments shown may be detected using Southern blotting and a suitable probe.

site for this information. This can be found at URL [http://www.ncbi.nlm.nih.gov/Omim/]. Many of these diseases have already been studied extensively using the retrospective technique of **pedigree analysis**. However, to generate useful genetic map data, it is often not necessary to be able to trace the actual gene responsible for the phenotypic effect. If a polymorphic marker can be identified that almost always segregates with the target gene, this can be just as useful. These are called **neutral molecular polymorphisms**. In the early 1980s a marker of this group called a **restriction fragment length polymorphism** (RFLP) began to be used to map genes. RFLPs (sometimes referred to as 'rifflips') are differences in the lengths of specific restriction fragments generated when DNA is digested with a particular enzyme (Fig. 10.9). They are produced when there is a variation

Mendelian Inheritance in Man (MIM) was started by Victor McKusick, who is often called 'the father of medical genetics'. MIM is a major resource for collating and distributing information about genetically based diseases. It is of course now available online as OMIM.

in DNA that alters either the recognition sequence or the location of a restriction enzyme recognition site. Thus, a point mutation might abolish a particular restriction site (or create a new one), whereas an insertion or deletion would alter the relative positions of restriction sites. If the RFLP lies within (or close to) the locus of a gene that causes a particular disease, it is often possible to trace the defective gene by looking for the RFLP, using the Southern-blotting technique in conjunction with a probe that hybridises to the region of interest. This approach is extremely powerful and enabled many genes to be mapped to their chromosomal locations before high-resolution genetic and physical maps became available. Examples include the genes for Huntington disease (chromosome 4), cystic fibrosis (chromosome 7), sickle-cell anaemia (chromosome 11), retinoblastoma (chromosome 13), and Alzheimer disease (chromosome 21).

Using RFLPs as markers, a genetic map of the human genome was available by 1987. However, this approach was limited in terms of the degree of polymorphism (a restriction site can only be present or absent) and in the level of resolution. The use of **minisatellites** and **microsatellites** enabled more detailed genetic maps to be constructed. Minisatellites are made up of tandem repeats of short (10–100 base pair) sequences. The number of elements in a minisatellite region can vary, and thus these are also known as **variable number tandem repeats** (**VNTR**s, Fig. 10.10). These can be used in mapping studies and also formed the basis of **genetic fingerprinting** (now more commonly referred to as **DNA profiling**, discussed in Chapter 12). One drawback with VNTRs is that they are not evenly distributed in the genome and tend to be located at the ends of chromosomes. Microsatellites have been used to overcome this difficulty. These are much shorter repeats, and the two base-pair CA repeat has been used as the standard microsatellite in mapping studies. Regions of CA repeats can be amplified using PCR, with primers that flank the repeated elements. Because the primers are derived from unique-sequence regions, this essentially means that microsatellites amplified in this way are a type of STS (see Table 10.1) and can, therefore, be used to link genetic and physical maps.

Development of physical mapping techniques along with genetic mapping enabled more detailed analysis of the genome in the latter half of the 1990s. A physical map with over 15 000 STS markers was published in 1995, and by 1998 this had been extended to 30 000 physical markers. Along with this increasing level of coverage and resolution, various types of software were developed to enable the map data to be correlated and viewed sensibly, and to be integrated with the emerging sequence data. The end result of all this activity was that, in a relatively short period, mapping of the genome had been moved on much more than many people had ever dared to hope.

> Genetic mapping of the human genome was an essential part of the preparative work that eventually resulted in the completion of the genome sequencing project.

10.4.3   Deriving and assembling the sequence

Sequence determination is perhaps the least troublesome area of modern genome analysis – particularly now that automated sequencing is the norm. However, dealing with the 3 billion As, Gs, Ts,

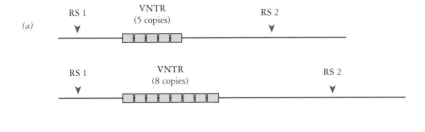

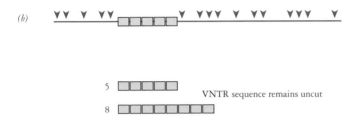

Fig. 10.10 Variable number tandem repeats (VNTRs) can generate what are essentially RFLPs, as shown in Fig 10.9(d) and in this figure (a). Here restriction sites 1 and 2 (RS1 and RS2) are separated by five and eight copies of a repeat element for illustration. If fragments are produced by cutting at the recognition sites for RS1 and RS2, different fragment lengths will be produced. (b) An alternative method of analysing VNTRs. If a restriction enzyme is used that cuts the flanking DNA frequently (shown by arrowheads) but does NOT cut within the VNTR, the VNTR sequence is effectively isolated, trimmed, and left as a marker that can be used in DNA profiling (fingerprinting).

and Cs of the human genome is not trivial! As with physical mapping, progress was more rapid than many thought possible, and reliable methods for generating large amounts of sequence information quickly and accurately were established by the mid 1990s.

Sequencing the genome was carried out using a number of different techniques, although all had similar aims. One key aspect of genome sequencing is the production of what are called **sequence-ready clones**. In theory, obtaining the human genome sequence could be done using a straightforward **shotgun sequencing** approach. However, the limiting step in this method is the task of putting the sequences together, which is very difficult for a genome as large as the human genome, in which there are many repetitive sequences. A **directed shotgun method**, in which there is some attempt to link the shotgun sequences to map data, is better but is still likely to generate some anomalies.

One favoured method for large-scale sequencing is based on the **clone contig method**. In this, sequences from one clone are used to identify similar sequences in contiguous genome regions from other clones, either by hybridisation or by PCR amplification. Clones with minimum (but unambiguous) overlap can be selected for further processing. This method is the one most likely to produce an accurate sequence in finished form, as it links clones together and can therefore be checked easily by multiple sequencing of different clones where this is required. The use of BAC vectors is a popular and efficient method of generating the DNA fragments; some 300 000 of these could in theory represent the entire human genome. By determining the sequences of the ends of the clones (to assist with clone ordering), and then assembling each clone sequence by a small-scale shotgun approach, sequence data can be generated in a controlled and accurate way.

The 'race' between the public and private sequencing groups provided an interesting personal and political dimension to the genome project, which had until then been mainly seen as a technical and scientific challenge.

One interesting aspect of the genome project has been the perceived 'race' between the public consortium of sequencing centres and a private company, Celera Genomics, established by Craig Venter in 1998 with the Perkin-Elmer Corporation. Venter, with a far-sighted vision, claimed that he could sequence the genome much more efficiently than the public consortium, by using a **whole genome shotgun** procedure. There was a good deal of what euphemistically might be called 'debate' when Venter put forward his ideas, particularly as he was going to use the freely available public consortium sequence data to assist his own efforts, which would result in Celera releasing its own data on a commercial basis. The whole matter became rather public and somewhat heated at times. However, it is undoubtedly true that Venter's appearance on the scene added to the urgency of the project (hence the so-called 'race' to finish the sequence). The fact that Venter, along with Francis Collins of the public consortium, shared in the U.S. announcement of the first draft of the sequence in June 2000 shows that Celera Genomics had made a significant input and that their contribution was at least recognised if not fully applauded or appreciated.

10.4.4 Presentation and interrogation of the sequence

In Chapter 9 we looked at how bioinformatics is reshaping the way that biologists deal with information. In particular, Section 9.4.4 used an example from the HGP to illustrate the general principles behind the presentation of database information. In this section we will look a little more closely at how gene sequence information can be presented and used.

Let's look again at one site that presents genome information – the Wellcome Trust Sanger Institute in Cambridge, UK. The best way to follow this example is to have the website active in front of you as you read the text. Go to the URL [http://www.sanger.ac.uk] and you should see the Institute's home page appear. You now have access to a set of resources that would have been unimaginable just a few years ago, and just a few clicks of your computer mouse takes you into the genome sequence itself. We will look at one example of a gene entry to illustrate the scope of the presentation of genome information.

Presentation interfaces and annotation schemes turn the raw sequence data into a resource of almost unimaginable scope.

The protein we will look at is **p53**, which is a **tumour-suppressor** (sometimes known as TP53) with a key role to play in the control of cell division. It is particularly important in the cell, as mutations in the gene can lead to loss of control of cell division and, hence, to cancers. From the home page of the Sanger Institute website, select Ensembl from the Database Resources section of the menu bar. This takes you to a list of genomes that are available using the Ensembl presentation interface. Select *Homo sapiens* to get into the human genome. The front page of the human genome section should appear, with some interesting facts and figures about the genome. There is a search box in the top right-hand corner of the page – select the box with the drop-down list, and type TP53 into the empty box. Click 'Go' or press the Enter key. In a few moments a listing of several genes will appear – select the one with identifier ENSG00000141510. This takes you into the gene report for this gene, which lists the types of information and links that can be accessed from this page. We will consider three journeys from this page to illustrate the sort of information that can be found easily.

In the Genomic Location section, click on the link that takes you to chromosome 17 – the locator base-pair figures are 7 512 464–7 531 642. This takes you to a screen showing the p53 gene as being on the short arm of chromosome 17 at band p13.1. In the detailed view and base-pair view sections, the location of the gene and its transcripts can be viewed at various magnifications. When you have had a look, get back to the gene report screen and, in the Description section, click on the Uniprot/SWISSPROT link. This takes you to the entry for p53 in the UniProt Knowledgebase. Scrolling down the entry gives you some idea of the type of information available through this interface, with the amino acid sequence data presented at the end of the listing. Go back to the gene report screen, and this time we want to scroll down to the Similarity Matches section, showing links to other database identifiers (note that you can access the p53 UniProt entry from this section by clicking the UniProt/Swiss-Prot link). However, scroll down until you come to MIM (Mendelian Inheritance in Man). Two sets of links are found here – the MIM disease links take you to entries in the online version of MIM that describe disease states associated with the gene, whilst the MIM gene entry (191170) takes you to the entry itself in MIM.

The searches just outlined illustrate the complexity and scope of the human genome information that can be accessed using Ensembl. With the addition of a few links to other resources, we have accessed the chromosomal location, gene sequence, information about the protein and the amino acid sequence, and information about the role of the protein in disease. If you have a look through some of the other resources available for p53, I think you will agree that the use of IT-based presentation tools has opened up access to information in a simple and user-friendly way.

10.5 | 'Otheromes'

The molecular biology era has been responsible for presenting a whole raft of new disciplines under the 'omics' unbrella. However, we should remember that the ecological term **biome** has in fact been in use for a long time, being first used in 1916. More recently, it was probably inevitable that once the genome and genomics had become

Over the past few years there has been a significant expansion in 'omes' and 'omics' – from originating in the term *biome*, the suffix is now used to describe a range of concepts and applications.

established as useful concepts, the terminology would be adopted for other purposes; **proteome** and **transcriptome** were early additions to the list. These terms are used to represent the total complement of proteins in a particular cell (the proteome) and the set of transcripts produced at any given point in time (the transcriptome). However, there are now many additional 'omes' that are not only terms to describe particular aspects of cell function but are also active research areas at the forefront of modern cell and molecular biology. One source [http://www.genomicglossaries.com/content/omes.asp] currently lists over 100 different variations of 'ome' and 'omics', and the renowned journal *Nature* has a website called the 'Omics Gateway' [http://www.nature.com/omics/index.html]. We will consider some aspects of the emerging field of 'omics' in the final sections of this chapter.

10.5.1   The transcriptome

The genome represents the entire collection of genes, regulatory regions, repetitive sequences, and intergenic regions. However, in any particular cell only a subset of genes in the genome will be actively expressed at any one time. In Section 2.5.3 we saw that some genes have general 'housekeeping' functions and are not highly regulated, whilst others can be adaptively or developmentally regulated depending on the type of cell and the environment it finds itself in. The set of transcripts produced from genes active in a cell is called the transcriptome, and obviously the analysis of the transcriptome has a key part to play in the investigation of gene expression. One way of achieving this is to use DNA microarray technology.

DNA microarray technology has revolutionised the way that we carry out analysis of the transcriptome.

There are several different ways that microarray technology can be used for transcriptome analysis. One typical application is to examine the pattern of transcripts produced by two different cell samples, as shown in Fig. 10.11. The samples could be cells grown under different conditions, or cells from different tissues. A microarray is constructed with either cDNAs or oligonucleotides and used to investigate the pattern of transcript binding in each of the two cell types. In this way it is possible to identify gene sequences that are expressed in one sample but not in the other.

A second approach to transcript analysis, often used to analyse normal and disease states (such as in cancer cells), is shown in Fig. 10.12. In this case the microarray or DNA chip is challenged with a mixture of cDNAs produced from samples of mRNA taken from normal and disease cells. The mRNAs from each cell are used to produce fluorescent-tagged cDNAs, with the normal and disease samples being tagged with different coloured dyes. The samples are mixed and incubated with the microarray. The cDNAs will bind to their complementary sequences on the microarray, and the pattern produced can be used to identify genes that are expressed differentially in normal and cancer cells.

Whilst the transcriptome represents the set of transcripts that exist at any given point, the term **expressome** is sometimes used

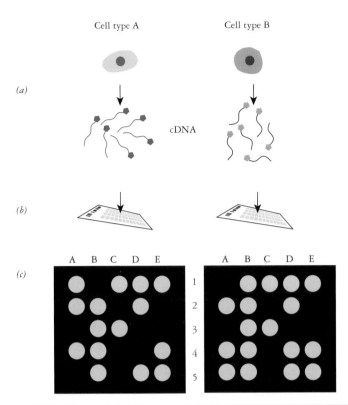

(a)

(b)

(c)

cDNA

Fig. 10.11 Transcriptome analysis in two cell types using a microarray. An array is generated with the sequences of interest attached to the support matrix. Samples of DNA are generated from the two cell types. Often this is achieved by synthesising cDNA, which is tagged with a fluorescent dye (*a*). The cDNA samples are hybridised separately to the microarray (*b*) and the signals analysed. Results are compared as shown in (*c*). A black spot indicates no expression of the gene represented by the field. In this example a positive result is shown by the blue circles. Comparison of the results shows where genes are differentially expressed. Thus, the gene in position A1 is expressed in cell A but not in B, whilst B1 is the reverse. Some genes (e.g. C2) are not expressed in either of the cell types, and some (e.g. E5) are expressed in both.

to indicate the entire complement of products of gene expression, including transcripts, proteins, and other components of cells. This term is also sometimes applied to tissues, organisms, and even species.

10.5.2   The proteome

Genes themselves are of course only the starting point for the study of how genetics enables cells to perform all the various functions that are required. The complementary area of how proteins function is now being re-examined in the light of increased knowledge about genomes. The term proteome is used to describe the set of proteins encoded by the genetic information in a cell, and great advances are being made in this area of research. The concept of the cell as a **molecular machine** is one that goes some way to describing the emphasis of modern molecular biology, in which genes are just part of

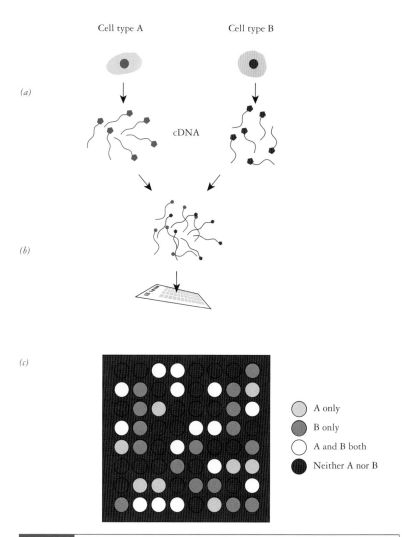

Cell type A   Cell type B

(a)

cDNA

(b)

(c)

A only

B only

A and B both

Neither A nor B

Fig. 10.12 Transcriptome analysis using pooled cDNAs to investigate differential gene expression. In this example, (a) mRNA from the two cells is used to prepare cDNA that is labelled with two different fluorescent dyes. (b) The cDNAs are mixed and hybridised to the array. The pattern of signal detected is shown in (c). The various colours produced indicate the levels of transcript in each of the cells. This can give an indication of differential expression of genes using a single microarray.

the overall picture. The biochemical interactions between the various proteins of the cell are just as important as the information in the genes themselves, and the ultimate aim of cell and molecular biology is to understand cells fully at the molecular level. In many ways this is the biologist's holy grail, similar in scope to the search in physics for the unifying theory that will link the various branches of the discipline together.

Some indication of the complexity of the proteome in a cell can be obtained by using basic techniques such as polyacrylamide gel electrophoresis (PAGE). Although the standard one-dimensional PAGE technique is useful, it is limited in terms of the number of proteins that can be resolved. Two-dimensional electrophoresis (2D PAGE) gives much greater resolution. In this method, proteins are first separated in one direction. The proteins are then separated in a second direction, often using a different gel system. One common technique uses isoelectric focusing in the first dimension followed by SDS-PAGE in the second. In this way several hundred proteins can usually be identified, although the technique can be limited by the sensitivity of the detection methods used. Additional methods such as mass spectrometry, X-ray crystallography, and nuclear magnetic resonance techniques can add valuable information about protein structure.

Several different techniques are required to move towards a full understanding of the structure and function of the many proteins that make up the proteome.

The advent of DNA microarrays enabled researchers to produce arrays using different types of molecules, which has facilitated studies on the proteome using this technology. One way of achieving this is to fix antibodies to the support matrix and use this to monitor the binding of proteins to their respective antibodies. This technology can also be used for other applications in proteomics research, which is now well established as a discipline and has given rise to the Human Proteome Organisation. This was established in 2001 to provide a focus for the study of the proteome. More information can be found at http://www.hupo.org/.

10.5.3 Metabolomes and interactomes

Extension of the 'ome' concept to other contents of the cell can be useful. If the aim of modern cell biology is to attempt to understand the way that a cell works, then all the reactions and interactions need to be considered. Two additional terms that can be used to describe facets of cell function are the **metabolome** and the **interactome**. The metabolome describes the small molecules that are found in the cell and includes the various metabolites that are essential for cell function. Thus, with the genome, transcriptome, proteome, and metabolome, we have essentially a way of describing the components of cells. The next step towards a full understanding of cell function is to see how these components interact with each other. This gave rise to the concept of what became known as the **interactome**. Analysis of the interactome can be carried out by constructing a map of protein–protein interactions, which provides some insight into how a cell functions. The principle of mapping interactions between proteins is shown in Fig. 10.13. As with research into the transcriptome and the proteome, analysis of cellular interactions is supported by a wide range of online resources, including a collaboration between the **European Bioinformatics Institute**, **Cold Spring Harbour Laboratory**, and the **Gene Ontology Consortium**. This is known as **Reactome** [see http://www.reactome.org/]. The collaboration's aims are to track information across a range of areas of research and to collate and curate data on interactions. Release 18 (August 2006) showed information from 23 organisms, spanning 26 areas of interest. Entries for *Homo sapiens* listed 1452 proteins and 1338 complexes, covering 1784 reactions and 686 metabolic pathways. Whilst perhaps not

Assembling data for all the various interactions that happen in cells is opening up a new level of understanding of how cells work.

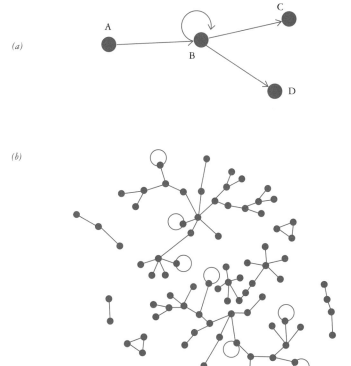

Fig. 10.13 Mapping interactions between proteins. (*a*) Four proteins are shown. Protein A interacts with B, which interacts with C and D. Protein B also interacts with itself (circular path). Using this type of analysis, a picture of interactions can be established as shown in (*b*). In this case the direction arrows have been omitted for clarity. There are several variations of computer software packages that can be used for the construction of interactome maps.

numerically in the same league as the 3 billion base pairs of the genome sequence, this aspect of research is likely to become the critical area for developing our understanding of how cells work.

10.6 | Life in the post-genomic era

The emphasis in genome research is now not so much on what the sequence *is*, but on what it *does*. Over the past few years, interpretation of the genome sequence has enabled a start to be made on the integration of genomics and proteomics and has led to the developments in other 'omic' areas discussed earlier. We are beginning to get a greater understanding of how cells and tissues work at the molecular level, although this is likely to take some time to develop. One obvious question to ask is, 'what impact will this have in the years ahead?'

Undoubtedly a major focus will be on the medical advances that will arise as a result of genome research. The wealth of information that already exists for genetically based diseases (see Online Mendelian Inheritance in Man, as noted earlier) will be supplemented with more detailed analysis at the molecular level, and this will facilitate new treatments in areas such as drug design and gene therapy (discussed more fully in Chapter 12).

A development of mapping that adds a further level of detail to genetic and physical maps is the identification of **single nucleotide polymorphisms** (SNPs). These are, as the name suggests, regions in which a single base is different between individuals, and they represent the major source of sequence variation. Thus, one sequence might read AGTTCGATGCG, and in another person might be AGTTAGATGCG, with the C at position 5 changed to an A. SNPs are (by definition) so small that they have not been subjected to the usual pressures of evolution, as most of these changes will not affect the reproductive fitness of an individual. Thus, they have become scattered evenly across the genome, one every few hundred base pairs. Whilst many will perhaps not be directly attributable as the cause of a particular disease, they will be useful markers for bringing an increased level of subtlety to genome-based diagnosis.

In addition to sequence data and SNPs, human genome research has recently uncovered a feature of the genome that was unexpected and has profound implications for the analysis of disease. In November 2006 an international team of researchers published an analysis of 270 genomes from a range of different groups of people. The research was aimed at looking for variations in the number of copies of genes and DNA sequences, rather than at small sequence differences such as SNPs. The striking finding was that there are many more differences between individual genomes than was first thought to be the case, and that these are caused by what are known as **copy number variations** (CNVs). One particularly exciting aspect of the work was the correlation between some of the CNVs found and the occurrence of disease traits as listed in the Online Mendelian Inheritance in Man database. This development shows how genome research is already having a major impact in a wider context than sequence analysis alone could provide.

The use of microarrays and DNA chips has already had a profound effect on the analysis of genome structure and expression, and development of new applications in this area is likely to expand as the technology becomes more accessible. This happened with DNA sequencing itself, which over a period of a few years moved from small-scale laboratory-based sequencing to centres of excellence where automated large-scale sequencing is carried out. The fast-developing fields of proteomics and analysis of intracellular interactions will also have major impacts in the area of cell and molecular biology.

Despite all the positive developments, concerns about the misuse of genome information have been around for as long as the project has been underway. There are very real **ethical** concerns. These range

An exciting extension to the analysis of SNPs and DNA sequences is the discovery of copy number variations (CNVs) in the genome; this is likely to be of major importance in helping us understand the puzzle of differential susceptibility to various diseases.

from the dilemma of whether or not to tell someone of a latent genetic condition that will appear in later life, to the use of genome information to discriminate against someone in areas such as life assurance or career path. There are also **legal** aspects that are difficult – the patenting of gene sequence information being one particularly troublesome area. However, these aspects are not peculiar to the genome project, as ethical dilemmas occur in many areas of science and technology. With common sense and a solid regulatory framework, many of the concerns can be alleviated. The whole aspect of ethics in genome analysis and gene manipulation generally is sometimes called **ELSI** (ethical, legal, and social implications) and will be discussed further in Chapter 15.

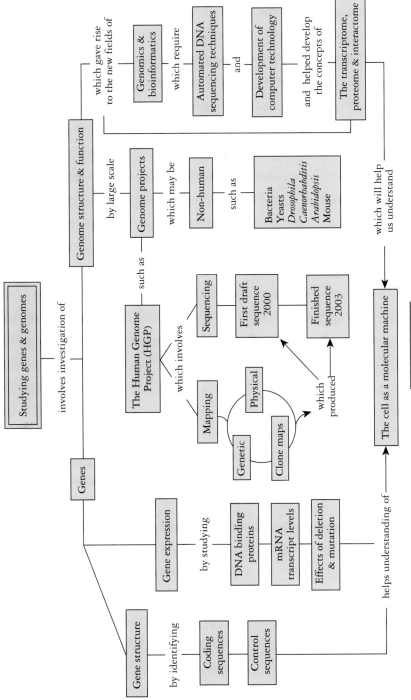

Concept map 10

Chapter 11 summary

Aims

- To outline the range and scope of biotechnology
- To illustrate the impact of rDNA technology on biotechnological applications
- To describe how recombinant DNA is used to produce proteins
- To outline the issues surrounding scale-up to commercial production capacity
- To describe a range of rDNA-based biotechnological applications

Chapter summary/learning outcomes

When you have completed this chapter you will have knowledge of:

- The impact that gene manipulation is having on biotechnology
- The production of proteins using rDNA technology
- The aims of and techniques used for protein engineering
- Funding and developing a biotechnology campany
- Issues involved in scale-up from laboratory to production plant
- The production and applications of a range of rDNA-generated products

Key words

Biotechnology, post-translational modification (PTM), process engineering, native proteins, fusion proteins, reading frame, *Saccharomyces cerevisiae*, *Schizosaccharomyces pombe*, *Pichia pastoris*, *Hansela polymorpha*, *Kluyveromyces lactis*, *Yarrowia lipolytica*, alcohol oxidase, baculovirus, polyhedra, polyhedrin, transfer vector, downstream processing (DSP), protein engineering, muteins, rational design, directed evolution, mutagenesis *in vitro*, oligonucleotide-directed mutagenesis, site-directed mutagenesis, error-prone PCR, DNA shuffling, small to medium entreprise (SME), seedcorn funding, venture capital, process economics, separation, concentration, purification, formulation, filtration, centrifugation, sedimentation, precipitation, evaporation, adsorption, chromatography, crystallisation, freeze-drying, rennet, chymosin, *Rhizomucor miehei*, *Endothia parasitica*, *Rhizomucor pusillus*, *Aspergillus niger*, genetically modified organism (GMO), proteases, lipases, Lipolase™, recombinant bovine somatotropin (rBST), technology transfer, bovine growth hormone, Food and Drug Administration (FDA), Posilac™, insulin-like growth factor (IGF-1), medical diagnostics, replacement, supplementation, specific disease therapy, recombinant vaccines, diabetes mellitus (DM), insulin, type I DM, insulin-dependent DM (IDDM), type II DM, non-insulin-dependent DN (NIDDM), injection, infusion, inhalation, A-chain, B-chain, C-chain, proinsulin, Humulin™, tissue plasminogen activator (TPA), plasminogen, plasmin, fibrin, Activase™, vaccine, hepatitis B, transgenic plants.

Chapter 11

Genetic engineering
and biotechnology

Biotechnology is one of those difficult terms that can mean different things to different people. In its broadest context, it is essentially the use of a biological system or process to improve the lot of humankind. More specifically, it can involve using a cell or organism (maybe a mammalian cell, or a microorganism) or a biologically derived substance (usually an enzyme) in a production or conversion process. Biotechnology has a very long history, having been applied in a societal context long before any elements of scientific development of the field became established. Thus, brewing and wine-making, and the production of bread and cheese, are processes that have been used for centuries. More highly developed food-processing and manufacturing techniques came on the scene much later, and biotechnology reached full scientific maturity when the biological and biochemical aspects of the topic began to be understood more fully. This led to the diverse range of modern applications, which (in addition to continued food and beverage production) include the production of specialist chemicals and biochemicals, pharmaceutical and therapeutic products, and environmental applications such as the treatment of sewage and pollutants.

> Biotechnology has a very long history and is not dependent on gene manipulation technology. However, the advent of rDNA techniques has helped the development of biotechnology in a way that would not have been possible otherwise.

In many biotechnological applications the organism or enzyme is used in its natural form and is not modified, apart from perhaps having been subjected to selection methods to enable the best strain or type of enzyme to be used for a particular application. Thus, whilst there is no *requirement* for genetic engineering to be associated with the discipline, modern biotechnology is often linked with the use of genetically modified systems. In this chapter we will consider the impact that gene manipulation technology has had on some biotechnological applications, with particular reference to the production of useful proteins.

The products of biotechnological processes are destined for use in a variety of fields such as medicine, agriculture, and scientific research. It is perhaps an arbitrary distinction to separate the production of a therapeutic protein from its clinical application, as both could be considered as 'biotechnology' in its broadest sense. In a similar way, the developing area of transgenic plants and animals is also part of biotechnology, and the information provided by genome sequencing will give rise to many more diverse biotechnological applications. The impact of gene manipulation in medicine and transgenics will be considered in more detail in Chapters 12 and 13.

The production of recombinant proteins is now a well-developed area of research and development, and there is a bewildering range of different vector/host combinations. Whilst many workers may still wish to develop their own vectors with specific characteristics, it is now possible to buy vector/host combinations that suit most common applications. As with basic cloning vectors, the commercial opportunities presented by increasing demand for sophisticated expression systems have been exploited by many suppliers, and a look through supplier's catalogues or websites is a good way to get an overview of the current state of the technology.

> The production of proteins using rDNA technology was one of the earliest applications of gene manipulation in the biotechnology industry.

11.1 | Making proteins

The synthesis and purification of proteins from cloned genes is one of the most important aspects of genetic manipulation, particularly where valuable therapeutic proteins are concerned. Many such proteins have already been produced by recombinant DNA (rDNA) techniques and are already in widespread use; we will consider some examples later in this chapter. In many cases a bacterial host cell can be used for the expression of cloned genes, but often a eukaryotic host is required for particular purposes. Eukaryotic proteins are often subjected to **post-translational modification (PTM)** *in vivo*, and it is important that any modifications are achieved in an expression system if a functional protein is to be produced.

In protein production there are two aspects that require optimisation: (1) the biology of the system and (2) the production process itself. Careful design of both aspects is required if the overall process is to be commercially viable, which is necessary if large-scale production and marketing of the protein is the aim. Thus, biotechnological applications require both a biological input and a **process engineering** input if success is to be achieved, and one of the key challenges in establishing a particular application is the scale-up from laboratory to production plant. We will consider this aspect more fully in Section 11.3.

> In biotechnology, the biological aspects of the application have to be complemented by the development of suitable engineering processes if commercial success is to be realised.

11.1.1 Native and fusion proteins

For efficient expression of cloned DNA, the gene must be inserted into a vector that has a suitable promoter (see Table 6.3 for some early examples of inducible promoters) and which can be introduced into an appropriate host such as *E. coli*. Although this organism is not ideal for expressing eukaryotic genes, many of the problems of using *E. coli* can be overcome by constructing the recombinant so that the expression signals are recognised by the host cell. Such signals include promoters and terminators for transcription, and ribosome binding sites

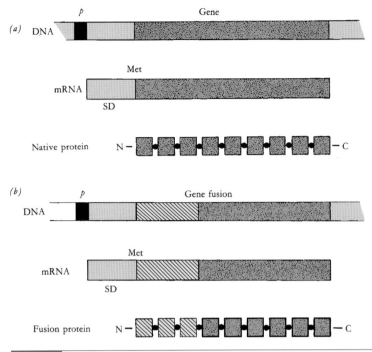

Fig. 11.1 Native and fusion proteins. (*a*) The coding sequence for the cloned gene (shaded) is not preceded by bacterial coding sequence; thus, the mRNA encodes only insert-specified amino acid residues. This produces a native protein, synthesised from its own N terminus. (*b*) The gene fusion contains bacterial codons (hatched); therefore, the protein contains part of the bacterial protein. In this example the first three N-terminal amino acid residues are of bacterial origin (hatched). The ribosome-binding site, or Shine–Dalgarno sequence, is marked SD.

(Shine–Dalgarno sequences) for translation. Alternatively, a eukaryotic host such as the yeast *S. cerevisiae*, or mammalian cells in tissue culture, may be more suitable for certain proteins.

For eukaryotic proteins, the coding sequence is usually derived from a cDNA clone of the mRNA. This is particularly important if the gene contains introns, as these will not be processed out of the primary transcript in a prokaryotic host. When the cDNA has been obtained, a suitable vector must be chosen. Although there is a very wide variety of expression vectors, there are two main categories, which produce either **native proteins** or **fusion proteins** (Fig. 11.1). Native proteins are synthesised directly from the N terminus of the cDNA, whereas fusion proteins contain short, N-terminal amino acid sequences encoded by the vector. In some cases these may be important for protein stability or secretion and are thus not necessarily a problem. However, such sequences can be removed if the recombinant is constructed so that the fusion protein contains a methionine residue at the point of fusion. The chemical cyanogen bromide (CNBr) can be used to cleave the protein at the methionine residue, thus releasing the desired peptide. A major problem with this

Two main classes of proteins that can be generated using rDNA techniques are native and fusion proteins.

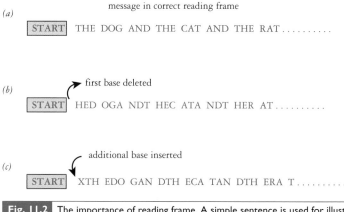

(a)

message in correct reading frame

START THE DOG AND THE CAT AND THE RAT . . . .

(b)

first base deleted

START HED OGA NDT HEC ATA NDT HER AT . . . .

(c)

additional base inserted

START XTH EDO GAN DTH ECA TAN DTH ERA T . . . .

Fig. 11.2 The importance of reading frame. A simple sentence is used for illustration. (*a*) The message has been 'cloned' downstream from the start site and is readable, as it is in the correct reading frame. (*b*) A deletion of one base at the start is enough to knock out the sense completely. Addition of an extra base also causes problems, as shown in (*c*).

approach occurs if the protein contains one or more internal methionine residues, as this will result in unwanted cleavage by CNBr.

When constructing a recombinant for the synthesis of a fusion protein, it is important that the cDNA sequence is inserted into the vector in a position that maintains the correct **reading frame**. The addition or deletion of one or two base pairs at the vector/insert junction may be necessary to ensure this, although there are vectors that have been constructed so that all three potential reading frames are represented for a particular vector/insert combination. Thus, by using the three variants of the vector, the correct in-frame fusion can be obtained. The importance of reading frame is shown in Fig. 11.2.

> When constructing a recombinant to express a protein, it is vital that the correct reading frame is maintained; otherwise the protein may not be synthesised correctly.

11.1.2 Yeast expression systems

As noted in Chapter 5, the yeast *Saccharomyces cerevisiae* has been the favoured microbial eukaryote in the development of recombinant DNA technology. Whilst *S. cerevisiae* is still useful for gene expression studies and protein production, other yeasts may offer advantages in terms of growth characteristics, yield of heterologous protein, and large-scale fermentation characteristics. Species that are used include *Schizosaccharomyces pombe*, *Pichia pastoris*, *Hansela polymorpha*, *Kluyveromyces lactis*, and *Yarrowia lipolytica*. These demonstrate many of the characteristics of bacteria with respect to ease of use – they grow rapidly on relatively inexpensive media, and a range of different mutant strains and vectors is available for various applications. In some cases scale-up fermentations present some difficulties compared to bacteria, but these can usually be overcome by careful design and process monitoring. Yields of heterologous proteins of around 12 g L^{-1} (10–100 times more than in *S. cerevisiae*) have been obtained using *P. pastoris*, which can be grown on methanol as sole carbon source. In this situation growth is regulated by the enzyme **alcohol oxidase**, which has a low specific activity and is

> There are several different yeasts that can be used in biotechnological processes, each of which has its own particular characteristics, which may suit a range of applications.

consequently overproduced in these cells, making up around 30% of total soluble protein. By placing heterologous genes downstream from the alcohol oxidase promoter (AOX1), high levels of expression are achieved.

One of the advantages of using yeast as opposed to bacterial hosts is that proteins are subjected to PTMs such as glycosylation. In addition, there is usually a higher degree of 'authenticity' with respect to three-dimensional conformation and the immunogenic properties of the protein. Thus, in a situation where the biological properties of the protein are critical, yeasts may provide a better product than prokaryotic hosts.

11.1.3   The baculovirus expression system

Baculoviruses infect insects and do not appear to infect mammalian cells. Thus, any system based on such viruses has the immediate attraction of a low risk of human infection. During normal infection of insect cells, virus particles are packaged within **polyhedra**, which are nuclear inclusion bodies composed mostly of the protein **polyhedrin**. This is synthesised late in the virus infection cycle and can represent as much as 50% of infected cell protein when fully expressed. Whilst polyhedra are required for infection of insects themselves, they are not required to maintain infection of cultured cells. Thus, the polyhedrin gene is an obvious candidate for construction of an expression vector, as it encodes a late-expressed dispensable protein that is synthesised in large amounts.

The baculovirus genome is a circular double-stranded DNA molecule. The genome size is from 88 to 200 kb, depending on the particular virus, and the genome is therefore too large to be directly manipulated. Thus, insertion of foreign DNA into the vector has to be accomplished by using an intermediate known as a **transfer vector**. These are based on *E. coli* plasmids and carry the promoter for the polyhedrin gene (or for another viral gene) and any other essential expression signals. The cloned gene for expression is inserted into the transfer vector, and the recombinant is used to co-transfect insect cells with non-recombinant viral DNA. Homologous recombination between the viral DNA and the transfer vector results in the generation of recombinant viral genomes, which can be selected for and used to produce the protein of interest. Systems based on baculoviruses demonstrate transient gene expression, in which the protein of interest is synthesised as part of the infection cycle of the viral-based vector system. More recently, stable insect-cell expression systems have also been developed, in which the cells can be used for continuous expression of protein.

One disadvantage of using insect cells as opposed to bacteria or yeast is that they require more complex growth media for maintenance and production. The cells are also less robust than the microbial cells and, thus, require careful handling if success is to be achieved. However, there are advantages, such as a greater degree

The baculovirus expression system offers an alternative to microbial systems and can be useful for expressing certain types of eukaryotic protein.

of fidelity of expression and PTMs that are more likely to reflect the situation *in vivo*.

11.1.4 Mammalian cell lines

Insect and mammalian cells require more complex media and are generally less robust than microbial cells such as bacteria and yeasts.

Where the expression of recombinant human proteins is concerned, it might seem obvious that a mammalian host cell would be a better system than bacteria, eukaryotic microbes, or insect cells. However, the use of such cell lines in protein production presents some problems. As with insect cell lines, the media required to sustain growth of mammalian cells are complex and expensive, and the cells are relatively fragile when compared with microbial cells, particularly where large-scale fermentation is involved. There may also be difficulties in the processing of the products (often the term **downstream processing (DSP)** is used to describe the operations needed to purify a protein from a fermentation process). Despite these difficulties, many vectors are now available for protein expression in mammalian cells. They exhibit characteristics that will by now be familiar – often based on a viral system, vectors utilise selectable markers (often drug-resistance markers) and have promoters that enable expression of the cloned gene sequence. Common promoters are based on simian virus (SV40) or cytomegalovirus. Some examples were presented in Table 5.3.

11.2 | Protein engineering

Altering the characteristcs of proteins by modifying gene sequences is a powerful and widely used technique that can greatly speed up the identification of novel protein variants.

One of the most exciting applications of gene manipulation lies in the field of **protein engineering**. This involves altering the structure of proteins *via* alterations to the gene sequence and has become possible because of the availability of a range of techniques, as well as a deeper understanding of the structural and functional characteristics of proteins. This has enabled workers to pinpoint the essential amino acid residues in a protein sequence; thus, alterations can be carried out at these positions and their effects studied. The desired effect might be alteration of the catalytic activity of an enzyme by modification of the residues around the active site, an improvement in the nutritional status of a storage protein, or an improvement in the stability of a protein used in industry or medicine. Proteins that have been engineered by the incorporation of mutational changes have become known as **muteins**.

There are two types of approach that can be used to engineer proteins. These are sometimes called **rational design** and **directed evolution**. We will consider these separately, although recent developments have begun to bring together the two approaches in novel procedures that make best use of the characteristics of both.

11.2.1 Rational design

The use of a rational design protocol depends on some detailed information about the protein being available. Typically the target protein may have been characterised biochemically, and its gene cloned

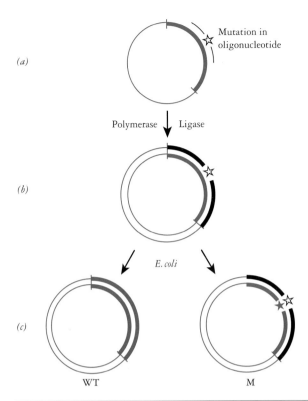

(a)

Mutation in oligonucleotide

Polymerase | Ligase

(b)

E. coli

(c)

WT

M

Fig. 11.3 Oligonucleotide-directed mutagenesis. (*a*) The requirement for mutagenesis *in vitro* is a single-stranded DNA template containing a cloned gene (heavy line). An oligonucleotide is synthesised that is complementary to the part of the gene that is to be mutated (but which incorporates the desired mutation). This is annealed to the template (the mutation is shown as an open star). (*b*) The molecule is made double-stranded in a reaction using DNA polymerase and ligase, which produces a hybrid wild-type/mutant DNA molecule with a mismatch in the mutated region. (*c*) On introduction into *E. coli* the molecule is replicated, thus producing double-stranded copies of the wild-type (WT) and mutant (M) forms. The mutant carries the original mutation and its complementary base or sequence (filled star).

and sequenced. Thus, the mRNA coding sequence, predicted or actual amino acid sequence, three-dimensional structure, folding characteristics, and so on may be available. This information is used to predict what the effect of changing part of the protein might be; the change can then be made and the altered protein tested to see if the desired changes have been incorporated. A procedure known as **mutagenesis *in vitro*** enables specific mutations to be introduced into a gene sequence. One variant of this technique is called **oligonucleotide-directed** or **site-directed mutagenesis** and is elegantly simple in concept (Fig. 11.3). The requirements are a single-stranded (ss) template containing the gene to be altered, and an oligonucleotide (usually 15–30 nucleotides in length) that is complementary to the region of interest. The oligonucleotide is synthesised with the desired mutation as part of the sequence. The ss template is often produced using the M13 cloning system, which produces ss DNA. The template and

Mutagenesis *in vitro* is an elegant technique that is used to introduce defined mutations into a cloned DNA sequence to alter the amino acid sequence of the protein that it encodes.

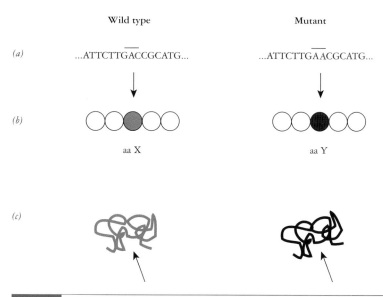

Wild type   Mutant

(a)   ...ATTCTTGACCGCATG...   ...ATTCTTGAACGCATG...

(b)

aa X   aa Y

(c)

Fig. 11.4 Protein engineering by the rational design method. In this procedure, some knowledge of the gene and protein sequence is required. A change is then introduced into the gene, as shown in (a). In this case the codon GAC is altered to GAA by changing the third base in the codon DNA sequence. This causes a change in the amino acid sequence, shown in (b) as a change from aa X to aa Y. This causes a change in the way that the protein folds, shown in (c). In this example the active site (shown by the arrows) becomes slightly larger in the altered molecule.

oligonucleotide are annealed (the mutation site will mismatch, but the flanking sequences will confer stability), and the template is then copied using DNA polymerase. This gives rise to a double-stranded DNA. When this is replicated it will generate two daughter molecules, one of which will contain the desired mutation.

Identification of the mutated DNA can be carried out by hybridisation with the mutating oligonucleotide sequence, which is radiolabelled. Non-mutated DNA will retain the original mismatch, whereas the mutant will match perfectly. By washing filters of the suspected mutants at high stringency, all imperfect matches can be removed and the mutants detected by autoradiography. Even a single base-pair change can be picked up using this technique. The mutant can then be sequenced to confirm its identity.

Having altered a gene by mutagenesis, the protein is produced using an expression system. Often a vector incorporating the *lac* promoter is used, so that transcription can be controlled by the addition of IPTG. Alternatively, the λP_L promoter can be used with a temperature-sensitive λcI repressor, so that expression of the mutant gene is repressed at 30°C but is permitted at 42°C. Analysis of the mutant protein is carried out by comparison with the wild-type protein. In this way, proteins can be 'engineered' by incorporating subtle structural changes that alter their functional characteristics. An overview of the rational design concept is shown in Fig. 11.4.

11.2.2 Directed evolution

One of the main disadvantages of the rational design approach to protein engineering is that a very large range of potential structures can be derived by modification of a protein sequence in various ways. Thus, it is very difficult to be sure that the modification that is being incorporated will have the desired effect – and the process is labour-intensive. Directed evolution is a recent development that has increased the range and scope of producing new protein variants. As the name suggests, the technique is more like an evloutionary process, rather than the incorporation of a specific alteration in a defined part of the protein. The increase in potential benefit comes not from the fact that any *particular* structural alteration is created, but that a large number of different alterations are generated and the desired variant selected by a process that mimics natural selection. Thus, the need for predictive structural alteration is removed, as the system itself enables large numbers of changes to be generated and screened efficiently.

The process generates a library of recombinants that encode the protein of interest, with random mutagenesis (often using techniques such as **error-prone PCR**) applied to generate the variants. Thus, a large number of different sequences can be generated, some of which will produce the desired effect in the protein. These can be selected by expressing the gene and analysing the protein using a suitable assay system to select the desired variants. Additional rounds of mutagenesis and selection can be applied if necessary. An extension to this technique called **DNA shuffling** can be used to mix pieces of DNA from variants that show desired characteristcs. This mimics the effect of recombination that would occur *in vivo* and can be an effective way of 'fast-tracking' the directed evolution technique.

The technique of directed evolution removes the need for predictive alterations to DNA and protein sequences and is potentially more powerful than site-directed methods.

11.3 | From laboratory to production plant

As stated in the introduction to this chapter, biotechnology covers a range of disciplines and is not easily defined. In this section we will take a wide-ranging view and consider that the field covers any biological system, product, or process that is developed at a technological level and exploited on a commercial basis. Some examples could be the production of amino acids or enzymes by traditional fermentation technology methods, drug discovery and production, products for use in other industries, food additives, and healthcare products. On this basis we would exclude any application that is not significantly high-tech; thus, bulk gathering and processing of materials would not be classed as a biotechnological process even though it may have a significant commercial impact (*e.g.* food production and processing, salmon farming, and other agricultural applications).

In this section we will look at how the transition from laboratory to production plant can be achieved. This will be of necessity a brief treatment of a complex topic but will allow us to consider some of

The range of outputs generated from biotechnology applications is extensive and covers a wide spectrum of different types of product for use in areas as diverse as food processing, agriculture, healthcare, and scientific research.

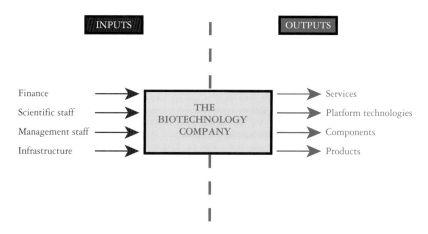

Finance → THE BIOTECHNOLOGY COMPANY → Services

Scientific staff → → Platform technologies

Management staff → → Components

Infrastructure → → Products

Fig. 11.5 The anatomy of a biotechnology company. The requirements are shown as inputs on the left-hand side of the diagram and include finance, staff, and infrastructure. The outputs are shown on the right-hand side. The output may be a single product or service, or might be a range of related products. The critical aspect that will determine how successful a new company will be is of course what happens inside the box – a good product, with good management of the staff and cost-effective production processes, may lead to sustainable production and ultimately profit and return for its investors.

the traps and hurdles that need to be negotiated successfully if a profitable biotechnology company is to emerge.

11.3.1 Thinking big – the biotechnology industry

Despite the range of applications and disciplines that make up the biotechnology sector, there are certain similarities across all areas. At the heart of any biotechnology process there has to be first-class science, and many applications are developed from fundamental research carried out in universities and research institutes. Often 'spin-off' companies are associated with the academic institution that employs the scientists who have developed the idea. Science parks have grown up over the past 20 years or so, with the aim of ensuring that appropriate location, infrastructure, and access to expertise is available to companies who locate to the area.

The anatomy of a biotechnology company can be summarised as a set of basic requirements or inputs, and a potential set of outputs. This is shown in Fig. 11.5. At the outset a company will usually have a particular output in mind – it may be a high-value product, development of a cloning technology, or perhaps some other service provision. Sometimes a company may be able to offer more than one output, but care must be taken to ensure that resources are not overstretched.

A critical aspect of converting an idea into a marketable product or service is financing of the business. This is particularly crucial in the early stages of development, as many potentially sound business ideas have failed to get further than the initial stages because of a drying up

Although biotechnology companies may be very different in terms of what they do, there are certain key elements (inputs and outputs) that define their basic operational characteristics.

of available funds. It is at this stage that a 'new start' small company (often known as an **SME** or **small to medium entreprise**) is at its most vulnerable. Companies that are extensions to the business of large multinational parent companies can often avoid these threats in the early stages.

Raising finance for a new biotechnology venture is not an easy process, and it can involve several stages. Usually there will have already been a significant investment in an idea before it ever gets to the biotechnological application stage – this is the investment in the basic science that led to the proposal for a biotech company. It is impossible to quantify this level of funding in most cases, as it can involve several years of effort with many research staff in universities or research institutes. Once an idea has been developed to the point where it can be exploited, finance is required to develop the core idea. This is sometimes known as **seedcorn funding** and may be supplied by government agencies (such as local enterprise companies), corporate lenders, or private investors. Seedcorn funding provides the cashflow that is essential to get a company up and running but is usually not sufficient to enable more than establishing the company and initial development of the idea. The second phase of funding is often provided by **venture capital (VC)** sources. This stage of funding often marks the transition from 'good idea' to a viable product and involves the 'proof-of-concept' stage and development of production capability. This is usually the most exposed stage for investment input, as it carries the high level of risk (and potential high reward). Investors who manage VC funds will not necessarily be expecting a short-term return on investment, but they will want to investigate the company, its business plan, and its staff before providing support. It is at this stage that the quality of the staff is most important, as there may be only a few people involved at this stage and their commitment and tenacity may be what marks the difference between success and failure. In the early stages the scientists who are involved with the project are often the major part of the management team, but this may change as the company begins to grow and there is a need for more specialised and formalised management structures.

Having survived the proof-of-concept stage, there is still no certainty that the product will make it to market and be profitable. At this stage there is often the need for another large injection of funds, perhaps by second-stage VC funding or by developing links with a larger corporate partner. This period can often be one of the most tricky in a company, as it may involve clinical trials or steering the product through regulatory procedures in the countries in which it will be marketed. It is only when all aspects of the product development and approval are in place that commercial production and distribution can begin. An illustration of funding levels and timescales for a biotechnology company in the early stages of development is shown in Fig. 11.6.

> Although a sound and commercially viable process or product is obviously essential, the most important factor in enabling a company to achieve success is the provision of realistic levels of funding for each of the stages of its development.

> As a company grows and develops, the roles of staff will change as responsibilities become more specific and demanding.

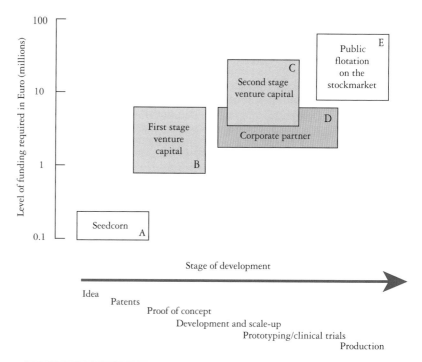

Fig. 11.6 Funding for the intial phases of a biotechnology company. The Y axis shows the level of funding in Euro (millions). The X axis is a timeline showing the various stages of development. At start-up seedcorn funding is required; levels and duration are indicated by box A. This is usually followed by the first major input of investment (box B). When the company is in the developmental stage, perhaps with a patented product that is going on to the trial stage, additional funds are usually essential and are provided by either second-stage venture capital funding or perhaps by a corporate partner. The final stage may be a flotation on the stockmarket and operational production. Modified from Bains and Evans (2001), The business of biotechnology. In *Basic Biotechnology*, 2nd Edition, Eds. Ratledge and Kristiansen. Cambridge University Press, Cambridge. Reproduced with permission.

11.3.2   Production systems

In addition to securing financial backing for a new biotechnology company, there are of course many aspects of the *science* to be considered if success is to be achieved. The fact that an idea has reached the stage of potential commercial exploitation usually means that the basic science has been tested and perhaps a patent obtained. During early development the emphasis is usually on getting the processes up and running, with little thought for efficiency or cost of materials. However, when considering the development to production scale, these aspects become critical, as controlling costs will have a major impact on the selling price and, thus, on profitability and potential market share.

One of the first considerations is the system to be used for production. Usually this is fairly clear-cut and is determined by the process itself; this will have a number of specific requirements, which in turn will suggest a range of operational possibilities. Production of a recombinant protein by fermentation of a microbial culture will require a different type of system than production of monoclonal antibodies. Even established and reliable systems may require fine tuning for a particular application, with initial testing and development often carried out at laboratory-scale levels in the first instance.

In addition to securing appropriate financing for a new company, the start-up staff must ensure that the product or process is developed from initial concept to commercial production capability.

11.3.3 Scale-up considerations

Having determined the process that is most suitable for the application, the next stage is to scale this up to a level where commercial production is realistic and cost-effective. There are several major aspects that have to be considered at this point. First the *physical requirements* of the production plant need to be established. This can include the type of process, bioreactor design, control and monitoring systems, infrastructure, and the disposal of waste products. In parallel with this the *biological* aspects have to be considered – will the process work most efficiently at 100, 1000, or 10000 litres? Is there a contamination issue that is best resolved by using a relatively small-scale system that is easy to sterilise, even if this does not provide the most efficient production? Finally, there are aspects of **process economics** that have to be considered, with careful analysis of input costs and management of the process to reduce wastage. Both capital costs (setup and recurrent) and operational costs are important in these calculations.

The scale-up stage is critical in that an efficient and cost-effective process will have a much higher chance of realising a profit than a process in which there is significant waste or inefficiency. Given the number of things that need to be considered, evaluated, designed, and controlled, it is clear that this is one of the most complex aspects of the development of a biotechnology company.

11.3.4 Downstream processing

Having reached the stage of actually making something, with (hopefully) a well-designed production plant operating at optimum efficiency, the next problem is how to process, package, market, and distribute the product. The first of these stages is often called downstream processing (DSP) and involves separation of the product from any co-products or wastes, and bringing it to a stage where the formulation is suitable for distribution. Further packaging and distribution may be carried out onsite, or the product may be transported in bulk to a separate packaging and distribution site. Often the volume and value of a product will have a bearing on its post-production fate, with high-value/low-volume products easily packaged at or close to the site of production.

When a product has been made, there are usually several additional steps required to purify, concentrate, formulate, package, and distribute the material.

DSP encompasses a range of techniques, many of which are well established and have been developed for use in the chemical engineering industry. Often the product will be suspended or dissolved in a liquid, perhaps the output of a batch fermentation process. It may be secreted from cells into the medium, or it may remain in the cells

Downstream processing usually involves four key stages, each of which may require a number of procedures, before a product is ready for packaging and distribution.

and may need to be released from them. Cells may be relatively fragile (*e.g.* mammalian cell cultures) or may be tougher yeast or bacterial cells. The product may be heat-sensitive or susceptible to chemical denaturation (*e.g.* a protein), or it may be resistant to many forms of chemical attack. All of these aspects have to be considered when designing a DSP procedure, although in most cases there are four key stages: **separation**, **concentration**, **purification**, and **formulation**.

Separation includes processes such as **filtration**, **centrifugation**, and **sedimentation**. The aim of such a procedure is usually to clarify a suspension. The product may remain in the clarified filtrate or supernatant or may be pelleted or collected on the filter if it is an intracellular product. If in the supernatant, concentration will usually be required, using processes such as **precipitation**, **evaporation**, or **adsorption**. At this point the product has reached the final stages of preparation, and final purification and formulation can involve **chromatography**, **crystallisation**, or **freeze-drying**.

As with any production system operated at a commercial level, the final packaged product hides a whole range of techniques and procedures that began with a set of raw materials and a process. An interesting exercise to illustrate the scope of biotechnology is to select a type of product, pick a brand name, and carry out a websearch for information on how it is produced. This gives some insight into the complexity of the production process for a single product. When this is multiplied up to cover the range of biotechnology products available, it is clear that the field has now matured into a significant international sector with major economic impact.

11.4 Examples of biotechnological applications of rDNA technology

In the final part of this chapter we will consider some examples of the types of products that can be produced using rDNA technology in biotechnological processes. This is a rapidly developing area for many biotechnology companies, with large-scale investment in both basic research and in development to production status. This aspect of gene manipulation technology is likely to become increasingly important in the future, particularly in medicine and general healthcare, with many diverse products being brought to market.

11.4.1 Production of enzymes

The commercial production and use of enzymes is already a well-established part of the biotechnology industry. Enzymes are used in brewing, food processing, textile manufacture, the leather industry, washing powders, medical applications, and basic scientific research, to name just a few examples. In many cases the enzymes are prepared from natural sources, but in recent years there has been a move towards the use of enzymes produced by rDNA methods, where this is possible. In addition to the scientific problems of producing a recombinant-derived enzyme, there are economic factors to take into account, and in many cases the cost-benefit analysis makes the use of a recombinant enzyme unattractive. Broadly speaking, enzymes are either high-volume/low-cost preparations for use in industrial-scale operations or are low-volume/high-value products that may have a very specific and relatively limited market.

There is a nice twist to the gene manipulation story in that some of the enzymes used in the procedures are now themselves produced using rDNA methods. Many of the commercial suppliers list recombinant variants of the common enzymes, such as polymerases (particularly for PCR) and others. Recombinant enzymes can sometimes be engineered so that their characteristics fit the criteria for a particular process better than the natural enzyme, which increases the fidelity and efficiency of the process.

In the food industry, one area that has involved the use of recombinant enzyme is the production of cheese. In cheese manufacture, **rennet** (also known as rennin, chymase, or **chymosin**) has been used as part of the process. Chymosin is a protease that is involved in the coagulation of milk casein following fermentation by lactic acid bacteria. It was traditionally prepared from animal (bovine or pig) or fungal sources. In the 1960s the Food and Agriculture Organisation of the United Nations predicted that a shortage of calf rennet would develop as more calves were reared to maturity to satisfy increasing demands for meat and meat products. Today there are six sources for natural chymosin – veal calves, adult cows and pigs, and the fungi *Rhizomucor miehei*, *Endothia parasitica*, and *Rhizomucor pusillus*. Chymosin is now also available as a recombinant-derived preparation from *E. coli*, *Kluyveromyces lactis*, and *Aspergillus niger*. Recombinant chymosin was first developed in 1981, approved in 1988, and is now used to prepare around 90% of hard cheeses in the UK.

Although the public acceptance of what is loosely called 'GM cheese' has not presented as many problems as has been the case with other areas of gene manipulation of foodstuffs, there are still concerns that need to be addressed. In cheese manufacture, there are three possible objections that can be raised by those who are concerned about GM foods. First, milk could have been produced from cows treated with recombinant growth hormone (see Section 11.4.2). Second, the cows could have been fed with animal feeds containing GM soya or maize. The third concern is the use of recombinant-derived chymosin. Despite these fears, many consumers are content that cheese is not itself a **genetically modified organism (GMO)**, but is the *product* of a *product* of a GMO.

A final example of recombinant-derived proteins in consumer products is the use of enzymes in washing powder. **Proteases** and **lipases** are commonly used to assist cleaning by degradation of protein and lipid-based staining. A recombinant lipase was developed in 1988 by Novo Nordisk A/V (now known as Novozymes). The company is the largest supplier of enzymes for commercial use in cleaning applications. Their recombinant lipase was known as **Lipolase™**, which was the first commercial enzyme developed using rDNA technology

The preparation of enzymes is a central part of biotechnology and ranges from the production of large amounts of low-cost preparations for bulk applications to highly specialised enzymes for use in diagnostics or other molecular biology techniques.

Some aspects of biotechnology, such as the use of GMOs in food preparation or modification, may lead to public concern about the potential impact on health. This is an important aspect that we all have a role in debating.

and the first lipase used in detergents. A further development involved an engineered variant of Lipolase called Lipolase Ultra, which gives enhanced fat removal at low wash temperatures.

11.4.2 The BST story

Not all rDNA biotechnology projects have a smooth passage from inception to commercial success. In Section 11.3 we considered some of the requirements for a biotechnology-based process to be developed. The story of **recombinant bovine somatotropin (rBST)** illustrates some of the problems that may be encountered once the scientific part of the process has been achieved. In bringing a recombinant product such as rBST (and the examples outlined earlier) to market, many aspects have to be considered. The basic science has to be carried out, followed by **technology transfer** to get the process to a commercially viable stage. Approval by regulatory bodies may be required, and finally (and most critical from a commercial standpoint) the product has to gain market acceptance and establish a consumer base. We can find all of these aspects in the BST story.

BST is also known as **bovine growth hormone** and is a naturally occurring protein that acts as a growth promoter in cattle. Milk production can be increased substantially by administering BST and, thus, it was an attractive target for cloning and production for use in the dairy industry. The basic science of rBST was relatively straightforward, and scientists were already working on this in the early 1980s. The BST gene was in fact one of the first mammalian genes to be cloned and expressed, using bacterial cells for production of the protein. Thus, the production of rBST at a commercial level, involving the basic science and technology transfer stages, was achieved without too much difficulty. A summary of the process is shown in Fig. 11.7.

With respect to approval of new rDNA products, each country has its own system. In the USA, the **Food and Drug Administration (FDA)** is the central regulatory body, and in 1994 approval was given for the commercial distribution of rBST, marketed by Monsanto under the trade name **Posilac™**. At that time the European Union did not approve the product, but this was partly for socioeconomic reasons (increasing milk production was not necessary) rather than for any concerns about the science. Evaluation of evidence at that time suggested that milk from rBST-treated cows was identical to normal untreated milk, and it was therefore unlikely that any negative effects would be seen in consumers.

The effects of rBST must be considered in three different contexts – the effect on milk production, the effects on the animals themselves, and the possible effects on the consumer. Milk production is usually increased by around 10–15% in treated cows, although yield increases of much more than this have been reported. Thus, from a dairy herd management viewpoint, use of rBST would seem to be beneficial. However, as is usually the case with any new development that is aimed at 'improving' what we eat or drink, public concern grew along with the technology. The concerns fuel a debate that is still ongoing and is

BST was one of the early successes of biotechnology, in that recombinant BST was the result of achieving the aim of producing a useful protein by expressing cloned DNA in a bacterial host.

(a)

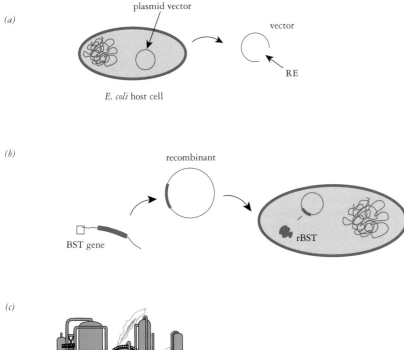

(b)

(c)

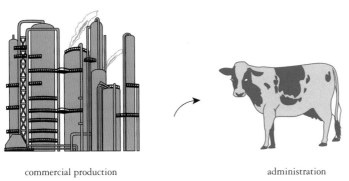

commercial production administration

Fig. 11.7 Production of recombinant bovine growth hormone (rBST). (a) A plasmid vector is prepared from *E. coli* and cut with a restriction enzyme. (b) The BST gene coding sequence is ligated into the plasmid to generate the recombinant, which produces rBST protein in the cell following transformation. Scale-up to commercial production is shown in (c) and, with product approval granted, administration can begin. The whole process from basic science to market usually takes several/many years from start to finish, with a large amount of investment capital required. From Nicholl (2000), *Cell & Molecular Biology*, Advanced Higher Monograph Series, Learning and Teaching Scotland. Reproduced with permission.

at times emotive. One area that is hotly debated is the effect of rBST on the cows themselves. Administering rBST can produce localised swelling at the site of injection and can exacerbate problems with foot infections, mastitis, and reproduction. The counter-argument is that many of these problems occur anyway, even in herds that are rBST-free. On balance, however, the evidence does suggest that animal welfare is compromised to some extent when rBST is used.

The possible effects of rBST use on human health is another area of great concern and debate. The natural hormone (and therefore the recombinant version also) affects milk production by increasing the levels of **insulin-like growth factor (IGF-1)**, which causes increased milk production. Administration of rBST generates elevated levels of IGF-1, and there is evidence that IGF-1 can stimulate the growth of cancer cells. Thus, the concern is that using rBST could pose a risk to health. The counter-argument in this case is that the levels of IGF-1 in the early stages of lactation are higher than those generated by the use of rBST in cows 100 days after lactation begins, which is often when it is administered. This arguably means that milk from early lactating cows should not be drunk at all if there are any concerns about IGF-1. Those who oppose the use of rBST point out that, unlike a therapeutic protein that would be used for a limited number of patients, milk is consumed by most people, and any inherent risk, no matter how small, is therefore unacceptable. On the basis of this uncertainty, many countries have banned the use of rBST, citing both the animal welfare issue and the potential risk to health as reasons. The arguments look set to continue into the future as commercial, animal welfare, and human health interests clash. Following the debate provides an interesting illustration of the problems surrounding the use of gene technology and of the need for objective assessment of risks and the avoidance of emotive judgements.

Concerns about a biotechnology product or process are often multifaceted and can generate emotive debate; it is sometimes difficult to separate evidence from speculation.

11.4.3   Therapeutic products for use in human healthcare

Although the production of recombinant-derived proteins for use in medical applications does raise some ethical concerns, there is little serious criticism aimed at this area of biotechnology. The reason is largely that therapeutic products and strategies are designed to alleviate suffering or to improve the quality of life for those who have a treatable medical condition. In addition, the products are used under medical supervision, and there is a perception that the corporate interests that tend to be highlighted in the food debate have less of an impact in the diagnosis and treatment of disease. In fact, there is just as much competition and investment risk associated with the medical products field as is the case in agricultural applications; there does, however, seem to be less emotive debate in this area, so use in medical applications is therefore apparently much more acceptable to the public. In addition to the actual *treatment* of conditions, the area of **medical diagnostics** is a large and fast-growing sector of the biotechnology market, with rDNA technology involved in many aspects.

Recombinant DNA products for use in medical therapy can be divided into three main categories. First, protein products may be used for **replacement** or **supplementation** of human proteins that may be absent or ineffective in patients with a particular illness. Second, proteins can be used in **specific disease therapy**, to alleviate a disease state by intervention. Third, the production of **recombinant vaccines** is an area that is developing rapidly and that offers great promise. Some examples of therapeutic proteins produced using rDNA technology are listed in Table 11.1. We will consider examples

Recombinant DNA products for medical applications are often more easily accepted by the public than is the case for GMOs used in food production.

Table 11.1. Selected rDNA-derived therapeutic products for use in humans

Product	Type	Trade name	FDA App.	Company	Use
Insulin	R/S	Humulin	1982	Eli Lilly	Diabetes treatment
Growth hormone	R/S	Protropin	1985	Genentech	Growth hormone deficiency in children
α-Interferon	SDT	Intron A	1986	Schering-Plough	Hairy cell leukaemia
Hepatitis B vaccine	V	Recombivax HB	1986	Merck & Co.	Hepatitis B prevention
Tissue plasminogen activator	SDT	Activase	1987	Genentech	Myocardial infarction
Growth hormone	R/S	Humatrope	1987	Eli Lilly	Growth hormone deficiency in children
Hepatitis B vaccine	V	Engerix-B	1989	SmithKline Beecham	Hepatitis B prevention
Factor VIII	R/S	Recombinate rAHF (antihaemophiliac factor)	1992	Baxter Healthcare	Treatment of haemophilia
Factor VIII	R/S	Kogenate	1993	Bayer	Treatment of haemophilia
DNase	SDT	Pulmozyme	1993	Genentech	Treatment of cystic fibrosis symptoms
Erythropoietin	R/S	Epogen	1993	Amgen	Anaemia
Imiglucerase	R/S	Cerezyme	1994	Genezyme	Type I Gaucher disease
Insulin	R/S	Humalog	1996	Eli Lilly	Diabetes treatment
Factor IX	R/S	BeneFix	1997	Wyeth	Treatment of haemophilia
Factor VIIa	R/S	NovoSeven	1999	Novo Nordisk	Treatment of haemophilia bleeding episodes
Insulin analogue	R/S	NovoLog	2000	Novo Nordisk	Diabetes treatment
Insulin analogue	R/S	Lantus	2000	Aventis	Diabetes treatment
Hepatitis A & B vaccine	V	Twinrix	2001	SmithKline Beecham	Mixed vaccine for hepatitis A & B
Laronidase	R/S	Aldurazyme	2003	BioMarin Pharmaceuticals & Genzyme	Enzyme replacement therapy for mucopolysaccharidosis I
Galsulphase	R/S	Naglazyme	2005	BioMarin Pharmaceuticals	Enzyme replacement therapy for mucopolysaccharidosis VI
Hyaluronidase	–	Hylenex	2005	Halozyme Therapeutics	Adjuvant for use with other drugs
Insulin	R/S	Exubera	2006	Pfizer	Inhalable form of insulin for diabetes treatment

Note: Type refers to replacement/supplementation (R/S), specific disease therapy (SDT), or vaccine (V). Dates in column 4 refer to year of first approval by the FDA for use in the USA. Subsequent approvals for additional uses or modifications are not shown. Products are listed chronologically with respect to year of first approval. Trade names are registered trademarks of the companies involved; company names are as given in the approval, and may have changed because of corporate policy, merger, and so forth. Further information can be found on the FDA website at http://www.fda.gov. A useful collated list of therapeutic products can be found at http://www.bio.org/speeches/pubs/er/approveddrugs.asp.

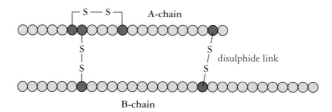

Fig. 11.8 The structure of insulin. Amino acids are represented by circles. The A-chain (21 amino acids) and B-chain (30 amino acids) are held together by disulphide linkages between cysteine residues (filled circles).

from each of these three areas to illustrate the type of approach taken in developing a therapeutic protein.

The widespread condition **diabetes mellitus (DM)** is usually caused either by β-cells in the islets of Langerhans in the pancreas failing to produce adequate amounts of the hormone **insulin**, or by target cells not being able to respond to the hormone. Many millions of people worldwide are affected by DM, and the World Health Organisation estimates that the global incidence will double by 2025. Some 3–6% of people in the UK and the USA have diabetes, although there is thought to be significant underdiagnosis. Sufferers are classed as having either **type I DM** (formerly known as **insulin-dependent DM or IDDM**) or **type II DM** (formerly **non-insulin-dependent DM or NIDDM**). Some 10% of patients have type I DM, with around 90% having type II. There are also some other variants of the disease that are much less common. Type I patients obviously require the hormone, but many type II patients also use insulin for satisfactory control of their condition. Delivery of insulin is achieved by **injection** (traditional syringe or 'pen'-type devices), **infusion** using a small pump and catheter, or **inhalation** of powdered insulin.

Insulin is composed of two amino acid chains, the **A-chain** (acidic, 21 amino acids) and **B-chain** (basic, 30 amino acids). When synthesised naturally, these chains are linked by a further 30–amino acid peptide called the **C-chain**. This 81–amino acid precursor molecule is known as **proinsulin**. The A- and B-chains are linked together by disulphide bonds between cysteine residues, and the proinsulin is cleaved by a protease to produce the active hormone shown in Fig. 11.8. Insulin was the first protein to be sequenced – by Frederick Sanger in the mid 1950s.

As DM is caused by a problem with a normal body constituent (insulin), therapy falls into the category of replacement or supplementation. Banting and Best developed the use of insulin therapy in 1921, and for the next 60 or so years diabetics were dependent on natural sources of insulin, with the attendant problems of supply and quality. In the late 1970s and early 1980s rDNA technology enabled scientists to synthesise insulin in bacteria, with the first approvals granted by 1982. Recombinant-derived insulin is now available in several forms and has a major impact on diabetes therapy.

Recombinant insulin for use in the treatment of diabetes is one of the major success stories of rDNA-based biotechnology in that its availability has had a major impact on the lives of millions of people.

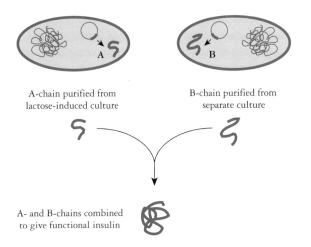

A-chain purified from
lactose-induced culture

B-chain purified from
separate culture

A- and B-chains combined
to give functional insulin

Fig. 11.9 Production of
recombinant-derived insulin by
separate fermentations for A- and
B-chains. Lactose is used to induce
transcription of the cloned gene
sequences from the *lac* promoter.
Following translation, the products
are purified to give A- and B-chains
that are then combined chemically
to give the final product.

One of the most widely used forms is marketed under the name
Humulin™ by the Eli Lilly Company.

In an early method for the production of recombinant insulin,
the insulin A- and B-chains were synthesised separately in two bacterial strains. The insulin A and B genes were placed under the control
of the *lac* promoter, so that expression of the cloned genes could be
switched on by using lactose as the inducer. Following purification
of the A- and B-chains, they were linked together by a chemical process to produce the final insulin molecule. The process is shown in
Fig. 11.9. A development of this method involves the synthesis of the
entire proinsulin polypeptide (shown in Fig. 11.10) from a single gene
sequence. The product is converted to insulin enzymatically.

There are many recombinant proteins for use in specific disease
therapy. One example of this type of protein is **tissue plasminogen
activator** (TPA). This is a protease that occurs naturally and functions
in breaking down blood clots. TPA acts on an inactive precursor protease called **plasminogen**, which is converted to the active form called

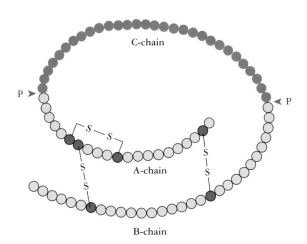

C-chain

P

P

S — S

S

S

A-chain

S

S

B-chain

Fig. 11.10 Proinsulin. This
molecular precursor of insulin is
synthesised as an 81–amino acid
polypeptide. The C peptide
sequence is then removed by a
protease (P) to leave the A- and
B-chains to form the final insulin
molecule. Proinsulin is now
synthesised intact during
rDNA-based production of insulin.

Tissue plasminogen activator is another example of a valuable therapeutic protein that is produced by rDNA technology.

plasmin. This protease attacks the clot by breaking up **fibrin**, the protein that is involved in clot formation. TPA is used as a treatment for heart attack victims. If administered soon after an attack, it can help reduce the damage caused by coronary thrombosis.

Recombinant TPA was produced in the early 1980s by the company Genentech using cDNA technology. It was licensed in the USA in 1987, under the trade name **Activase**, for use in treatment of acute myocardial infarction. It was the first recombinant-derived therapeutic protein to be produced from cultured mammalian cells, which secrete rTPA when grown under appropriate conditions. The amount of rTPA produced in this way was sufficient for therapeutic use; thus, a major advance in coronary care was achieved. Further uses were approved in 1990 (for acute massive pulmonary embolism) and 1996 (for acute ischaemic stroke).

The final group of recombinant-derived products are **vaccines**. There are now many vaccines available for animals, and the development of human vaccines is also beginning to have an impact in healthcare programmes. One vaccine that has been produced by rDNA methods is the **hepatitis B** vaccine. The yeast *S. cerevisiae* is used to express the surface antigen of the hepatitis B virus (HBsAg), under the control of the alcohol dehydrogenase promoter. The protein can then be purified from the fermentation culture and used for inoculation. This removes the possibility of contamination of the vaccine by blood-borne viruses or toxins, which is a risk if natural sources are used for vaccine production.

Vaccine delivery by incorporation into foods is an elegant concept that has the potential to benefit millions of people in countries where large-scale vaccination by traditional methods is difficult.

A further development in vaccine technology involves using **transgenic plants** as a delivery mechanism. This area of research and development has tremendous potential, particularly for vaccine delivery in underdeveloped countries where traditional methods of vaccination may not be fully effective because of cost and distribution problems. The attraction of having a vaccine-containing banana or tomato is clear, and development and trials are currently under way for a variety of plant vaccines.

The use of gene manipulation techniques in the biotechnology industry is a major developing area of applied science. In addition to the scientific and engineering aspects of the work, the financing of biotechnology companies is an area that presents its own risks and potential rewards – for example, a new drug may take 10–15 years to develop, at a cost of several hundreds of millions of pounds. The stakes are therefore high, and many fledgling companies fail to survive their first few years of operation. Even established and well-financed companies are not immune to the risks associated with the development of a new and untried product. The next few years will certainly be interesting for this sector of the applied science industry.

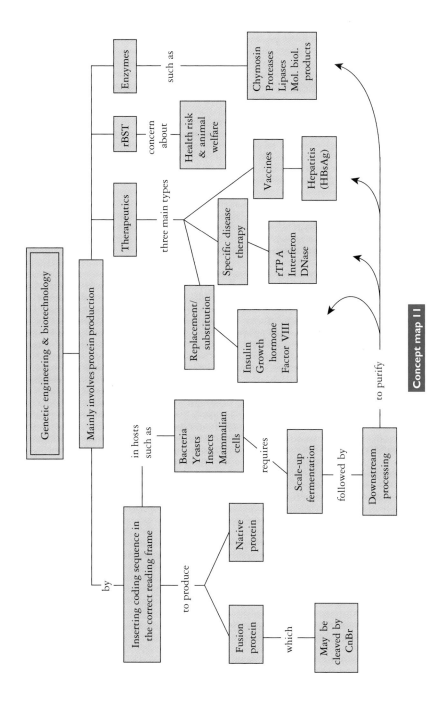

Concept map 11

Chapter 12 summary

Aims

- To outline the impact of gene manipulation in medicine and forensic science
- To describe the form and patterns of inheritance for some genetically based diseases
- To evaluate the potential of gene therapy and RNA interference in therapeutics
- To describe the use of gene manipulation technology in an investigative context

Chapter summary/learning outcomes

When you have completed this chapter you will have knowledge of:

- The diagnosis and characterisation of medical conditions
- The main patterns of inheritance of genetic conditions
- Genetically based diseases
- Gene therapy
- RNA interference and its potential in therapeutic applications
- DNA profiling and its applications in the analysis of genomes

Key words

Congenital, aetiology, human immunodeficiency virus (HIV), acquired immune deficiency syndrome (AIDS), enzyme-linked immunosorbent assay (ELISA), enzyme immunoassay, Western blot, indirect immunofluorescence assay (IFA), false negative, tuberculosis, human papilloma virus, Lyme disease, diploid, autosomes, sex chromosomes, haploid, gametes, meiosis, zygote, monogenic, polygenic, alleles, dominant, recessive, homozygous, heterozygous, autosomal dominant, autosomal recessive, X-linked, maternal pattern of inheritance, penetrance, expressivity, incomplete penetrance, variable expressivity, multiple alleles, incomplete dominance, co-dominance, partial dominance, multifactorial, Online Mendelian Inheritance in Man (OMIM), chromosomal abnormalities (aberrations), gene mutations, ploidy number, triploid, tetraploid, aneuploidy, monosomic, trisomic, Down syndrome, non-disjunction, gene mutations, cystic fibrosis (CF), pancreatic exocrine deficiency, cystic fibrosis transmembrane conductance regulator (CFTR), positional cloning, chromosome walking, chromosome jumping, phenylalanine, allele-specific oligoonucleotides (ASOs), Huntington disease (HD), trinucleotide repeat, muscular dystrophy (MD), Duchenne muscular dystrophy (DMD), Becker muscular dystrophy (BMD), dystrophin, transgene, gene therapy, somatic cell, germ line cell, gene replacement therapy, gene addition therapy, antisense mRNA, *ex vivo* gene therapy, *in vivo* gene therapy, vehicle, retrovirus, adenovirus, adeno-associated virus, liposome, lipoplex, adenosine deaminase deficiency, severe combined immunodeficiency syndrome, enzyme replacement therapy, animal model, ornithine transcarbamylase deficiency, X-linked severe combined immunodeficiency syndrome, enhancement gene therapy, RNA interference (RNAi), down-regulation, knockdown, gene silencing, post-transcriptional gene silencing, dicer, slicer, short interfering RNAs (siRNAs), RNA induced silencing complex (RISC), micro RNAs (miRNAs), age-related macular degeneration (AMD), macula, vascular endothelial growth factor, monozygotic twins, DNA fingerprinting, DNA profiling, minisatellite, variable number tandem repeat (VNTR), multi-locus probe, single-locus probe, short tandem repeat (STR), microsatellite, European Network of Forensic Science Institutes, Combined DNA Index System, molecular paleontology, molecular ecology.

Chapter 12

Medical and forensic applications of gene manipulation

The diagnosis and treatment of human disease is one area in which genetic manipulation is beginning to have a considerable effect. As outlined in Chapter 11, many therapeutic proteins are now made by recombinant DNA (rDNA) methods, and the number available is increasing steadily. Thus, the treatment of conditions by recombinant-derived products is already well established. In this chapter we will look at how the techniques of gene manipulation impact more directly on medical diagnosis and treatment, and we will also examine the use of rDNA technology in forensic science. Recent progress in both of these areas is of course closely linked to our increasing knowledge of the human genome, and new developments in medical and forensic applications will undoubtedly appear as we continue to decipher the genome.

12.1 | Diagnosis and characterisation of medical conditions

Genetically based diseases (often called simply 'genetic diseases') represent one of the most important classes of disease, particularly in children. A disorder present at birth is termed a **congenital** abnormality, and around 5% of newborn babies will suffer from a serious medical problem of this type. In most of these cases there will be a significant genetic component in the **aetiology** (cause) of the disease. It is estimated that about a third of primary admissions to paediatric hospitals are due to genetically based problems, whilst some 70% of cases presenting more than once are due to genetic defects. In addition to genetic problems appearing at birth or in childhood, it seems that a large proportion of diseases presenting in later life also have a genetic cause or predisposition. Thus, medical genetics, in its traditional non-recombinant form, has already had a major impact on the diagnosis of disease and abnormality. The development of molecular genetics and rDNA technology has not only broadened the range of techniques available for diagnosis, but has also opened up the possibility of novel gene-based treatments for certain conditions.

As many disease conditions have a major genetic component, gene manipulation technology has provided new tools for investigating and treating what are sometimes called 'genetic diseases'.

12.1.1 Diagnosis of infection

In addition to genetic conditions that affect the individual, rDNA technology is also important in the diagnosis of certain types of infection. Normally, bacterial infection is relatively simple to diagnose, once it has taken hold. Thus, the prescription of antibiotics may follow a simple investigation by a general practitioner. A more specific characterisation of the infectious agent may be carried out using microbiological culturing techniques, and this is often necessary when the infection does not respond well to treatment. Viral infections may be more difficult to diagnose, although conditions such as *Herpes* infections are usually obvious.

Despite traditional methods being applied in many cases, there may be times when these methods are not appropriate. Infection by the **human immunodeficiency virus** (**HIV**) is one case in point. The virus is the causative agent of **acquired immune deficiency syndrome** (**AIDS**). The standard test for HIV infection requires immunological detection of anti-HIV antibodies, using techniques such as **ELISA** (**enzyme linked inmmunosorbent assay**, sometimes known as the **enzyme immunoassay**), **Western blot**, and **IFA** (**indirect immunofluorescence assay**). However, these antibodies may not be detectable in an infected person until weeks after initial infection, by which time others may have been infected. A test such as this, where no positive result is obtained even though the individual is infected, is a **false negative**. The use of DNA probes and PCR technology circumvents this problem by assaying for nucleic acid of viral origin in the T-lymphocytes of the patient, thus permitting a diagnosis before the antibodies are detectable.

Other examples of the use of rDNA technology in diagnosing infections include **tuberculosis** (caused by the bacterium *Mycobacterium tuberculosis*), **human papilloma virus** infection, and **Lyme disease** (caused by the spirochaete *Borrelia burgdorferi*).

> In some cases, viral infections can be diagnosed by using rDNA techniques (such as PCR) to identify viral DNA before antibodies have reached detectable levels.

12.1.2 Patterns of inheritance

Although diagnosis of infection is an important use of rDNA technology, it is in the characterisation of genetic disease that the technology has perhaps been most applied in medicine to date. Before dealing with some specific diseases in more detail, it may be useful to review the basic features of transmission genetics, and outline the range of factors that may determine how a particular disease state presents in a patient.

Since it was rediscovered in 1900, the work of Gregor Mendel has formed the basis for our understanding of how genetic characteristics are passed on from one generation to the next. We have already seen that the human genome is made up of some 3 billion base pairs of information. This is organised as a **diploid** set of 46 chromosomes, arranged as 22 pairs of **autosomes** and one pair of **sex chromosomes**. Prior to reproduction, the **haploid** male and female **gametes** (sperm and oocyte, respectively) are formed by the reduction division of **meiosis**, which reduces the chromosome number to 23. On

> Transmission genetics, the principles of which were first established by Gregor Mendel, is still an important part of modern medicine. Molecular genetics complements transmission genetics to provide a powerful range of methods for genetic analysis.

fertilisation of the oocyte by the sperm, diploid status is restored, with the **zygote** receiving one member of each chromosome pair from the father and one from the mother. In males the sex chromosomes are X and Y, in females XX, and thus it is the father that determines the sex of the child.

Traits may be controlled by single genes, or by many genes acting in concert. Single-gene disease traits are known as **monogenic** disorders, whilst those involving many genes are **polygenic**. Inheritance of a monogenic disease trait usually follows a basic Mendelian pattern and can therefore often be traced in family histories by pedigree analysis. A gene may have **alleles** (different forms) that may be **dominant** (exhibited when the allele is present) or **recessive** (the effect is masked by a dominant allele). With respect to a particular gene, individuals are said to be either **homozygous** (both alleles the same) or **heterozygous** (the alleles are different, perhaps one dominant and one recessive). Patterns of inheritance of monogenic traits can be associated with the autosomes, as either **autosomal dominant** or **autosomal recessive**, or may be sex-linked (usually with the X chromosome, thus showing **X-linked** inheritance). The Mendelian patterns and ratios for these types of inheritance are shown in Fig. 12.1. In addition to the nuclear chromosomes, mutated genes associated with the mitochondrial genome can cause disease. As the mitochondria are inherited along with the egg, these traits show **maternal patterns of inheritance**. We will consider specific examples of the patterns of inheritance in the next section.

> Genetic traits can be transmitted from generation to generation in different ways. These patterns of inheritance follow set 'rules' and can be useful in the diagnosis and tracing of disease patterns in families.

The effect of a gene depends not only on its allelic form and character but also on how it is expressed. The terms **penetrance** and **expressivity** are used to describe this aspect. Penetrance is usually quoted as the percentage of individuals carrying a particular allele who demonstrate the associated phenotype. Expressivity refers to the degree to which the associated phenotype is presented (the *severity* of the phenotype is one way to think of this). Thus, alleles showing **incomplete penetrance** and/or **variable expressivity** can greatly affect the range of phenotypes derived from what is actually a simple Mendelian pattern of inheritance. Further complications arise when **multiple alleles** are involved in determining traits, or when alleles demonstrate **incomplete dominance**, **co-dominance**, or **partial dominance**. In many cases the route from genotype to phenotype also involves one or more environmental factors, when traits are said to be of a **multifactorial** nature.

Despite the complexities of transmission patterns and outcomes, there are many cases where the defect can be traced with reasonable certainty. As stated in Chapter 10, data for transmission of disease traits are collated in **Online Mendelian Inheritance in Man (OMIM)**, which now runs to over 17000 entries in various categories. The database is the electronic version of the text *Mendelian Inheritance in Man*, by Victor McKusick of Johns Hopkins University, who published the first edition in 1966. McKusick is rightly considered to be the father of medical genetics.

> Online Mendelian Inheritance in Man is another good example of how the availability of powerful desktop computers and the Internet has transformed the way we deal with complex data sets.

(a) Autosomal dominant

♂ Genotype: **Aa**
  Phenotype: *Affected male*

X

♀ Genotype: **aa**
  Phenotype: *Normal female*

Genotype ratio is 1:1
Aa to **aa**
Thus 50% affected
     50% normal

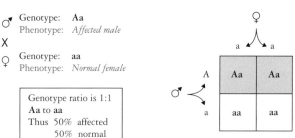

(b) Autosomal recessive

♂ Genotype: **Bb**
  Phenotype: *Carrier male*

X

♀ Genotype: **Bb**
  Phenotype: *Carrier female*

Genotype ratio is 1:2:1
BB/Bb/bb
Phenotype ratio is 3:1
Thus 25% affected
     75% normal

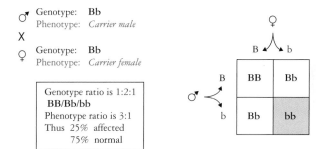

(c) X-linked

♂ Genotype: **XY**
  Phenotype: *Normal male*

X

♀ Genotype: **X^{c}X**
  Phenotype: *Carrier female*

Genotype ratio is 1:1:1:1
XX/XY/X^{c}X/X^{c}Y
Thus 50% of male
children are affected

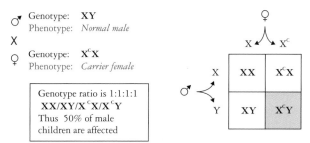

Fig. 12.1 Patterns of inheritance. (*a*) An autosomal dominant disease allele is designated **A**, normal form **a**. Half of the gametes from an affected individual (in this case the male) will carry the disease allele. On mating (box diagram) the gametes can mix in the combinations shown. The result is that half the offspring will be heterozygous and, therefore, have the disease (shown by shaded boxes). (*b*) An autosomal recessive pattern. The disease-causing allele is designated **b**, the normal variant **B**. On a mating between two carriers heterozygous for the defective allele, there is a one-in-four chance of having an affected child. (*c*) An X-linked pattern for a disease allele designated **c**. In this case a recessive allele is shown. Half of the male children will be affected, as there is only one X chromosome and, thus, no dominant allele to mask the effect. No female children are affected. However, in the case of an X-linked dominant allele, females are also affected.

12.1.3 Genetically based disease conditions

Genetic problems may arise from either **chromosomal abnormalities (aberrations)** or **gene mutations**. An abnormal chromosome complement can involve whole chromosome sets (variation in the **ploidy number**, such as **triploid**, **tetraploid**, *etc.*) or individual chromosomes

Table 12.1. Examples of types of chromosomal aberrations in humans

Condition	Chromosome designation	Syndrome	Frequency per live births
Autosomal			
Trisomy-13	47, 13+	Patau syndrome	1:12 500–1:22 000
Trisomy-18	47, 18+	Edwards syndrome	1:6 000 -1:10 000
Trisomy-21	47, 21+	Down syndrome	1:800
Sex chromosome variation			
Missing Y	45, X	Turner syndrome	1:3 000 female births
Additional X	47, XXX	Triplo-X	1:1 200 female births
Additional X	47, XXY	Klinefelter syndrome	1:500 male births
Additional Y	47, XYY	Jacobs syndrome	1:1 000 male births
Structural defects	**Cause**		
Deletion	Part of chromosome deleted e.g. ABCDEFGH → ABFGH		
Duplication	Part of chromosome duplicated e.g. ABCDEFGH → ABCDBCDEFGH		
Inversion	Part of chromosome inverted e.g. ABCDEFGH → ABCFEDGH		
Translocation	Fragment moved to different chromosome e.g. ABCDEFGH → PQRDEFSTUV		
Fragile-X syndrome	Region of X-chromosome susceptible to breakage; known as Martin–Bell syndrome, presenting as 1:1 250 male births and 1:2 500 female births		

Note: Chromosome designation lists the total number of chromosomes, followed by the specific defect. Thus, 47, 13+ indicates an additional chromosome 13, and 47, XXY a male with an additional X chromosome. The syndrome is usually named after the person who first described it; the possessive (*e.g. Down's* syndrome) is sometimes still used, but the modern convention is to use the non-possessive (*e.g. Down* syndrome).

(**aneuploidy**). Any such variation usually has very serious consequences, often resulting in spontaneous abortion, as gene dosage is affected and many genes are involved. Multiple chromosome sets are rare in most animals but are quite often found in plants. As gamete formation involves meiotic cell division in which homologous chromosomes separate during the reduction division, even-numbered multiple sets are most commonly found in polyploid plant species that remain stable.

Aneuploidy is a much more common form of chromosomal variation in humans, but is still relatively rare in terms of live-birth presentations. A missing chromosome gives rise to a **monosomic** condition, which is usually so severe that the foetus fails to develop fully. An additional chromosome gives a **trisomic** condition, which is more likely to persist to term. Monosomy and trisomy can affect both autosomes and sex chromosomes, with several recognised syndromes such as **Down syndrome** (trisomy-21). Most cases involving changes to chromosome number are caused by **non-disjunction** at meiosis during gamete formation. In addition to variation in chromosome number, structural changes can affect parts of chromosomes and can cause a range of conditions. Some examples of chromosomal aberrations in humans are shown in Table 12.1.

Chromosomal abnormalities often have very serious effects on the organism, as the disruption to normal genetic balance is usually severe.

| Table 12.2. | Selected monogenic traits in humans |

Inheritance pattern/disease	Frequency per live births	Features of the disease condition
Autosomal recessive		
Cystic fibrosis	1:2000–1:2500 in Western Caucasians	Ion transport defects; lung infection and pancreatic dysfunction result
Tay-Sachs disease	1:3000 in Ashkenazi Jews	Neurological degeneration, blindness, and paralysis
Sickle-cell anaemia	1:50–1:100 in African populations where malaria is endemic	Sickle-cell disease affects red blood cells; heterozygous genotype confers a level of resistance to malaria
Phenylketonuria	1:2000–1:5000	Mental retardation due to accumulation of phenylalanine
α_1-antitrypsin deficiency	1:5000–1:10000	Lung tissue damage and liver failure
Autosomal dominant		
Huntington disease	1:5000–1:10000	Late onset motor defects, dementia
Familial hypercholesterolaemia	1:500	Premature susceptibility to heart disease
Breast cancer genes BRCA1 and 2	1:800 (1:100 in Ashkenazi Jews)	Susceptibility to early onset breast and ovarian cancers
Familial retinoblastoma	1:14000	Tumours of the retina
X-linked		
Duchenne muscular dystrophy	1:3000–1:4000	Muscle wastage, teenage onset
Haemophilia A/B	1:10000	Defective blood clotting mechanism
Mitochondrial		
Leber hereditary optic neuropathy (LHON)	Mitochondrial defect, maternally inherited/late onset thus difficult to estimate	Optic nerve damage, may lead to blindness, but complex penetrance of the defective gene due to mitochondrial pattern of inheritance

Although chromosomal abnormalities are a very important type of genetic defect, it is in the characterisation of **gene mutations** that molecular genetics has had the most impact. Many diseases have now been almost completely characterised, with their mode of transmission and action defined at both the chromosomal and molecular levels. Table 12.2 lists some of the more common forms of monogenic disorder that affect humans. We will consider some of these in more detail to outline how a disease can be characterised in terms of the effects of a mutated gene.

Cystic fibrosis (CF) is the most common genetically based disease found in Western Caucasians, appearing with a frequency of around 1 in 2000–2500 live births. It is transmitted as an autosomal recessive characteristic and, therefore, the birth of an affected child may be the first sign that there is a problem in the family. The carrier frequency for the CF defective allele is around 1 in 20–25 people. The disease presents with various symptoms, the most serious of which is the clogging of respiratory passageways with thick, sticky mucus. This is too thick to be moved by the cilia that line the air passages, and the patient is likely to suffer persistent and repeated infections. Lung function is therefore compromised in CF patients, and even with improved treatments the life expectancy is only around 30 years. The pancreatic duct may also be affected by CF, resulting in **pancreatic exocrine deficiency**, which causes problems with digestion.

Cystic fibrosis can be traced in European folklore, from which the following puzzling statement comes: '*Woe to that child which when kissed on the forehead tastes salty. He is bewitched and soon must die*'. The condition was first described clinically in 1938, although characterisation of the disease at the molecular level was not achieved until the gene responsible was cloned in 1989. The defect responsible for CF affects a membrane protein involved in chloride ion transport, which results in epithelial cell sheets having insufficient surface hydration – hence the sticky mucus. There is also an increase in the salt content of sweat – hence the statement quoted earlier. The gene/protein responsible for CF is called the **cystic fibrosis transmembrane conductance regulator (CFTR)**. So how was the gene cloned and characterised?

The hunt for the CF gene is a good example of how the technique of **positional cloning** can be used to find a gene for which the protein product is unknown, and for which there is little cytogenetic or linkage information available. Positional cloning, as the name suggests, involves identifying a gene by virtue of its position – essentially by deciphering the molecular connection between phenotype and genotype. In 1985, a linkage marker (called *met*) was found that localised the CF gene on the long arm of chromosome 7. The search for other markers uncovered two that showed no recombination with the CF locus – thus, they were much closer to the CF gene. These markers were used as the start points for a trawl through some 280 kbp of DNA, looking for potential CF genes. This was done using the techniques of **chromosome walking** and **chromosome jumping** to search for contiguous DNA sequences from clone banks (this was before yeast and bacterial artificial chromosome vectors enabled large fragments to be cloned). The basis of chromosome walking and jumping is shown in Fig. 12.2. Using these methods, four candidate genes were identified from coding sequence information, and by tracing patterns of expression of each of these in CF patients, the search uncovered the 5′ end of a large gene that was expressed in the appropriate tissues. This became known as the CFTR gene. A summary of the hunt for CFTR is shown in Fig. 12.3.

Having identified the CFTR gene, more detailed characterisation of its normal gene product, and the basis of the disease state, could begin. The gene is some 250 kbp in size and encodes 27 exons that produce a protein of 1480 amino acids. The protein is similar to the ATP-binding cassette family of membrane transporter proteins. When the gene was being characterised, it was noted that around 70% of CF

Cystic fibrosis is an example of a serious disease that has been studied from the viewpoint of molecular, transmission, and population genetics.

Location and identification of the CFTR gene was a major breakthrough in medical genetics.

(a) Chromosome walking

(b) Chromosome jumping

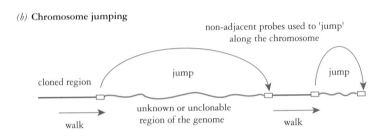

Fig. 12.2 Chromosome walking and jumping. Chromosome walking (a) uses probes derived from the ends of overlapping clones to enable a 'walk' along the sequence. Thus, a probe from clone 1 identifies the next clone, which then provides the probe for the next, and so on. In this way a long contiguous sequence can be assembled. In chromosome jumping (b), regions that are difficult to clone can be 'jumped'. The probes are prepared using a technique that enables fragments from distant sites to be isolated in a single clone by circularising a large fragment and isolating the region containing the original probe and the distant probe. This can then be used to isolate a clone containing sequences from the distant region. Often a combination of walks and jumps is needed to move from a marker (such as an RFLP) to a gene sequence. From Nicholl (2000), *Cell & Molecular Biology*, Advanced Higher Monograph Series, Learning and Teaching Scotland. Reproduced with permission.

cases appeared to have a similar defective region in the sequence – a three-base-pair deletion in exon 10. This causes the amino acid **phenylalanine** to be deleted from the protein sequence. This mutation is called the ΔF508 mutation (Δ for deletion, F is a single-letter abbreviation for phenylalanine, and 508 is the position in the primary sequence of the protein). It affects the folding of the CFTR protein, which means that it cannot be processed and inserted into the membrane correctly after translation. Thus, patients who carry two ΔF508 alleles do not produce any functional CFTR, with the associated disease phenotype arising as a consequence of this. The ΔF508 mutation is summarised in Fig. 12.4.

Molecular characterisation of a gene opens up the possibility of accurate diagnosis of disease alleles. Although screening for CF traditionally involved the 'sweat test', there is now a range of molecular techniques that can be used to confirm the presence of a defective CFTR allele, which can enable heterozygous carriers to be identified with certainty. Two of these involve the use of PCR to amplify a

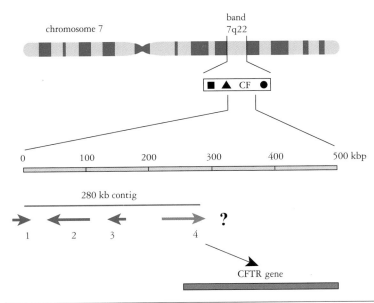

band
7q22

chromosome 7

■ ▲ CF ●

0   100   200   300   400   500 kbp

280 kb contig

1   2   3   4   ?

CFTR gene

Fig. 12.3 The hunt for the cystic fibrosis gene. (a) Mapping studies placed the gene on the long arm of chromosome 7, at band position 7q22. (b) Markers associated with this region (square, triangle, and circle) were mapped in relation to the CF gene. (c) A region of some 500 kbp was examined and a contiguous sequence (clone 'contig') of 280 kbp was identified. This region contained 4 candidate gene sequences or open reading frames (ORFs, labelled 1–4). Further analysis of mRNA transcripts and DNA sequences eventually identified ORF 4 as the start of the 'CF gene', which was named the cystic fibrosis transmembrane conductance regulator (CFTR) gene.

fragment around the ΔF508 region to identify the 3 bp deletion, and the use of **allele-specific oligonucleotides** (ASOs) in hybridisation tests. The use of these techniques is shown in Fig. 12.5.

Although the ΔF508 mutation is the most common cause of CF, to date around 1500 mutations have been identified in the CFTR gene. A database of these is maintained by staff at the Hospital for Sick Children in Toronto, where the gene was discovered, and can be found at http://www.genet.sickkids.on.ca/cftr/app. Many types of mutation have been characterised, including promoter mutations, frameshifts, amino acid replacements, defects in splicing, and deletions. With more sophisticated diagnosis, patients are being diagnosed with milder presentations of CF, which may not appear as early or be as severe as the ΔF508-based disease. Thus, the CF story provides a good illustration of the scope of molecular biology in medical diagnosis, as it has enabled the common form of the disease to be characterised and has also extended our knowledge of how highly polymorphic loci can influence the range of effects that may be caused by mutation.

In the area around Lake Maracaibo in Venezuela, there is a large family group of people who are descended from a woman who had migrated from Europe in the 1800s. Members of this group share a common ailment. They begin to exhibit peculiar involuntary

Many different mutations of the CFTR gene have been identified, although the most prevalent is the absence of phenylalanine at position 508 in the protein.

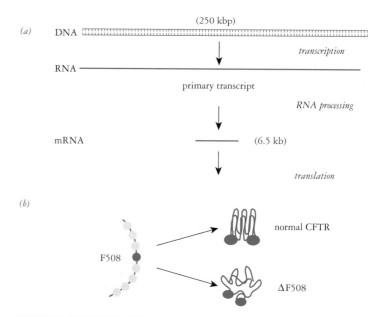

(a) DNA (250 kbp)

transcription

RNA

primary transcript

RNA processing

mRNA ——— (6.5 kb)

translation

(b)

normal CFTR

F508

ΔF508

Fig. 12.4 The cystic fibrosis ΔF508 mutation. (a) The CF gene. Transcription produces the primary RNA transcript that is converted into the functional 6.5 kb mRNA by removal of intervening sequences. On translation the transmembrane conductance regulator protein (CFTR) is produced. (b) The normal and mutant proteins are shown. Normal CFTR has phenylalanine (F) at position 508. In the mutant ΔF508 protein this is deleted, causing the protein to fold incorrectly, which prevents it reaching the site of incorporation into the membrane. From Nicholl (2000), *Cell & Molecular Biology*, *Advanced Higher Monograph Series*, Learning and Teaching Scotland. Reproduced with permission.

movements and also suffer from dementia and depression. Time of onset is usually around the age of 40–50. Their children, who were born when their parents were healthy, also develop the symptoms of this distressing condition, which is known as **Huntington disease** (**HD**; previously known as Huntington's chorea, which describes the *choreiform* movements of sufferers).

A clinical psychologist named Nancy Wexler has made a long-term study of thousands of HD sufferers from the Lake Maracaibo population, by carrying out an extensive pedigree analysis. This confirmed that HD follows an autosomal dominant pattern of inheritance, where the presence of a single defective allele is enough to trigger the disease state. Thus, children of an affected parent have a 50% chance of inheriting the condition. As the disease presents with late onset (relative to childbearing age), many people would wish to know if they carried the defective allele, so that informed choices could be made about having a family. The search for the gene responsible for HD involved tracing a restriction fragment length polymorphism (RFLP) that is closely linked to the HD locus. The RFLP, named G8, was identified in 1983. It segregates with the HD gene in 97% of cases. The HD gene itself was finally identified in 1993, located near the end of the

Pedigree analysis can be an invaluable tool for tracing the pattern of inheritance of a trait in a population.

(*a*) Normal and mutant F508 sequences

Ile   Ile   Phe   Gly

Normal gene sequence   5' - GAA AAT ATC AT **C TT**T GGT GTT TCC - 3'

Mutant gene sequence   5' - GAA AAT ATC ATT GGT GTT TCC - 3'

Ile   Ile   Gly

(*b*) PCR amplification of deleted region

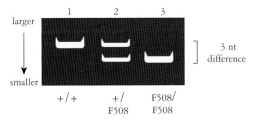

(*c*) Using allele-specific oligonucleotide probes

ASO 1 (normal)   3' - CTTTTATAGTAGAAACCACAAAGG - 5'

ASO 2 (mutant)   3' - CTTTTATAGTAACCACAAAGG - 5'

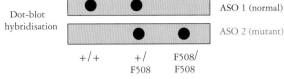

Fig. 12.5   Diagnosis of ΔF508 CF allele. (*a*) The normal and mutant gene sequences around position 508 (Phe) are shown. The deleted 3 base pairs are shaded in the normal sequence. This causes loss of phenylalanine. (*b*) A PCR-based test. A 100-base-pair region around the deletion is amplified using PCR, and the products run on a gel that will discriminate between the normal fragment and the mutant fragment, which will be 3 nucleotides smaller. Lanes 1, 2, and 3 show patterns obtained for homozygous normal (+/+), heterozygous carrier (+/ΔF508), and homozygous recessive CF patient (ΔF508/ΔF508). (c) A similar pattern is seen with the use of allele-specific oligonucleotide probes (ASOs). The probe sequence is shown, derived from the gene sequences shown in (*a*). By amplifying DNA samples from patients using PCR and performing a dot-blot hybridisation with the radiolabelled ASOs, simple diagnosis is possible. In this example hybridisation with each probe separately enables the three genotypes to be determined by examining an autoradiograph.

short arm of chromosome 4. The defect involves a relatively unusual form of mutation called a **trinucleotide repeat**. The HD gene has multiple repeats of the sequence CAG, which codes for glutamine. In normal individuals, the gene carries up to 34 of these repeats. In HD alleles, more than 42 of the repeats indicates that the disease condition will appear. There is also a correlation between the number of repeats and the age of onset of the disease, which appears earlier in cases where larger numbers of repeats are present. As with CF, the availability of the gene sequence enables diagnostic tests to be developed for HD. Using PCR, the repeat region can be amplified and the products separated by gel electrophoresis to determine the number of repeats and, thus, the genetic fate of the individual with respect to HD.

Most X-linked gene disorders are recessive. However, their pattern of inheritance means that they are *effectively* dominant in males (XY) as there is no second allele present as would be the case for females (XX), or in an autosomal diploid situation. Thus X-linked diseases are often most serious in boys, as is the case for **muscular dystrophy (MD)**. This is a muscle wasting disease that is progressive, usually from a teenage onset, which causes premature death. The severe form of the disease is called **Duchenne muscular dystrophy (DMD)**, although there is a milder form called **Becker muscular dystrophy (BMD)**. Both these defects map to the same location on the X chromosome. The MD gene was isolated in 1987 using positional cloning techniques. It is extraordinarily large, covering 2.4 Mb of the X chromosome (that's 2 400 kbp, or around 2% of the total!). The 79 exons in the MD gene produce a transcript of 14 kb, which encodes a protein of 3 685 amino acids called **dystrophin**. Its function is to link the cytoskeleton of muscle cells to the sarcolemma (membrane).

The dystrophin gene, which is involved in muscular dystrophy, is vast – some 2.4 Mb in length.

The MD gene shows a much higher rate of mutation than is usual – some two orders of magnitude higher than other X-linked loci. This may simply be due to the extreme size of the gene, which therefore presents an 'easy target' for mutation. Most of the mutations characterised so far are deletions. Those that affect reading frame generally cause the severe DMD, whilst deletions that leave reading frame intact tend to be associated with BMD.

12.2 | Treatment using rDNA technology – gene therapy

Once genetic defects have been identified and characterised, the possibility of treating the patient arises. If the defective gene can be replaced with a functional copy (sometimes called the **transgene**, as in transgenic) that is expressed correctly, the disease caused by the defect can be prevented. This approach is known as **gene therapy**. Although it has not yet fulfilled its early expectations, it remains one of the most promising aspects of the use of gene technology in medicine. There are two possible approaches to gene therapy: (1) introduction of the transgene gene into the **somatic cells** of the affected tissue or (2) introduction into the reproductive cells (**germ line cells**). These two approaches have markedly different ethical implications. Most scientists and clinicians consider somatic cell gene therapy an acceptable practice – no more morally troublesome than taking an aspirin. However, tinkering with the reproductive cells, with the probability of germ line transmission, is akin to altering the gene pool of the human species, which is regarded as unacceptable by most people.

Gene therapy holds great promise that has not yet been fully realised.

Thus, genetic engineering of germ cells is an area that is likely to remain off limits at present.

There are several requirements for a gene therapy protocol to be effective. First, the gene defect itself will have been characterised, and the gene cloned and available in a form suitable for use in a clinical programme. Second, there must be a system available for getting the gene into the correct site in the patient. Essentially these are vector systems that are functionally equivalent to vectors in a standard gene cloning protocol – their function is to carry the DNA sequence into the target cells. This also requires a mechanism for physical delivery to the target, which may involve inhalation, injection, or other similar methods. Finally, if these requirements can be satisfied, the inserted gene must be expressed in the target cells if a non-functional gene is to be 'corrected'. Ideally, the faulty gene would be replaced by a functional copy. This is known as **gene replacement therapy** and requires recombination between the defective gene and the inserted functional copy. Because of technical difficulties in achieving this reliably in target cells, the alternative is to use **gene addition therapy**. In addition therapy there is no absolute requirement for reciprocal exchange of the gene sequences, and the inserted gene functions alongside the defective gene. This approach is useful only if the gene defect is not dominant, in that a dominant allele will still produce the defective protein, which may overcome any effect of the transgene. Therapy for dominant conditions could be devised using **antisense mRNA**, in which a reversed copy of the gene is used to produce mRNA in the antisense configuration. This can bind to the mRNA from the defective allele and effectively prevent its translation. Antisense technology will be discussed in more detail in Chapter 13.

> Although the concept of gene replacement or substitution therapy is elegantly simple, it is much more difficult to achieve in reality.

A further complication in gene therapy is the target cell or tissue system itself. In some situations it may be possible to remove cells from a patient and manipulate them outside the body. The altered cells are then replaced, with function restored. This approach is known as *ex vivo* **gene therapy**. It is mostly suitable for diseases that affect the blood system. It is not suitable for tissue-based diseases such as DMD or CF, in which the problem lies in dispersed and extensive tissue such as the lungs and pancreas (CF) or the skeletal muscles (DMD). It is difficult to see how these conditions could be treated by *ex vivo* therapy; therefore, the technique of treating these conditions at their locations is used. This is known as *in vivo* **gene therapy**. Features of these two types of gene therapy are illustrated in Fig. 12.6, with both approaches having been used with some success.

12.2.1 Getting transgenes into patients

Before looking at two examples of gene therapy procedures, it is worth reviewing the key methods available for getting the transgene into the cells of the patient. As we have seen, there are two aspects to this.

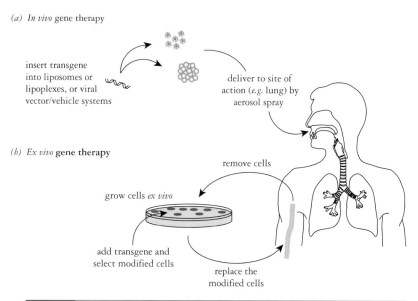

(a) *In vivo* gene therapy

insert transgene
into liposomes or
lipoplexes, or viral
vector/vehicle systems

deliver to site of
action (*e.g.* lung) by
aerosol spray

(b) *Ex vivo* **gene therapy**

remove cells

grow cells *ex vivo*

add transgene and
select modified cells

replace the
modified cells

Fig. 12.6 *In vivo* and *ex vivo* routes for gene therapy. The *in vivo* approach is shown in (*a*). The gene is inserted into a vector or liposome/lipoplex and introduced into target tissue of the patient. In this case the lung is the target, and an aerosol can be used to deliver the transgene. Such an approach can be used with cystic fibrosis therapy. (*b*) The *ex vivo* route. Cells (*e.g.* from blood or bone marrow) are removed from the patient and grown in culture medium. The transgene is therefore introduced into the cells outside the body. Modified cells can be selected and amplified (as in a typical gene cloning protocol with mammalian cells) before they are injected back into the patient.

The *biology* of the system must be established and evaluated, and then the *physical* method for getting the gene to the site of action has to be considered. Deciding on the best method for addressing these two aspects of a therapeutic procedure is one important part of the strategy.

As with vectors for use in cloning procedures, viruses are an attractive option for delivering genes into human cells. We can use the term **vector** in its cloning context, as a piece of DNA into which the transgene is inserted. The viral particle itself is often called the **vehicle** for delivery of the transgene, although some authors describe the whole system simply as a vector system. The main viral systems that have been developed for gene therapy protocols are based on **retroviruses**, **adenoviruses**, and **adeno-associated viruses**. The advantage of viral systems is that they provide a specific and efficient way of getting DNA into the target cells. However, care must be taken to ensure that viable virus particles are not generated during the therapy procedure, as this would potentially be detrimental to the patient.

Many of the problems associated with gene therapy have been due to the viral vectors used for delivery of the treatment gene.

In addition to viral-based systems, DNA can be delivered to target cells by non-viral methods. Naked DNA can be used directly, although this is not an efficient method. Alternatively, the DNA can be encapsulated in a lipid micelle called a **liposome**. Development of this

Table 12.3.	Vector/vehicle systems for gene therapy
System	Features
Viral-based	
Retroviruses	RNA genome, usually used with cDNA, requires proliferating cells for incorporation of the transgene into the nuclear material. Not specific for one cell type and can activate cellular oncogenes.
Adenoviruses	Double-stranded DNA genome, virus infects respiratory and gastrointestinal tract cells, thus effective in non- or slowly dividing cells. Generally provokes a strong immune response.
Adeno-associated viruses	Replication-defective, thus requires helper virus. Some benefits over adenoviral systems; may show chromosome-specific integration of transgene.
Non-viral	
Liposomes	System based on lipid micelles that encapsulate the DNA. Some problems with size, as micelles are generally small and may restrict the amount of DNA encapsulated. Inefficient compared to viral systems.
Lipoplexes	Benefits over liposomes include increased efficiency due to charged groups present on the constituent lipids. Non-immunogenic, so benefits compared to viral systems.
Naked DNA	Inefficient uptake but may be useful in certain cases.

technique produced more complex structures that resemble viral particles, and these were given the name **lipoplexes** to distinguish them from liposomes. Some features of selected delivery systems are shown in Table 12.3.

When a delivery system is available, the patient can be exposed to the virus in a number of ways. Delivery into the lungs by aerosol inhalation is one method appropriate to *in vivo* therapy for CF, as this is the main target tissue. Injection or infusion are other methods that may be useful, particularly if an *ex vivo* protocol has been used.

12.2.2 Gene therapy for adenosine deaminase deficiency

The first human gene therapy treatment was administered in September 1990 to a four-year-old girl named Ashanti DaSilva, who received her own genetically altered white blood cells. Ashanti suffered from a recessive defect known as **adenosine deaminase (ADA) deficiency**, which causes the disease **severe combined immunodeficiency syndrome**. Although a rare condition, this proved to be a suitable target for first steps in gene therapy in that the gene defect was known (the 32 kbp gene for ADA is located on chromosome 20), and an *ex vivo* strategy could be employed. Before gene therapy was available, patients could be treated by **enzyme replacement therapy**. A major development in this area was preparing the ADA enzyme with polyethylene glycol (the main component of antifreeze!) to stabilise delivery of the enzyme. The treatment is still important as an additional supplement to gene therapy, the response to which can be variable in different patients.

> Gene therapy for ADA deficiency was the first successful demonstration that the process could improve the condition.

For the first ADA treatments, lymphocytes were removed from the patients and exposed to recombinant retroviral vectors to deliver the functional ADA gene into the cells. The lymphocytes were then replaced in the patients. Further developments came when bone marrow cells were used for the modification. The stem cells that produce T-lymphocytes are present in bone marrow; thus, altering these progenitor cells should improve the effect of the ADA transgene, particularly with respect to the duration of the effect. The problem is that T-lymphocyte stem cells are present as only a tiny fraction of the bone marrow cells and, thus, efficient delivery of the transgene is difficult. Umbilical cord blood is a more plentiful source of target cells, and this method has been used to effect ADA gene therapy in newborn infants diagnosed with the defect.

12.2.3 Gene therapy for cystic fibrosis

CF is an obvious target for gene therapy, as it presents much more frequently than ADA deficiency and is a major health problem. Drug therapies can help to alleviate some of the symptoms of CF by digestive enzyme supplementation and the use of antibiotics to counter infection. However, as with ADA, enzyme replacement therapy is an attack on the *symptoms* of the disease, rather than on the *cause*. As outlined in Section 12.1.3, the defective gene/protein involved in CF has been defined and characterised. As CF is a recessive condition cause by a faulty protein, if a functional copy of the CFTR gene could be inserted into the appropriate tissue (chiefly the lung) then normal CFTR protein could be synthesised by the cells and, thus, restore the normal salt transport mechanism. Early indications that this could be achieved came from experiments that demonstrated that normal CFTR could be expressed in cell lines to restore defective CFTR function, thus opening up the real possibility of using this approach in patients.

Development of a suitable therapy for a disease such as CF usually involves developing an **animal model** for the disease, so that research can be carried out to mimic the therapy in a model system before it reaches clinical trials. In CF, the model was developed using transgenic mice that lack CFTR function. Adenovirus-based vector/vehicle

Table 12.4.	Vectors and delivery methods used in gene therapy trials since 1989	
Vector/delivery method	Number of trials	% of total
Adenovirus	318	25.2
Retrovirus	290	23
Naked/plasmid DNA	229	18.2
Lipofection	99	7.9
Vaccinia virus	63	5
Poxvirus	60	4.8
Adeno-associated virus	46	3.7
Herpes simplex virus	43	3.4
Poxvirus + vaccinia virus	25	2
RNA transfer	16	1.3
Lentivirus	8	0.6
Other vectors/methods	43	3.4
Unknown	36	2.9
Total	**1 260**	

Note: Data show the number of gene therapy trials using different types of vector or delivery methods. In total, 1 260 trials have been recorded from 1989 to 2007.

Source: Data selected from *Gene Therapy Clinical Trials Worldwide*, provided by *The Journal of Gene Medicine*, copyright John Wiley & Sons Ltd (2004, 2007). URL [http://www.wiley.co.uk/genmed/clinical/]. Data manager M. Edelstein. Reproduced with permission.

systems were used, and these were shown to be effective. Thus, the system seemed to be effective, and human trials could begin. In moving into a human clinical perspective, there are several things that need to be taken into account in addition to the science of the gene and its delivery system. For example, how can the efficacy of the technique be measured? As CF therapy involved cells deep in the lung, it is difficult to access these cells to investigate the expression of the normal CFTR transgene. Using nasal tissue can give some indications, but this is not completely reliable. Also, how effective must the transgene delivery/expression be in order to produce a clinically significant effect? Do all the affected cells in the lung have to be 'repaired', or will a certain percentage of them enable restoration of near-normal levels of ion transport?

Despite the problems associated with devising, applying, and monitoring gene therapy for CF, over 1 200 clinical trials have been instigated from 1989 to date (see Table 12.4), with some successes achieved. Both viral-based and liposome/lipoplex delivery systems have been used, although the application of CF gene therapy by a widely available, robust, and effective method is still relatively distant. However, steady progress is being made, and many scientists believe that an effective therapy for CF is within reach.

The availability of gene therapy 'medicine' in an off-the-shelf, reliable, and tested form for a variety of diseases is still a long way off, despite steady progress.

12.2.4   What does the future hold for gene therapy?

As stated earlier, gene therapy has not yet realised its full potential in effective clinical applications. Whilst this is dissapointing, the very rapid pace of developments in modern genetics sometimes leads to the expectation of 'instant success'. Unfortunately something like gene therapy presents a number of very complex challenges and, thus, we should maybe not be *too* surprised when progress is not as rapid as we would wish. At present gene therapy is still an experimental application, and, as the technology is not fully developed, at times there may be problems. In the 10-year period from 1990, several thousand patients were treated by gene therapy, mostly without long-term success.

Lack of success is of course dissapointing (and often devastating) from the patient's perspective, but more distressing are cases involving the deaths of patients. Most of the difficulties tend to be associated with the use of viral vectors. In 1999 a young man named Jesse Gelsinger was undergoing gene therapy for **ornithine transcarbamylase deficiency**. Sadly, he suffered an adverse reaction to the vector and died a few days after treatment. Another setback was the development of a leukaemia-like disease in patients in a French trial that had initially delivered a positive outcome for the treatment of **X-linked severe combined immunodeficiency syndrome**. Following any major setbacks like these, it often takes a number of years before public confidence is restored.

In addition to the obvious scientific and medical challenges involved, there are very real ethical concerns around gene therapy. As already mentioned, the difference between somatic cell therapy and germ line therapy is a clear distinction that is accepted by most scientists. However, as we move on in terms of societal ethics, this distinction may become blurred and the position may even be questioned. An additional ethical problem arises when we consider gene therapy in the context of enhancing characteristics rather than treating disease. The potential use of **enhancement gene therapy**, whereby certain traits in an individual could be enhanced by the technique, raises very difficult ethical issues, and in some senses really does bring the potential for 'designer babies' closer.

> Any novel process will run into difficulties; in medical applications the consequences are often distressing and can prove to be serious setbacks.

> The potential for enhancing an individual's characteristics by gene therapy raises difficult ethical questions.

12.3 | RNA interference – a recent discovery with great potential

As we saw in Chapter 7 when we considered the PCR, from time to time a new discovery appears that enables a step change in a discipline. The discovery of **RNA interference** (**RNAi**) is one recent example of this, for which Andrew Fire and Craig Mello were awarded the 2006 Nobel prize in Physiology or Medicine. RNAi is a fascinating and complex topic, which we will not be able to do justice to in this book. However, there are potentially very significant applications in

> Many scientists think that RNA interference has the potential to be a major therapeutic tool in combating a wide range of diseases.

both basic science and in therapeutics, so it is important to at least introduce the basics of what RNAi is and what it can be used for.

12.3.1 What is RNAi?

RNAi was discovered in 1998 in the nematode *Caenorhabditis elegans*, although earlier work with *Petunia* pigmentation genes had raised the question of how gene addition could apparently lead to reduction in expression. The process of RNAi is thought to have arisen as a defence against viruses and transposons and is triggered by the presence of double-stranded RNA molecules (dsRNAs). The term *interference* gives some clue as to the functioning of the RNAi system. It refers to what is sometimes called **down-regulation** (or **knockdown**) of gene expression, or **gene silencing**, which in this case is mediated *via* short RNA molecules that enable specific regulation of particular mRNAs. In plants the process is called **post-transcriptional gene silencing**. The control of gene silencing by RNAi is complex, but in essence it works by facilitating the degradation of mRNA following a sequence-specific recognition event. Other mechanisms may prevent translation by antisense RNA binding, or may shut off transcription by methylation of bases in the promoter sequence of the gene.

> RNAi works by down-regulating or 'silencing' gene expression by acting after the mRNA has been transcribed. It is therefore sometimes called post-transcriptional gene silencing.

The mechanism of mRNA degradation involves two enzymes, called **dicer** and **slicer**. In the cell, dsRNA is recognised by the dicer enzyme, which processively degrades the dsRNA into fragments around 21 nucleotides in length. These are called **short interfering RNAs (siRNAs)**. These associate with a protein complex termed **RNA induced silencing complex (RISC)**, which contains a nuclease called slicer. RISC is activated when the dsRNA fragment is converted to single-strand form. The RISC complex contains the antisense RNA fragment, which binds to the sense strand mRNA molecule. The slicer nuclease then cuts the mRNA and the product is degraded by cellular nucleases. Thus, the expression of the gene is effectively neutralised by removal of mRNA transcripts. An outline of this mechanism of RNAi is shown in Fig. 12.7.

> The enzymes involved in generating short interfering RNA (siRNA) have been named 'dicer' and 'slicer' – appropriate given their roles in chopping up RNA molecules.

In some cases RNAi is effected by the synthesis of short RNAs from control genes. These are called **micro RNAs (miRNAs)** and are targeted by dicer to generate the response and silence the target gene by either degradation or preventing translation of the mRNA transcript.

12.3.2 Using RNAi as a tool for studying gene expression

As we outlined in Chapter 10, many of the methods for studying gene expression require some knowledge of the gene and its function and are labour-intensive. With the availability of large-scale DNA sequencing technology, the generation of genome sequence data became a reality. Whilst this has provided the sequence information, in most cases it gives little clue as to the function of a particular (or potential) gene sequence.

> Genome sequencing provides the information necessary to investigate gene expression using RNAi methods.

The discovery of RNAi presented genome scientists with exactly the tool that they had been seeking for years – the ability to switch

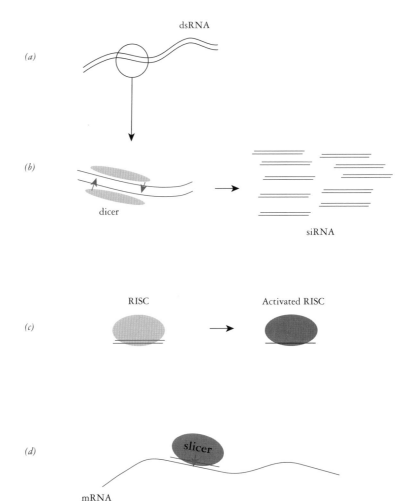

dsRNA

(a)

(b)

dicer

siRNA

RISC   Activated RISC

(c)

(d)

slicer

mRNA

Fig. 12.7 RNA interference. This is triggered by double-stranded RNA (dsRNA) as shown in (a). The RNA is cut into pieces about 21 bp in length by the enzyme *dicer*, as shown in (b). This generates short interfering RNA (siRNA), which binds to the RISC (c) and activates it. The complex binds to its complementary sequence in an mRNA molecule (d) and the *slicer* enzyme cuts the mRNA.

off any gene for which the DNA sequence was known. By generating dsRNAs that produce antisense RISCs for each gene, the effect of turning off expression of the gene can be investigated. This will enable researchers to investigate gene expression at the genome level in a way that was previously impossible, and many groups are already well on the way to systematically inactivating all the genes in their particular organism. International co-operation in the provision of the cloned sequences that generate the dsRNAs means that the future of RNAi in investigating gene expression seems assured for years to come and will provide much useful information on how cells work.

12.3.3 RNAi as a potential therapy

One exciting prospect for RNAi is in the area of therapeutic applications. Some scientists are a little cautious about expecting too much too soon, bearing in mind the problems around establishing gene therapy (as outlined earlier in Section 12.2, gene therapy has not yet delivered proven and reliable treatments). However, many more are convinced that RNAi technology can deliver therapies that can be adapted to almost any disease that involves expression of a gene or genes. If the gene can be identified and the sequence-specific dsRNA introduced into the target cells, then the RNAi system should switch on and result in the knockdown of that particular gene.

Despite some technical problems, typical of any developing technology, RNAi is elegantly simple in that the cell machinery takes over if the nucleic acid can be delivered to the correct cells.

As with gene therapy, there are several stages that are critical if a therapeutic effect is to be realised. The first stage is the preparation of the nucleic acid that is to be delivered to the target cells. In gene therapy this is usually a functional gene, which may be problematic in that genes are often large and complex. In RNAi, the dsRNA is much smaller and is likely to pose fewer problems. The next stage is delivering the nucleic acid to the target cells, which poses similar problems to those found in gene therapy. In some cases, direct injection of dsRNAs can be achieved (for accessible tissues), whilst the problems are much more complex for internal tissues.

Assuming the delivery of the dsRNA can be achieved, the final step is effecting the therapeutic action. This is where RNAi has great potential, in that the system should trigger the effect when the dsRNA enters the cell. Thus, there is no need for recombination or expression of a gene sequence, as is the case in gene therapy. A major disadvantage is that the effects of RNAi in response to a single delivery of dsRNA are transient; thus, repeated or continuous delivery of the therapy is likely to be needed if a long-term benefit is to be achieved.

The eye disease **age-related macular degeneration** (AMD) was an early target for the development of RNAi therapy. In this condition, overpromotion of blood vessel growth in the eye leads to reduction in visual function of the **macula** (a sensitive part of the retina). This is caused by too much of a protein called **vascular endothelial growth factor**, which stimulates blood vessel development. Thus, the elements needed for RNAi therapy are all present in AMD – a defined target protein that should be responsive to RNAi, a localised tissue area, and a straightforward delivery method (direct injection). Phase I clinical trials were started in 2004 by the American company Acuity Pharmaceuticals, which was the first time that RNAi had been delivered to patients as a potential threapy. Phase II results presented in 2006 appeared promising, and development of this area is continuing to lead the field of possible RNAi-based therapy.

Despite the problems associated with RNAi as a therapeutic technology, many in the field think that this is perhaps the best method that we are likely to have to down-regulate gene expression by a specific and targetted mechanism. In addition to macular degeneration, conditions such as diabetic retinopathy, cystic fibrosis, HIV infection, hepatitis C, respiratory infections, HD, and many others are potential targets for RNAi treatment. There are of course many problems to be overcome, and it may be difficult to devise effective therapies for polygenic traits, but optimism remains very high for the development of RNAi as perhaps the most important gene-based therapy in the years ahead.

12.4 | DNA profiling

Given the size of the human genome, and our knowledge of genome structure, it is relatively easy to calculate that each person's genome is unique, the only exceptions being **monozygotic twins** (twins derived from a single fertilised ovum). This provides the opportunity to use the genome as a unique identifier, if suitable techniques are available to generate robust and unambiguous results. The original technique was called **DNA fingerprinting**, but with improved technology the range of tests that can be carried out has increased, and today the more general term **DNA profiling** is preferred. The technique has found many applications in both criminal cases and in disputes over whether people are related or not (paternity disputes and immigration cases are the most common). The basis of all the techniques is that a sample of DNA from a suspect (or person in a paternity or immigration dispute) can be matched with that of the reference sample (from the victim of a crime, or a relative in a civil case). In scene-of-crime investigations, the technique can be limited by the small amount of DNA available in forensic samples. Modern techniques use the PCR to amplify and detect minute samples of DNA from bloodstains, body fluids, skin fragments, or hair roots.

DNA profiling techniques exploit the fact that each genome is unique (apart from the genomes of monozygotic twins).

12.4.1 The history of 'genetic fingerprinting'

The original DNA fingerprinting technique was devised in 1985 by Alec Jeffreys, who realised that the work he was doing on sequences within the myoglobin gene could have wider implications. The method is based on the fact that there are highly variable regions of the genome that are specific to each individual. These are **minisatellite** regions, which have a variable number of short repeated-sequence elements known as **variable number tandem repeats** (VNTRs, see Chapter 10 and Fig. 10.10). Within the VNTR there are core sequence motifs that can be identified in other polymorphic VNTR loci, and also sequences that are restricted to the particular VNTR. The arrangement of the VNTR sequences, and the choice of a suitable probe sequence, are the key elements that enable a unique 'genetic fingerprint' to be produced.

The first requirement is to isolate DNA and prepare restriction fragments for electrophoresis. As shown in Fig. 10.10, if an enzyme is used that does *not* cut the core sequence, but cuts frequently outside it, then the VNTR is effectively isolated. For human DNA the

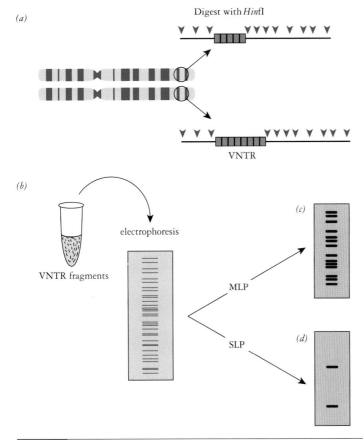

(a) Digest with HinfI

VNTR

(b)

electrophoresis

VNTR fragments

MLP

(c)

SLP

(d)

Fig. 12.8 Genetic fingerprinting of minisatellite DNA sequences. (a) A chromosome pair, with one minisatellite (VNTR) locus highlighted. In this case the locus is heterozygous for VNTR length. Cutting with HinfI effectively isolates the VNTR. (b) The VNTR fragments produced (from many loci) are separated by electrophoresis and blotted. Challenging with a multi-locus probe (MLP) produces the 'bar code' pattern shown in (c). If a single-locus probe (SLP) is used, the two alleles of the specific VNTR are identified as shown in (d).

enzyme HinfI is often used. By using a probe that hybridises to the core sequence, and carrying out the hybridisation under low stringency, polymorphic loci that bind the probe can be identified. The probe in this case is known as a **multi-locus probe**, as it binds to multiple sites. This generates a pattern of bands that is unique – the 'genetic fingerprint'. If probes with sequences that are specific for a particular VNTR are used (**single-locus probes**), a more restricted fingerprint is produced, as there will be two alleles of the sequence in each individual, one maternally derived and one paternally derived. An overview of the basis of the technique is shown in Fig. 12.8.

In forensic analysis, the original DNA profiling technique has now been largely replaced by a PCR-based method that amplifies parts of the DNA known as **short tandem repeats** (**STRs**, also known as **microsatellites**). These are repeats of 2, 3, 4, or 5 base pairs. A major

The original method of generating a DNA profile (sometimes called DNA fingerprinting) produces results in the now familiar 'bar code' format.

advantage over minisatellite (VNTR) repeats is that STRs are found throughout the genome; thus, better coverage is achieved than with minisatellites. The PCR overcomes any problems associated with the tiny amounts of sample that are often found at the crime scene. The reaction is set up to amplify the loci involved – usually 3 or 4 are sufficient if the loci are selected carefully to optimise the information generated. By using fluorescent labels and automated DNA detection equipment (similar to the genome sequencing equipment shown in Fig. 10.8) a DNA profile can be generated quickly and accurately.

12.4.2 DNA profiling and the law

The use of DNA profiling is now accepted as an important way of generating evidence in legal cases. In addition to the science itself, which may involve multi-locus or single-locus probes, or (more usually) STR amplification, there are several factors that must be considered if the evidence is to be sound. The techniques must be reliable and must be accessible to trained technical staff, who must be aware of the potential problems with the use of DNA profiling. To ensure that results from DNA profile analysis are admissible as evidence in legal cases, rigorous quality control measures must be in place. These include accurate recording of the samples as they arrive at the laboratory, and careful cross-checking of the procedures to make sure that the test is carried out properly and that the samples do not get mixed up. If PCR amplification is used as part of the procedure, great care must be taken to ensure that no trace of DNA contamination is present. A smear of the operator's sweat can often be enough to ruin a test, so strict operating procedures must be observed and laboratories must be inspected and authorised to conduct the tests. This is essential if public confidence in the technique is to be maintained.

> To be useful in a legal context, DNA profiling must be managed and regulated within an agreed framework so that the public can have confidence in the procedure.

As well as the detail of laboratory protocols, and the operational management of forensic laboratories, an important part of ensuring the acceptance of DNA evidence is the scientific and legal framework within which the procedures are carried out. The use of STR-based methods has resulted in the setting of accepted standard protocols and gene loci or target sequences. In Europe this is overseen by the **European Network of Forensic Science Institutes**, and in the USA by the American FBI's **Combined DNA Index System**.

An important consideration in legal cases is the likelihood of matching DNA profiles being generated by chance from two different individuals. This is obviously critical in cases where legal decisions are made on the strength of DNA fingerprint evidence, and perhaps custodial sentences passed. Although there is no dispute about the fact that we all have unique genomes, DNA profiling of course can only examine a small part of the genome. Thus, the odds of a chance match need to be calculated. The more bands present in a DNA profile, the less likely a non-related match will be found. The odds against a chance match for varying numbers of bands in a DNA profile are shown in Table 12.5. It is generally accepted that an approved DNA profiling agency, working under agreed conditions with standardised

| Table 12.5. | The odds against chance matches in a DNA fingerprint |

Number of bands in fingerprint	Odds against a chance match
4	250: 1
6	4 000: 1
8	65 000: 1
10	1 million: 1
12	17 million: 1
14	268 million: 1
16	4 300 million: 1
18	68 000 million: 1
20	1 million million: 1

Note: The more bands present in a DNA fingerprint, the less likely it is that any match is due to chance. However, allele frequencies for different genes may have to be taken into account. Allele frequencies can vary in different populations, and again this may be important in a legal situation. Generally, problems can be avoided by taking all known factors into account and assessing the risk of a chance match by taking the highest estimate.

Source: Data courtesy of Cellmark Diagnostics. Reproduced with permission.

protocols, will generate results that are valid and reliable. As case law has developed in forensic applications of DNA analysis, acceptance has increased and the methods are now seen as robust. In addition to dealing with new cases, DNA evidence is also being used to revisit so-called 'cold cases' and in many of these to either confirm or overturn original convictions.

An example of the use of multi-locus DNA profiling in a forensic case is shown in Fig. 12.9. In this example, blood from the victim is the reference sample. Samples from seven suspects were obtained and treated along with the sample from the victim. By matching the band patterns it is clear that suspect 5 is the guilty party.

Single-locus probes will bind to just one complementary sequence in the haploid genome. Thus, two bands will be visible in the resulting autoradiograph; one from the paternal chromosome and one from the maternal chromosome. This gives a simple profile that is often sufficient to demonstrate an unambiguous match between the suspect and the reference. The result of a paternity test using a single-locus probe is shown in Fig. 12.10. Sometimes two or more probes can be used to increase the number of bands in the profile. Single-locus probes are more sensitive than multi-locus probes and can detect much smaller amounts of DNA. Usually both single-locus and multi-locus probes are used in any given case, and the results combined.

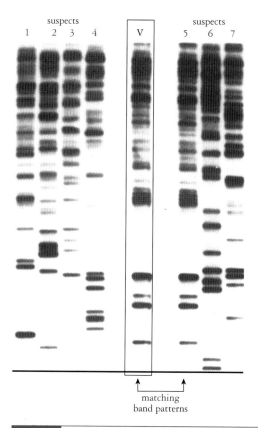

Fig. 12.9 A DNA profile prepared using a multi-locus probe. Samples of the suspect's DNA isolated from the victim (V; boxed) and seven candidate suspects (1–7) were cut with a restriction enzyme and separated on an agarose gel. The fragments were blotted onto a filter and challenged with a radioactive probe. The probe hybridises to the target sequences, producing a profile pattern when exposed to X-ray film. The band patterns from the victim's sample and suspect 5 match. Courtesy of Cellmark Diagnostics. Reproduced with permission.

12.4.3 Mysteries of the past revealed by genetic detectives

Another interesting development and application of rDNA technology has been in examining the past – a sort of 'genetic history' trail. Although the book (and subsequent film) *Jurassic Park* perhaps was a little too fanciful, it is surprising how some of the ideas portrayed in fiction have been used in the real world.

Identifying individuals using DNA profiling can go much further back in time than the immediate past, where perhaps a recently deceased individual is identified by DNA analysis. In the 1990s, DNA analysis enabled the identification of notable people such as **Joseph Mengele** (of Auschwitz notoriety) and **Czar Nicholas II** of Russia, thus solving long-running debates about their deaths. DNA has also been extracted from an Egyptian mummy (2400 years old) and a human bone that is 5500 years old! Thus, it is possible to use rDNA

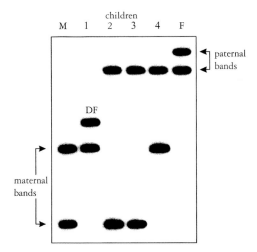

Fig. 12.10 A DNA profile prepared using a single-locus probe for paternity testing. Samples of DNA from the mother (M), four children (1–4), and the father (F) were prepared as in Fig. 12.9. A single-locus probe was used in this analysis. The band patterns therefore show two maternal bands and two paternal bands, one band from each homologous chromosome on which the target sequence is located. In the case of child 1, the paternal band is different from either of the two bands in lane F, indicating a different father (band labelled DF). This child was in fact born to the mother during a previous marriage. Courtesy of Cellmark Diagnostics. Reproduced with permission.

technology to study ancient DNA recovered from museum specimens or newly discovered archaeological material. Topics such as the migration of ancient populations, the degree of relatedness between different groups of animals, and the evolution of species can be addressed if there is access to DNA samples that are not too degraded. This area of work is sometimes called **molecular paleontology**. With DNA having been extracted from fossils as old as 65 million years, the 'genetics of the past' looks to provide evolutionary biologists, taxonomists, and paleontologists with much useful information in the future.

Use of human remains in the identification of disease-causing organisms can also be a fruitful area of research. For example, there was some debate as to the source of tuberculosis in the Americas – did it exist before the early explorers reached the New World, or was it a 'gift' from them? By analysing DNA from the lung tissue of a Peruvian mummy, researchers found DNA that corresponded to the tubercle bacillus *Mycobacterium tuberculosis*, thus proving that the disease was in fact endemic in the Americas prior to the arrival of the European settlers.

In addition to its use in forensic and legal procedures, and in tracing genetic history, DNA profiling is also a very powerful research tool that can be applied in many different contexts. Techniques such as RAPD analysis and genetic profiling are being used with many other organisms, such as cats, dogs, birds, and plants. Application of the

In addition to medical and forensic applications, the use of DNA profiling and identification techniques is proving to be of great value in many other areas of science.

technique in an ecological context enables problems that were previously studied by classical ecological methods to be investigated at the molecular level. This use of **molecular ecology** is likely to have a major impact on the study of organisms in their natural environments and (like molecular paleontology) is a good example of the coming together of branches of science that were traditionally treated as separate disciplines.

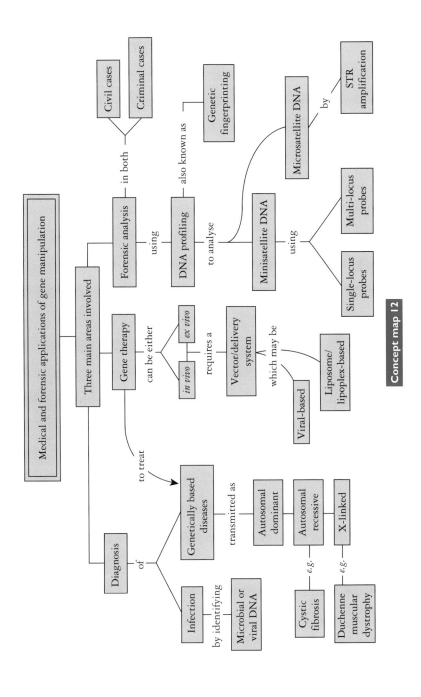

Concept map 12

Chapter 13 summary

Aims

- To outline the range and scope of transgenic technology
- To describe the methods used to produce transgenic plants and animals
- To discuss the current and potential uses of transgenic organisms
- To present the scientific, commercial, and ethical issues surrounding transgenic organism technology

Chapter summary/learning outcomes

When you have completed this chapter you will have knowledge of:

- The definition of the term 'transgenic'
- The range of methods used to generate tranegenic organisms
- The uses and potential applications of transgenic plants
- The uses and potential applications of transgenic animals
- Issues affecting the development of transgenics to commercial success
- Ethical issues surrounding the development and use of transgenic organisms

Key words

Transgenic, genetically modified foods, selective breeding, genetically modified organisms (GMOs), polygenic trait, calumoviruses, geminiviruses, *Agrobacterium tumefaciens*, crown gall disease, Ti plasmid, opines, octopine, nopaline, T-DNA, octopine synthase, tripartite (triparental) cross, cointegration, disarmed vector, binary vector, mini-Ti, 'Frankenfoods', ice-forming bacteria, *Pseudomonas syringae*, ice-minus bacteria, deliberate release, *Bacillis thuringiensis*, Bt plants, glyphosate, Roundup™, Tumbleweed™, EPSP synthase, Roundup- ready, Flavr Savr, antisense technology, polygalacturonase, iron deficiency, vitamin A deficiency, micronutrient, β-carotene, Golden rice, gene protection technology, technology protection system, terminator technology, genetic use restriction technology (GURT), genetic trait control technology, traitor technology, plant-made pharmaceuticals (PMPs), pharm animal, pharming, nuclear transfer, pronuclei, 'supermouse', mosaic, chimaera, oncomouse, *c-myc* oncogene, mouse mammary tuour (MMT) virus, prostate mouse, severe combined immunodeficiency syndrome (SCIDS), Alzheimer disease, knockout mouse, knockin mouse, Mouse Knockout and Mutation Database (MKMD), tissue plasminogen activator (TPA), blood coagulation factor IX, whey acid protein (WAP), β-lactoglobulin (BLG), α-1-antitrypsin, fibrinogen, xenotransplantation, green fluorescent protein (GFP).

Chapter 13

Transgenic plants and animals

The production of a **transgenic** organism involves altering the genome so that a permanent change is effected. This is different from somatic cell gene therapy, in which the effects of the transgene are restricted to the individual who receives the treatment. In fact, the whole point of generating a transgenic organism is to alter the germ line so that the genetic change is inherited in a stable pattern following reproduction. This is one area of genetic engineering that has caused great public concern, and there are many complex issues surrounding the development and use of transgenic organisms. In addition, the scientific and technical problems associated with genetic engineering in higher organisms are often difficult to overcome. This is partly due to the size and complexity of the genome, and partly due to the fact that the development of plants and animals is an extremely complex process that is still not yet fully understood at the molecular level. Despite these difficulties, methods for the generation of transgenic plants and animals are now well established, and the technology has already had a major impact in a range of different disciplines. In this chapter we will consider some aspects of the development and use of transgenic organisms.

13.1 Transgenic plants

All life on earth is dependent on the photosynthetic fixation of carbon dioxide by plants. We sometimes lose sight of this fact, as most people are removed from the actual process of generating our food, and the supermarket shelves have all sorts of exotic processed foods and pre-prepared meals that seem to swamp the vegetable section. Despite this, the generation of transgenic plants, particularly in the context of **genetically modified foods**, has produced an enormous public reaction to an extent that no one could have predicted. We will return to this aspect of the debate in Chapter 15. In this section, we will look at the science of transgenic plant production.

> It is often easy to forget that we are dependent on the photosynthetic reaction for our foods, and plants are therefore the most important part of our food supply chain.

13.1.1 Why transgenic plants?

For thousands of years humans have manipulated the genetic characteristics of plants by **selective breeding**. This approach has been extremely successful and will continue to play a major part in agriculture. However, classical plant breeding programmes rely on being able to carry out genetic crosses between individual plants. Such plants must be sexually compatible (which usually means that they have to be closely related); thus, it has not been possible to combine genetic traits from widely differing species. The advent of genetic engineering has removed this constraint and has given the agricultural scientist a very powerful way of incorporating defined genetic changes into plants. Such changes are often aimed at improving the productivity and 'efficiency' of crop plants, both of which are important to help feed and clothe the increasing world population.

There are many diverse areas of plant genetics, biochemistry, physiology, and pathology involved in the genetic manipulation of plants. Some of the prime targets for the improvement of crop plants are listed in Table 13.1. In many of these, success has already been achieved to some extent. However, many people are concerned about the possible ecological effects of the release of **genetically modified organisms (GMOs)** into the environment, and there is much debate about this aspect. The truth of the matter is that we simply do not know what the long-term consequences might be – a very small alteration to the balance of an ecosystem, caused by a more vigorous or disease-resistant plant, might have a considerable knock-on effect over an extended time scale.

> In agriculture, several aspects of plant growth are potential targets for improvement, either by traditional plant breeding methods or by gene manipulation.

There are two main requirements for the successful genetic manipulation of plants: (1) a method for introducing the manipulated gene into the target plant and (2) a detailed knowledge of the molecular genetics of the system that is being manipulated. In many cases the latter is the limiting factor, particularly where the characteristic under study involves many genes (a **polygenic trait**). However, despite the problems, plant genetic manipulation is already having a considerable impact on agriculture.

13.1.2 Ti plasmids as vectors for plant cells

Introducing cloned DNA into plant cells is now routine practice in many laboratories worldwide. A number of methods can be used to achieve this, including physical methods such as microinjection or biolistic DNA delivery. Alternatively a biological method can be used in which the cloned DNA is incorporated into the plant by a vector. Although plant viruses such as **calumoviruses** or **geminiviruses** may be attractive candidates for use as vectors, there are several problems with these systems. Currently the most widely used plant cell vectors are based on the Ti plasmid of *Agrobacterium tumefaciens*, which is a soil bacterium that is responsible for **crown gall disease**. The bacterium infects the plant through a wound in the stem, and a tumour of cancerous tissue develops at the crown of the plant.

> The *Agrobacterium* Ti plasmid-based vector is the most commonly used system for the introduction of recombinant DNA into plant cells.

Table 13.1. Possible targets for crop plant improvement

Target	Benefits

Resistance to:
  Diseases
  Herbicides
  Insects
  Viruses
	Improve productivity of crops and reduce losses due to biological agents

Tolerance of:
  Cold
  Drought
  Salt
	Permit growth of crops in areas that are physically unsuitable at present

Reduction of photorespiration	Increase efficiency of energy conversion
Nitrogen fixation	Capacity to fix atmospheric nitrogen extended to a wider range of species
Nutritional value	Improve nutritional value of storage proteins by protein engineering
Storage properties	Extend shelf-life of fruits and vegetables
Consumer appeal	Make fruits and vegetables more appealing with respect to colour, size, shape, etc.

The agent responsible for the formation of the crown gall tumour is not the bacterium itself, but a plasmid known as the **Ti plasmid** (Ti stands for tumour-inducing). Ti plasmids are large, with a size range in the region of 140 to 235 kb. In addition to the genes responsible for tumour formation, the Ti plasmids carry genes for virulence functions and for the synthesis and utilisation of unusual amino acid derivatives known as **opines**. Two main types of opine are commonly found, these being **octopine** and **nopaline**, and Ti plasmids can be characterised on this basis. A map of a nopaline Ti plasmid is shown in Fig. 13.1.

The region of the Ti plasmids responsible for tumour formation is known as the **T-DNA**. This is some 15–30 kb in size and also carries the gene for octopine or nopaline synthesis. On infection, the T-DNA becomes integrated into the plant cell genome and is therefore a possible avenue for the introduction of foreign DNA into the plant genome. Integration can occur at many different sites in the plant genome, apparently at random. Nopaline T-DNA is a single segment, wheras octopine DNA is arranged as two regions known as the left and right segments. The left segment is similar in structure to nopaline T-DNA, and the right is not necessary for tumour formation. The structure of nopaline T-DNA is shown in Fig. 13.2. Genes for tumour morphology are designated *tms* ('shooty' tumours), *tmr* ('rooty' tumours), and *tml* ('large' tumours). The gene for nopaline synthase is designated *nos* (in octopine T-DNA this is *ocs*, encoding **octopine**

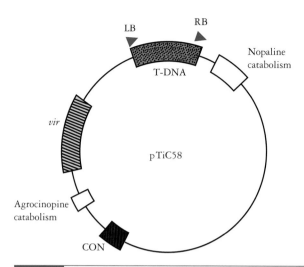

Fig. 13.1 Map of the nopaline plasmid pTiC58. Regions indicated are the T-DNA (shaded), which is bordered by left and right repeat sequences (LB and RB), the genes for nopaline and agrocinopine catabolism, and the genes specifying virulence (vir). The CON region is responsible for conjugative transfer. From Old and Primrose (1989), *Principles of Gene Manipulation*, Blackwell. Reproduced with permission.

synthase). The *nos* and *ocs* genes are eukaryotic in character, and their promoters have been used widely in the construction of vectors that express cloned sequences.

Ti plasmids are too large to be used directly as vectors, so smaller vectors have been constructed that are suitable for manipulation *in vitro*. These vectors do not contain all the genes required for Ti-mediated gene transfer and, thus, have to be used in conjunction with other plasmids to enable the cloned DNA to become integrated

> Ti plasmids are too large to be directly useful, and so they have been manipulated to have the desired characteristics of a good cloning vehicle.

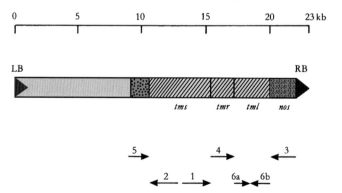

Fig. 13.2 Map of the nopaline T-DNA region. The left and right borders are indicated by LB and RB. Genes for nopaline synthase (nos) and tumour morphology (tms, tmr, and tml) are shown. The transcript map is shown below the T-DNA map. Transcripts 1 and 2 (tms) are involved in auxin production, transcript 4 (tmr) in cytokinin production. These specify either shooty or rooty tumours. Transcript 3 encodes nopaline synthase, and transcripts 5 and 6 encode products that appear to supress differentiation. From Old and Primrose (1989), *Principles of Gene Manipulation*, Blackwell. Reproduced with permission.

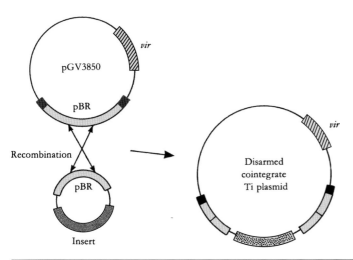

Fig. 13.3 Formation of a cointegrate Ti plasmid. Plasmid pGV3850 carries the *vir* genes but has had some of the T-DNA region replaced with pBR322 sequences (pBR). The left and right borders of the T-DNA are present (filled regions). An insert (shaded) cloned into a pBR322-based plasmid can be inserted into pGV3850 by homologous recombination between the pBR regions, producing a disarmed cointegrate vector.

into the plant cell genome. Often a **tripartite** or **triparental cross** is required, where the recombinant is present in one *E. coli* strain and a conjugation-proficient plasmid in another. A Ti plasmid derivative is present in *A. tumefaciens*. When the three strains are mixed, the conjugation-proficient 'helper' plasmid transfers to the strain carrying the recombinant plasmid, which is then mobilised and transferred to the *Agrobacterium*. Recombination then permits integration of the cloned DNA into the Ti plasmid, which can transfer this DNA to the plant genome on infection.

13.1.3   Making transgenic plants

In the development of transgenic plant methodology, two approaches using Ti-based plasmids were devised: (1) cointegration and (2) the binary vector system. In the **cointegration** method, a plasmid based on pBR322 is used to clone the gene of interest (Fig. 13.3). This plasmid is then integrated into a Ti-based vector such as pGV3850. This carries the *vir* region (which specifies virulence) and has the left and right borders of T-DNA, which are important for integration of the T-DNA region. However, most of the T-DNA has been replaced by a pBR322 sequence, which permits incorporation of the recombinant plasmid by homologous recombination. This generates a large plasmid that can facilitate integration of the cloned DNA sequence. Removal of the T-DNA has another important consequence, as cells infected with such constructs do not produce tumours and are subsequently much easier to regenerate into plants by tissue culture techniques. Ti-based plasmids lacking tumourigenic functions are known as **disarmed vectors**.

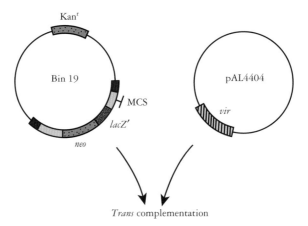

Fig. 13.4 Binary vector system based on Bin 19. The Bin 19 plasmid carries the gene sequence for the α-peptide of β-galactosidase (*lacZ'*), downstream from a polylinker (MCS) into which DNA can be cloned. The polylinker/*lacZ'*/*neo* region is flanked by the T-DNA border sequences (filled regions). In addition genes for neomycin phosphotransferase (*neo*) and kanamycin resistance (Kan^r) can be used as selectable markers. The plasmid is used in conjunction with pAL4404, which carries the *vir* genes but has no T-DNA. The two plasmids complement each other in *trans* to enable transfer of the cloned DNA into the plant genome. From Old and Primrose (1989), *Principles of Gene Manipulation*, Blackwell. Reproduced with permission.

The **binary vector** system uses separate plasmids to supply the disarmed T-DNA (**mini-Ti** plasmids) and the virulence functions. The mini-Ti plasmid is transferred to a strain of *A. tumefaciens* (which contains a compatible plasmid with the *vir* genes) by a triparental cross. Genes cloned into mini-Ti plasmids are incorporated into the plant cell genome by *trans* complementation, where the *vir* functions are supplied by the second plasmid (Fig. 13.4).

When a suitable strain of *A. tumefaciens* has been generated, containing a disarmed recombinant Ti-derived plasmid, infection of plant tissue can be carried out. This is often done using leaf discs, from which plants can be regenerated easily, and many genes have been transferred into plants by this method. The method is summarised in Fig. 13.5. The one disadvantage of the Ti system is that it does not normally infect monocotyledonous (monocot) plants such as cereals and grasses. As many of the prime target crops are monocots, this has hampered the development with these varieties. However, other methods (such as direct introduction or the use of biolistics) can be used to deliver recombinant DNA to the cells of monocots, thus avoiding the problem.

The regeneration of a functional and viable organism is a critical part of generating a transgenic plant and is often achieved using leaf discs.

13.1.4 Putting the technology to work

Transgenic plant technology has been used for a number of years, with varying degrees of success. One of the major problems we have already mentioned has been the public acceptance of transgenic plants. In fact, in some cases, this resistance from the public has been more of

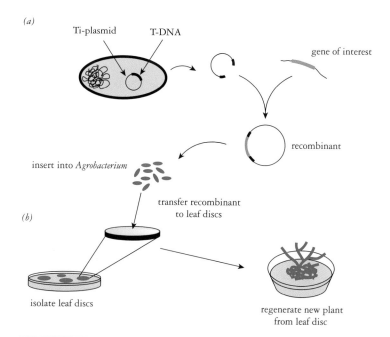

(a)

Ti-plasmid   T-DNA

gene of interest

recombinant

insert into *Agrobacterium*

transfer recombinant
to leaf discs

(b)

isolate leaf discs

regenerate new plant
from leaf disc

Fig. 13.5 Regeneration of transgenic plants from leaf discs. (*a*) The target gene is cloned into a vector based on the Ti plasmid. The recombinant plasmid is used to transform *Agrobacterium tumefaciens* (often a triparental cross is used as described in the text). The bacterium is then used to infect plant cells that have been grown as disc explants of leaf tissue, as shown in (*b*). The transgenic plant is regenerated from the leaf disc by propagation on an appropriate tissue culture medium.

a problem than the actual science. Towards the end of 1999 the backlash against so-called '**Frankenfoods**' had reached the point where companies involved were being adversely affected, either directly by way of action against field trials or indirectly in that consumers were not buying the transgenic products. We will examine this area in a little more depth in Chapter 15.

One of the first recombinant DNA experiments to be performed with plants did not in fact produce transgenic plants at all but involved the use of genetically modified bacteria. In nature, ice often forms at low temperatures by associating with proteins on the surface of so-called **ice-forming bacteria**, which are associated with many plant species. One of the most common ice-forming bacterial species is *Pseudomonas syringae*. In the late 1970s and early 1980s researchers removed the gene that is responsible for synthesising the ice-forming protein, producing what became known as **ice-minus bacteria**. Plans to spray the ice-minus strain onto plants in field trials were ready by 1982, but approval for this first **deliberate release** experiment was delayed as the issue was debated. Finally, approval was granted in 1987, and the field trial took place. Some success was achieved as the engineered bacteria reduced frost damage in the test treatments.

Public confidence in and acceptance of transgenic plants has been a difficult area in biotechnology, with many different views and interests. A rational and balanced debate is important if the technology is ultimately to be of widespread value.

The bacterium ***Bacillus thuringiensis*** has been used to produce transgenic plants known as **Bt plants**. The bacterium produces toxic crystals that kill caterpillar pests when they ingest the toxin. The bacterium itself has been used as an insecticide, sprayed directly onto crops. However, the gene for toxin production has been isolated and inserted into plants such as corn, cotton, soybean, and potato, with the first Bt crops planted in 1996. By 2000 over half of the soybean crop in the USA was planted with Bt-engineered plants, although there have been some problems with pests developing resistance to the Bt toxin.

One concern that has been highlighted by the planting of Bt corn is the potential risk to non-target species. In 1999 a report in *Nature* suggested that larvae of the Monarch butterfly, widely distributed in North America, could be harmed by exposure to Bt corn pollen, even though the regulatory process involved in approving the Bt corn has examined this possibility and found no significant risk. Risk is associated with both toxicity and exposure, and subsequent research has demonstrated that exposure levels are likely to be too low to pose a serious threat to the butterfly. However, the debate continues. This example illustrates the need for more extensive research in this area, if genetically modified crops are to gain full public acceptance.

Herbicide resistance is one area where a lot of effort has been directed. The theory is simple – if plants can be made herbicide-resistant, then weeds can be treated with a broad-spectrum herbicide without the crop plant being affected. One of the most common herbicides is **glyphosate**, which is available commercially as **Roundup**™ and **Tumbleweed**™. Glyphosate acts by inhibiting an amino acid biosynthetic enzyme called 5-enolpyruvylshikimate-3-phosphate synthase (**EPSP synthase** or EPSPS). Resistant plants have been produced by either increasing the synthesis of EPSPS by incorporating extra copies of the gene, or by using a bacterial EPSPS gene that is slightly different from the plant version and produces a protein that is resistant to the effects of glyphosate. Monsanto has produced various crop plants, such as soya, that are called **Roundup-ready**, in that they are resistant to the herbicide. Such plants are now used widely in the USA and some other countries, and herbicide resistance is the most commonly manipulated trait in genetically modified (GM) plants.

Since the first GM crops were planted commercially, there has been a steady growth in the area of GM plantings worldwide. Fig. 13.6 shows the total area of GM crops planted from 1996 to 2006. Two points are of interest here, in addition to the overall growth in area year-on-year. First, the symbolically significant 100 million hectare barrier was broken in 2006. Second, the developing countries of the world are embracing GM crop plants and are now catching up with the industrialised countries. This is also evident from Table 13.2, which shows data for areas planted in 2006 for each of the 22 countries growing GM crops commercially.

The four main GM crops are soybean, maize, cotton, and canola (oilseed rape). Table 13.3 shows areas planted with GM varieties for

Bt plants, in which a bacterial toxin is used to confer resistance to caterpillar pests, have been successfully established and are grown commercially in many countries.

Herbicide resistance is an area that has been exploited by plant biotechnologists to engineer plants that are resistant to common herbicides that are used to control weeds.

GM plants are now grown in some 22 countries, with over 100 million hectares being planted in 2006.

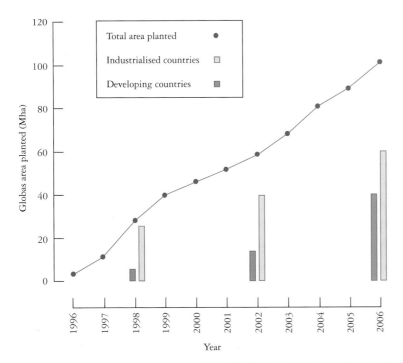

Fig. 13.6 The total area of GM crops planted worldwide is shown for the period 1996–2006. Areas are in millions of hectares (MHa). For 1998, 2002, and 2006 the balance between industrialised countries and developing countries is shown by the shaded bars. Data from James (2006), ISAAA Brief 35. (See www.isaaa.org for further information). Reproduced with permission.

each of these crops. The most commonly grown GM crop is soybean, which made up around 59% of the total of GM crops planted in 2006.

In addition to the 'big four' GM crops, other plant species have of course been genetically modified, with one example being the tomato. Tomatoes are usually picked green, so that they are able to withstand shipping and transportation without bruising. They are then ripened artificially by using ethylene gas, as ethylene is a key trigger for the ripening process. In trying to delay natural ripening, two approaches have been used. One is to target the production of ethylene itself, thus delaying the onset of the normal ripening mechanism. A second approach illustrates how a novel idea, utilising advanced gene technology to achieve an elegant solution to a defined problem, can still fail because of other considerations – this is the story of the **Flavr Savr** (*sic*) tomato.

The biotechnology company Calgene developed the Flavr Savr tomato using what became known as **antisense technology**. In this approach, a gene sequence is inserted in the opposite orientation, so that on transcription an mRNA that is complementary to the normal mRNA is produced. This antisense mRNA will therefore bind to the normal mRNA in the cell, inhibiting its translation and effectively shutting off expression of the gene. The principle of the method is

Table 13.2.	Areas of GM crops planted in 2006 by country

Country	Total area planted (MHa)
Countries planting > 50 000 Ha	
USA	54.6
Argentina	18.0
Brazil	11.5
Canada	6.1
India	3.8
China	3.5
Paraguay	2.0
South Africa	1.4
Uruguay	0.4
Philippines	0.2
Australia	0.2
Romania	0.1
Mexico	0.1
Spain	0.1
Countries planting < 50 000 Ha	
Czech Republic	–
France	–
Germany	–
Iran	–
Slovakia	–
Portugal	–
Honduras	–
Colombia	

Note: Areas of GM crops planted by country.

Source: Data from James (2006), ISAAA Brief 35. (See www.isaaa.org for further information). Reproduced with permission.

Table 13.3.	The four main GM crops

Year	1998		2002		2006	
	MHa	%	MHa	%	MHa	%
Soybean	14	(54)	36	(62)	59	(58)
Maize	8	(30)	12	(21)	25	(24)
Cotton	2	(8)	7	(12)	13	(13)
Canola (oilseed rape)	2	(8)	3	(5)	5	(5)
Total	26	(100)	58	(100)	102	(100)

Note: Approximate areas of the four main GM crops (in MHa), and as a % of the total GM crops grown, are shown for the years 1998, 2002, and 2006.

Source: Data from James (1998, 2002, 2006), ISAAA Briefs 8, 27, 35. (See www.isaaa.org for further information.) Reproduced with permission.

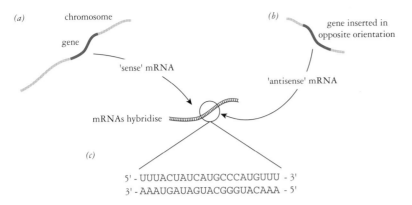

(a) chromosome

gene

'sense' mRNA

(b) gene inserted in opposite orientation

'antisense' mRNA

mRNAs hybridise

(c)

5' - UUUACUAUCAUGCCCAUGUUU - 3'
3' - AAAUGAUAGUACGGGUACAAA - 5'

Fig. 13.7 Antisense technology. The target gene is shown in (*a*). A copy of the gene is introduced into a separate site on the genome, but in the opposite orientation, as shown in (*b*). On transcription of the antisense gene, an antisense mRNA is produced. This binds to the normal mRNA, preventing translation. Part of the sequence is shown in (*c*) to illustrate.

shown in Fig. 13.7. In the Flavr Savr, the gene for the enzyme **polygalacturonase (PG)** is the target. This enzyme digests pectin in the cell wall and leads to fruit softening and the onset of rotting. The elegant theory is that inhibition of PG production should slow the decay process and the fruit should be easier to handle and transport after picking. It can also be left on the vine to mature longer than is usually the case, thus improving flavour. After much development, the Flavr Savr became the first genetically modified food to be approved for use in the USA, in 1994. The level of PG was reduced to something like 1% of the normal levels, and the product seemed to be set for commercial success. However, various problems with the characteristics affecting growth and picking of the crop led to the failure of the Flavr Savr in commercial terms. Calgene is now part of Monsanto, having been stretched too far by development of the Flavr Savr. Despite the failure of the Flavr Savr, the company has continued to produce innovations in biotechnology, including rapeseed oil (known as canola oil in the USA) with a high concentration of lauric acid, which is beneficial from a health perspective.

Attempts to improve the nutritional quality of crops are not restricted to the commercial or healthfood sectors. For many millions of people around the world, access to basic nutrition is a matter of survival rather than choice. Rice is the staple food of some 3 billion people, and about 10% of these suffer from health problems associated with vitamin A deficiency. It is estimated that around 1 million children die prematurely from this deficiency, with a further 350 000 going blind. Thus, rice has been one of the most intensively studied crop species with respect to improving quality of life for around half of the world's population. This led to the development of 'Miracle Rice', which was a product of the green revolution of the 1960s. However, widespread planting throughout Southeast Asia led to a rice

Sometimes an elegant and well-designed scientific solution to a particular problem does not guarantee commercial success for the produce, with many factors having an impact on the success or failure of a GM product.

Rice is of such importance as a staple food that it is an obvious target for GM technologists.

monoculture, with increased susceptibility to disease and pests, and the increased dependence on pesticides that this brings. Thus, as with genetically modified crops, there can be problems in adopting new variants of established crop species.

There are two main nutritional deficiency problems that are particularly prevalent in developing countries. These are **iron deficiency** and **vitamin A deficiency**, both of which result from a lack of these **micronutrients** in the diet. In rice, there is a low level of iron in the endosperm. There are also problems with iron re-absorption due to the presence of a chemical called phytate. Low levels of sulphur also contribute to the difficulties, as this is required for efficient absorption of iron in the intestine. The vitamin A problem is due to the failure of rice to synthesise β-carotene, which is required for the biosynthesis of vitamin A. In 1999 Ingo Potrykus, working in Zurich, succeeded in producing '**Golden rice**' with β-carotene in the grain endosperm, where it is not normally found. As β-carotene is a precursor of vitamin A, increasing the amount available by engineering rice in this way should help to alleviate some of the problems of vitamin A deficiency. This is obviously a positive development. However, corporate interests in patent rights to the technologies involved, and other non-scientific problems, had to be sorted out before agreement was reached that developing countries could access the technology freely. Development work and legal wrangles continue, but once again it is clear that it is difficult to strike a balance between commercial factors and the potential benefits of transgenic plant technology. This is a particularly sensitive topic when those who would benefit most may not have the means to afford the technology, and again raises some difficult ethical questions.

A further twist in the corporate *vs.* common good debate can be seen in the so-called **gene protection technology**. This is where companies design their systems so that their use can be controlled, by some sort of manipulation that is essentially separate from the actual transgenic technology that they are designed to deliver. This has caused great concern among public and pressure groups, and a vigorous debate has ensued between such groups and the companies who are developing the technology. Even the terms used to describe the various approaches reflect the strongly held views of proponents and opponents. Thus, one system is called a **technology protection system** by the corporate sector, and **terminator technology** by others. Another type of system is called **genetic use restriction technology (GURT)** or **genetic trait control technology**; this is also known more widely as **traitor technology**. So what do these emotive terms mean, and is there any need for concern?

Terminator technology is where plants are engineered to produce seeds that are essentially sterile, or do not germinate properly. Thus, the growers are prevented from gathering seeds from one year to plant the next season, and are effectively tied to the seed company, as they have to buy new seeds each year. Companies reasonably claim that they have to obtain returns on the considerable investment

The development of 'Golden Rice' is an example of good science and good intent, although as is often the case the political agenda may have a role to play in determining how benefits are shared and distributed.

The development and implementaion of protection technologies is a highly controversial and emotive area of plant genetic manipulation, with strong opinions on both sides of the argument.

required to develop a transgenic crop plant. Others, equally reasonably, insist that this constraint places poor farmers in developing countries at a considerable disadvantage, as they are unable to save seed from one year's crop to develop the next season's planting. Approximately half of the world's farmers are classed as 'poor' and cannot afford to buy new seed each season. They produce around 20% of world food output and feed some 1.4 billion people. Thus, there is a major ethical issue surrounding the prevention of seed-collecting from year to year, if terminator technology were to be applied widely. There is also great concern that any sterility-generating technology could transfer to other variants, species, or genera, thus having a devastating effect on third-world farming communities.

Traitor technology or GURT involves the use of a 'switch' (often controlled by a chemical additive) to permit or restrict a particular engineered trait. This is perhaps a little less contentious than terminator technology, as the aim is to regulate a particular modification rather than to prevent viable seed production. However, there are still many who are very concerned about the potential uses of this technology, which could again tie growers to a particular company if the 'switch' requires technology that only that company can supply.

Some of the major agricultural biotechnology companies have stated publicly that they will not develop terminator technology, which is seen as a partial success for the negative reaction of the public, farmers, and pressure groups. There is, however, still a lot of uncertainty in this area, with some groups claiming that traitor technology is being developed further, and that corporate mergers can negate promises made previously by one of the partners in the merger. The whole field of transgenic plant technology is therefore in some degree of turmoil, with many conflicting interests, views, and personalities involved. The debates are set to continue for a long time to come.

In addition to GM crops, transgenic plants also have the potential to make a significant impact on the biotechnology of therapeutic protein production. As we saw in Chapter 11, bacteria, yeast, and mammalian cells are commonly used for the production of high-value proteins by recombinant DNA technology. An obvious extension would be to use transgenic plants to produce such proteins. Plants are cheap to grow compared to the high-cost requirements of microbial or mammalian cells and, thus, cost reduction and potentially unlimited scale-up opportunities make transgenic plants an attractive option for producing high-value proteins. This is an area of active development at the moment, and the term **plant-made pharmaceuticals (PMPs)** has been coined to describe this aspect of transgenic plant technology.

> The use of plants as 'factories' for the synthesis of therapeutic proteins is an area that is currently being developed to enable the production of plant-made pharmaceuticals (PMPs).

13.2 | Transgenic animals

The generation of transgenic animals is one of the most complex aspects of genetic engineering, both in terms of technical difficulty

The debate around the use of GM technology becomes even more far-reaching when transgenic animals are concerned, and often raises animal welfare issues as well as the genetic modification aspects of the technology.

and in the ethical problems that arise. Many people, who accept that the genetic manipulation of bacterial, fungal, and plant species is beneficial, find difficulty in extending this acceptance when animals (particularly mammals) are involved. The need for sympathetic and objective discussion of this topic by the scientific community, the media, and the general public are likely to present one of the great challenges in scientific ethics over the next few years.

13.2.1 Why transgenic animals?

Genetic engineering has already had an enormous impact on the study of gene structure and expression in animal cells, and this is one area that will continue to develop. Cancer research is one obvious example, and current investigation into the molecular genetics of the disease requires extensive use of gene manipulation technology. In the field of protein production in biotechnology (discussed in Chapter 11), the synthesis of many mammalian-derived recombinant proteins is often best carried out using cultured mammalian cells, as these are sometimes the only hosts that will ensure the correct expression of such genes.

Cell-based applications such as those outlined above are an important part of genetic engineering in animals. However, the term 'transgenic' is usually reserved for whole organisms, and the generation of a transgenic animal is much more complex than working with cultured cells. Many of the problems have been overcome using a variety of animals, with early work involving amphibians, fish, mice, pigs, and sheep.

The term 'transgenic' is mostly used to describe whole organisms that have been modified to contain transgenes in a stable form that are inherited by transmission through the germ line.

Transgenics can be used for a variety of purposes, covering both basic research and biotechnological applications. The study of embryological development has been extended by the ability to introduce genes into eggs or early embryos, and there is scope for the manipulation of farm animals by the incorporation of desirable traits *via* transgenesis. The use of whole organisms for the production of recombinant protein is a further possibility, and this has already been achieved in some species. The term **pharm animal** or **pharming** (from *pharm*aceutical) is sometimes used when talking about the production of high-value therapeutic proteins using transgenic animal technology.

When considered on a global scale, the potential for exploitation of transgenic animals would appear to be almost unlimited. Achieving that potential is likely to be a long and difficult process in many cases, but the rewards are such that a considerable amount of money and effort has already been invested in this area.

13.2.2 Producing transgenic animals

There are several possible routes for the introduction of genes into embryos, each with its own advantages and disadvantages. Some of the methods are (1) direct transfection or retroviral infection of embryonic stem cells followed by introduction of these cells into an embryo at the blastocyst stage of development; (2) retroviral infection of early embryos; (3) direct microinjection of DNA into oocytes, zygotes, or early embryo cells; (4) sperm-mediated transfer; (5) transfer into unfertilised ova; and (6) physical techniques such as biolistics or electrofusion. In addition to these methods, the technique of **nuclear transfer** (used in organismal cloning, discussed in Chapter 14) is sometimes associated with a transgenesis protocol.

Early success was achieved by injecting DNA into one of the **pronuclei** of a fertilised egg, just prior to the fusion of the pronuclei (which produces the diploid zygote). This approach led to the production of the celebrated '**supermouse**' in the early 1980s, which represents one of the milestones of genetic engineering. The experiments that led to the 'supermouse' involved placing a copy of the rat growth hormone (GH) gene under the control of the mouse metallothionine (mMT) gene promoter. To create the 'supermouse', a linear fragment of the recombinant plasmid carrying the fused gene sequences (MGH) was injected into the male pronuclei of fertilised eggs (linear fragments appear to integrate into the genome more readily than circular sequences). The resulting fertilised eggs were implanted into the uteri of foster mothers, and some of the mice resulting from this expressed the GH gene. Such mice grew some 2–3 times faster than control mice and were up to twice the size of the controls. Pronuclear microinjection is summarised in Fig. 13.8.

> The production of 'supermouse' in the early 1980s represents one of the milestone achievements in genetic engineering.

In generating a transgenic animal, it is desirable that all the cells in the organism receive the transgene. The presence of the transgene in the germ cells of the organism will enable the gene to be passed on to succeeding generations, and this is essential if the organism is to be useful in the long term. Thus, introduction of genes has to be carried out at a very early stage of development, ideally at the single-cell zygote stage. If this cannot be achieved, there is the possibility that a **mosaic** embryo will develop, in which only some of the cells carry the transgene. Another example of this type of variation is where the embryo is generated from two distinct individuals, as is the case when embryonic stem cells are used. This results in a **chimaeric** organism. In practice this is not necessarily a problem, as the organism can be crossed to produce offspring that are homozygous for the transgene in all cells. A chimaeric organism that contains the transgene in its germ line cells will pass the gene on to its offspring, which will therefore be heterozygous for the transgene (assuming they have come from a mating with a homozygous non-transgenic). A further cross with a sibling will result in around 25% of the offspring being homozygous for the transgene. This procedure is outlined for the mouse in Fig. 13.9.

> Mosaics and chimaeras, where not all the cells of the animal carry the transgene, are variations that can arise when producing transgenic animals.

13.2.3 Applications of transgenic animal technology

Introduction of growth hormone genes into animal species has been carried out, notably in pigs, but in many cases there are undesirable side effects. Pigs with the bovine growth hormone gene show greater feed efficiency and have lower levels of subcutaneous fat than normal pigs. However, problems such as enlarged heart, high incidence of

Fig. 13.8 Production of 'supermouse'. (*a*) Fertilised eggs were removed from a female and (*b*) the DNA carrying the rat growth hormone gene/mouse metallothionein promoter construct (MGH) was injected into the male pronucleus. (*c*) The eggs were then implanted into a foster mother. (*d*) Some of the pups expressed the MGH construct (MGH$^+$) and were larger than the normal pups.

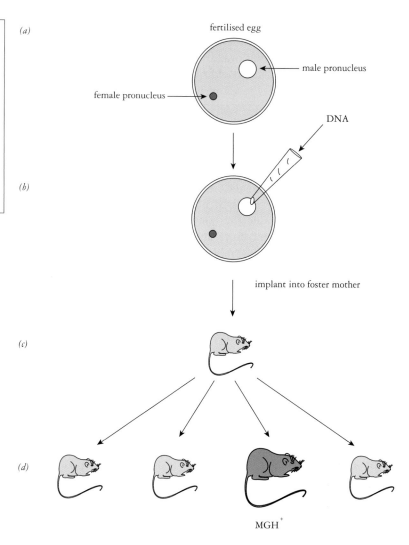

stomach ulcers, dermatitis, kidney disease, and arthritis have demonstrated that the production of healthy transgenic farm animals is a difficult undertaking. Although progress is being made, it is clear that much more work is required before genetic engineering has a major impact on animal husbandry.

The study of development is one area of transgenic research that is currently yielding much useful information. By implanting genes into embryos, features of development such as tissue-specific gene expression can be investigated. The cloning of genes from the fruit fly *Drosophila melanogaster*, coupled with the isolation and characterisation of transposable elements (P elements) that can be used as vectors, has enabled the production of stable transgenic *Drosophila* lines. Thus, the fruit fly, which has been a major contributor to the field of classical genetic analysis, is now being studied at the molecular level by employing the full range of gene manipulation techniques.

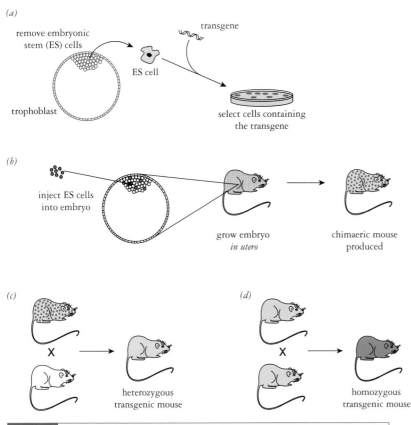

(a)

remove embryonic
stem (ES) cells

transgene

ES cell

trophoblast

select cells containing
the transgene

(b)

inject ES cells
into embryo

grow embryo
in utero

chimaeric mouse
produced

(c)

X

heterozygous
transgenic mouse

(d)

X

homozygous
transgenic mouse

Fig. 13.9 Production of transgenic mice using embryonic stem cell technology.
(a) Embryonic stem cells (ES cells) are removed from an early embryo and cultured. The
target transgene is inserted into the ES cells, which are grown on selective media.
(b) The ES cells containing the transgene are injected into the ES cells of another
embryo, where they are incorporated into the cell mass. The embryo is implanted into a
pregnant mother, and a chimaeric transgenic mouse is produced. By crossing the
chimaera with a normal mouse as shown in (c), some heterozygous transgenics will be
produced. If they are then self-crossed, homozygous transgenics will be produced in
about 25% of the offspring, as shown in (d).

In mammals, the mouse is proving to be one of the most useful
model systems for investigating embryological development, and the
expression of many transgenes has been studied in this organism. One
such application is shown in Fig. 13.10, which demonstrates the use
of the *lacZ* gene as a means of detecting tissue-specific gene expression. In this example the *lacZ* gene was placed under the control of
the weak thymidine kinase (TK) promoter from herpes simplex virus
(HSV), generating an HSV-TK–*lacZ* construct. This was used to probe for
active chromosomal domains in the developing embryos, with one
of the transgenic lines showing the brain-specific expression seen in
Fig. 13.10.

Although the use of transgenic organisms is providing many
insights into developmental processes, inserted genes may not always

The mouse has proved to be one
of the most useful model
organisms for transgenesis
research, with many different
variants having been produced
for a variety of purposes.

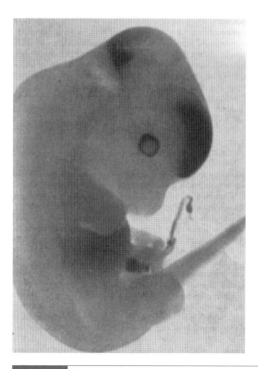

Fig. 13.10 Expression of a transgene in the mouse embryo. The β-galactosidase (*lacZ*) coding region was placed under the control of a thymidine kinase promoter from herpes simplex virus to produce the HSV-TK–*lacZ* gene construct. This was injected into male pronuclei and transgenic mice were produced. The example shows a 13-day foetus from a transgenic strain that expresses the transgene in brain tissue during gestation. Detection is by the blue colouration produced by the action of β-galactosidase on X-gal. Thus, the dark areas in the fore- and hindbrain are regions where the *lacZ* gene has been expressed. Photograph courtesy of Dr S. Hettle. From Allen *et al.* (1988), *Nature (London)* 333, 852–855. Copyright (1988) Macmillan Magazines Limited. Reproduced with permission.

be expressed in exactly the same way as would be the case in normal embryos. Thus, a good deal of caution is often required when interpreting results. Despite this potential problem, transgenesis is proving to be a powerful tool for the developmental biologist.

Mice have also been used widely as animal models for disease states. One celebrated example is the **oncomouse**, generated by Philip Leder and his colleagues at Harvard University. Mice were produced in which the *c-myc* **oncogene** and sections of the **mouse mammary tumour** (MMT) **virus** gave rise to breast cancer. The oncomouse has a place in history as the first complex animal to be granted a patent in the USA. Other transgenic mice with disease characteristics include the **prostate mouse** (prostate cancer), mice with **severe combined immunodeficiency syndrome** (SCIDS), and mice that show symptoms of **Alzheimer disease**.

Many new variants of transgenic mice have been produced and have become an essential part of research into many aspects of human disease. Increased knowledge of molecular genetics, and the

continued development of the techniques of transgenic animal production, have enabled mice to be generated in which specific genes
can be either activated or inactivated. Where a gene is inactivated or
replaced with a mutated version, a **knockout mouse** is produced. If
an additional gene function is established, this is sometimes called
a **knockin mouse**. The use of knockout mice in cystic fibrosis (CF)
research is one example of the technology being used in both basic
research and in developing gene therapy procedures. Mice have been
engineered to express mutant CF alleles, including the prevalent
ΔF508 mutation that is responsible for most serious CF presentations.
Having a mouse model enables researchers to carry out experiments
that would not be possible in humans, although (as with developmental studies) results may not be exactly the same as would be the case
in a human subject.

In 1995 a database was established to collate details of knockout
mice. This is known as the **Mouse Knockout and Mutation Database
(MKMD)** and can be found at [http://research.bmn.com/mkmd/]. At the
time of writing some 5 000 entries were listed, representing over 2 000
unique genes, with more being added daily. This demonstrates that
the mouse is proving to be one of the mainstays of modern transgenic
research.

> There are over 5 000 entries in the Mouse Knockout and Mutation Database, which is a major resource for those interested in using the mouse as a model system for the study of gene expression, development, and disease.

Early examples of protein production in transgenic animals
include expression of human **tissue plasminogen activator (TPA)** in
transgenic mice, and of human **blood coagulation factor IX** (FIX)
in transgenic sheep. In both cases the transgene protein product was
secreted into the milk of lactating organisms by virtue of being placed
under the control of a milk-protein gene promoter. In the mouse
example the construct consisted of the regulatory sequences of the
whey acid protein (WAP) gene, giving a WAP–TPA construct. Control
sequences from the **β-lactoglobulin (BLG)** gene were used to generate a BLG–FIX construct for expression in the transgenic sheep. Other
examples of transgenic animals acting as bioreactors include pigs that
express human haemoglobin, and cows that produce human lactofer-
rin.

Producing a therapeutic protein in milk provides an ideal way of
ensuring a reliable supply from lactating animals, and downstream
processing to obtain purified protein is relatively straightforward. This
approach was used by scientists from the Roslin Institute near Edinburgh, working in conjunction with the biotechnology company PPL
Therapeutics. In 1991 PPL's first transgenic sheep (called Tracy) was
born. Yields of human proteins of around 40 g L^{-1} were produced
from milk, demonstrating the great potential for this technology. PPL
continues to develop a range of products, such as **α-1-antitrypsin** (for
treating CF), **fibrinogen** (for use in medical procedures), and human
factor IX (for haemophilia B).

Using animals as bioreactors offers an alternative to the fermentation of bacteria or yeast that contain the target gene. The technology
is now becoming well established, with many biotechnology companies involved. As we have already seen, this area is sometimes called

> The use of transgenic animals as producers of high-value products is more fully developed than is the case for transgenic plants, although there are often scientific and commercial problems in making a success of the technology.

pharming, with the transgenic animals referred to as pharm animals. Given the problems in achieving correct expression and processing of some mammalian proteins in non-mammalian hosts, this is proving to be an important development of transgenic animal technology.

As we saw in Chapter 11, good science does not necessarily lead to a viable product, with many strategic and commercial aspects having an impact on whether or not a product is developed fully to market. One such case recently has been the potential use of chicken eggs to produce anticancer therapeutics. In a collaboration among the Roslin Institute, Oxford Biomedica, and Viragen, potentially useful proteins have been produced in the whites of eggs laid by a transgenic hen. Whilst the science is novel and impressive, the commercial aspects of the project led to an announcement by Viragen in June 2007 that it was not proceeding with further development of the system. Such setbacks are not unusual, and although this might seem unfortunate, the field as a whole will continue to progress, with successful commercially viable developments undoubtedly appearing more frequently in the future as the industry becomes more mature.

Transgenic animals also offer the potential to develop organs for **xenotransplantation**. The pig is the target species for this application, as the organs are of similar size to human organs. In developing this technology the key target is to alter the cell surface recognition properties of the donor organs, so that the transplant is not rejected by the human immune system. In addition to whole organs from mature animals, there is the possibility of growing tissue replacements as an additional part of a transgenic animal. Both these aspects of xenotransplantation are currently being investigated, with a lot of scientific and ethical problems still to be solved. Given the shortfall in organ donors, and the consequent loss of life or quality of life that results, many people feel that the objections to xenotransplantation must be discussed openly, and overcome, to enable the technology to be implemented when fully developed.

The use of transgenic technology to enable xenotransplantation is an area that could alleviate some of the problems of supply and demand that exist in the field of organ transplantation.

Our final word on transgenic animals brings the technology closer to humans. In January 2001 the birth of the first transgenic non-human primate, a rhesus monkey, was announced. This was developed by scientists in Portland, Oregon. He was named ANDi, which stands for inserted DNA (written backwards!). A marker gene from jellyfish, which produces **green fluorescent protein** (**GFP**), was used to confirm the transgenic status of a variety of ANDi's cells. ANDi is particularly significant in that he opens up the possibility of a near-human model organism for the evaluation of disease and therapy. He also brings the ethical questions surrounding transgenic research a little closer to the human situation, which some people are finding a little uncomfortable. It will certainly be interesting to follow the developments in transgenic technology over the next few years.

Transgenic organisms — also known as — **Genetically modified organisms (GMOs)**

Genetically modified organisms (GMOs)

can be — **Animals**

can be — **Plants**

Animals — e.g. — Mice / Sheep / Pigs / Cattle / Monkeys

Animals — produced using

Animals:
- **Pharm animals** — may offer — **Increased yields** — and produce — **Therapeutic products / Tissues for replacement or transplantation**
- **Knockouts** — as — **Disease models** — such as — **Cystic fibrosis mouse models** — to develop therapies for
- **Therapeutic products / Tissues for replacement or transplantation** — for use in — **Medical applications**

Transfection / Retroviral infection / Microinjection / Biolistics / Electroporation / Nuclear transfer — to introduce

Genetically modified organisms (GMOs) — aim is to — **Alter the organism's genetic makeup** — by incorporating — **Genes from a different organism** — to effect — **Beneficial changes** — useful for

Beneficial changes — useful for:
- **Medical applications**
- **Basic research**
- **Agricultural biotechnology**

Ti plasmid vectors of *Agrobacterium* / Microinjection / Biolistics / Electroporation / Tissue culture — to introduce

Plants — produced using

Plants — aim is to — **Alter the characteristics of crop plants** — to improve:
- **Tolerance** — e.g. — Drought / Salinity / Cold
- **Resistance** — e.g. — Pests / Herbicides
- **Qualities** — e.g. — Flavour / Texture / Shelf life / Nutrition

as part of — **Agricultural biotechnology**

Concept map 13

Chapter 14 summary

Aims

- To outline the range and scope of organismal cloning technology
- To describe the methods used to produce cloned plants and animals
- To discuss the current and potential uses of reproductive and therapeutic cloning
- To present the scientific, commercial, and ethical issues surrounding organismal cloning technology

Chapter summary/learning outcomes

When you have completed this chapter you will have knowledge of:

- The definition of the terms 'reproductive cloning' and 'therapeutic cloning'
- The history of organismal cloning and its development
- The concept of totipotent, pluripotent, and multipotent cells
- The methods used for cloning organisms
- The difference between nuclear transfer and embryo splitting techniques
- Ethical issues around the use of organismal cloning

Key words

Dolly, molecular cloning, organismal cloning, reproductive cloning, therapeutic cloning, development, preformationism, homunculus, epigenesis, temporal, spatial, nuclear transfer, somatic cell nuclear transfer (SNCT), embryo splitting, nuclear totipotency, pluripotent, multipotent, irreversibly differentiated, *Rana pipiens*, *Xenopus laevis*, cell cycle, G_0 phase, telomere, large offspring syndrome.

Chapter 14

The other sort of cloning

On 5 July 1996 a lamb was born at the Roslin Research Institute near Edinburgh. It was an apparently normal event, yet it marked the achievement of a milestone in biological science. The lamb was a clone, and she was named **Dolly**. She was the first organism to be cloned from adult differentiated cells, which is what makes the achievement such a ground-breaking event. In this chapter we will look briefly at this area of genetic technology.

In this book so far, we have been considering the topic of **molecular cloning**, where the aim of an experimental process is to isolate a gene sequence for further analysis and use. In **organismal cloning**, the aim is to generate an organism from a cell that carries a complete set of genetic instructions. We have looked at the methods for generating transgenic organisms in Chapter 13, and a discussion of organismal cloning is a natural extension of this, although transgenic organisms are not necessarily clones. In a similar way, a clone need not necessarily be transgenic. Thus, although not strictly part of gene manipulation technology, organismal cloning has become a major part of genetics in a broader sense. From a wider public perspective, organismal cloning is seen as an issue for concern and, thus, a discussion of the topic is essential even in a book where the primary goal is to illustrate the techniques of gene manipulation.

Organismal cloning can be further subdivided according to the purpose of the procedure. Where the function is to generate a 'copy' of the original organism, this is termed **reproductive cloning**. Recent advances in stem cell technology open up the possible use of cloning embryos to enable production of matched tissue types for use in reseach and potentially in the treatment of disease. This aspect of cloning is called **therapeutic cloning**.

> Organismal cloning can be divided into reproductive cloning and therapeutic cloning, each of which has a different purpose.

14.1 | Early thoughts and experiments

The announcement of the birth of Dolly in a paper in the journal *Nature* in February 1997 rocked the scientific community. However, the basic scientific principles that underpin organismal cloning have had a long history and can be traced back to the early days of experimental embryology in the late 1800s and early 1900s. The early embryologists were seeking answers to the central question of development – how does a complex multicellular organism develop from a single fertilised egg?

There are two contrasting theories of **development**. One is that everything is pre-formed in some way, and development is simply the unfolding of this already-existing pattern. This theory is called **preformationism**, and at its most extreme was thought of as a fully formed organism called a **homunculus** (a little man) sitting inside the sperm ready to 'grow' into a new individual during development. The alternative view is **epigenesis**, in which development is seen as an iterative process in which cells communicate with each other, and with their environment, as development proceeds. As with many opposing theories, there are aspects of each that can be considered valuable even today. Certainly the genetic information is already 'pre-formed': thus, the genome *could* be considered as a set of instructions that enable the unfolding of all structures in the embryo. However, with the growing appreciation of the importance of the proteins in the cell (the proteome), it became clear that development is indeed an interactive process that involves many factors, and that differential gene expression is the mechanism by which complexity is generated from the genetic information in the embryo. In addition to this control of gene expression, which essentially directs the process of embryogenesis, movement of cells and the formation of defined patterns enable structural complexity to arise. Overall, the process of development therefore involves a series of complex interactive steps in both a **temporal** and a **spatial** context. This is illustrated in Fig. 14.1.

Development of the foetus from a single cell is a complex process that involves many interactions between cells and their environment.

Differential gene expression, and interactions of various components in both temporal and spatial contexts, is required if the development of the embryo is to proceed normally.

14.1.1 First steps towards cloning

Our current knowledge of embryological development, as shown in Fig. 14.1, has been established over a long period. The first 'embryologist' appears to have been Aristotle, who is credited with establishing an early version of the theory of epigenesis. More recently, August Weismann attempted to explain development as a unidirectional process. In 1885 he proposed that the genetic information of cells diminishes with each cell division as development proceeds. This set a number of scientists to work trying to prove or disprove the theory. Results were somewhat contradictory, but in 1902 Hans Spemann managed to split a two-cell salamander embryo into two parts, using a hair taken from his baby son's head! Each half developed into a normal organism. Further work confirmed this result and also showed that in cases where a nucleus was separated from the embryo, and cytoplasm was retained to effectively give a complete cell, a normal organism developed. Thus, Weismann's idea of diminishing genetic

Early experiments in cloning used the technique of embryo splitting to generate cloned organisms.

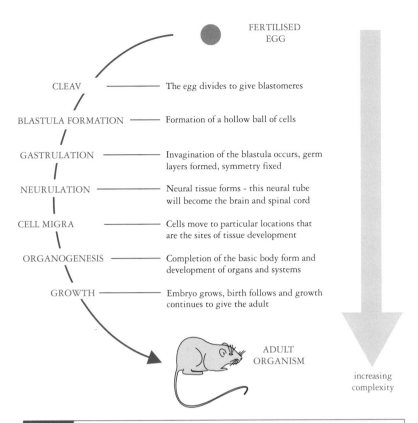

FERTILISED EGG

CLEAV ———————— The egg divides to give blastomeres

BLASTULA FORMATION ——— Formation of a hollow ball of cells

GASTRULATION ———————— Invagination of the blastula occurs, germ layers formed, symmetry fixed

NEURULATION ———————— Neural tissue forms - this neural tube will become the brain and spinal cord

CELL MIGRA ———————— Cells move to particular locations that are the sites of tissue development

ORGANOGENESIS ———————— Completion of the basic body form and development of organs and systems

GROWTH ———————— Embryo grows, birth follows and growth continues to give the adult

ADULT ORGANISM

increasing complexity

Fig. 14.1 Developmental sequence in the mouse. The journey from fertilised egg to adult organism involves cell division, cell differentiation, and the generation of complex patterns during embryogenesis. All these events must be coordinated and regulated in time and space if the process is to proceed successfully to completion. Modified after Nicholl (2000), *Cell & Molecular Biology*, Advanced Higher Monograph Series, Learning and Teaching Scotland.

resources was shown to be incorrect – all cells in developing embryos retained the ability to programme the entire course of development.

In 1938 Spemann published his book *Embryonic Development and Induction*, detailing his work. In this he proposed what he called 'a fantastical experiment', in which the nucleus would be removed from a cell and implanted into an egg from which the nucleus had been removed. Spemann could not see any way of achieving this, hence his caution by using the word 'fantastic'. However, he had proposed the technique that would later become known as cloning by **nuclear transfer** (more fully known as **somatic cell nuclear transfer** or **SCNT**). Unfortunately, he did not live to see this attempted; the first cloning success was not achieved until 1952, eleven years after his death.

The idea of transplanting an intact nucleus into an egg cell to generate a clone had been proposed as early as 1938. The technique would not be developed fully until much later, culminating in 1996 with the birth of Dolly.

14.1.2 Nuclear totipotency

The work of Spemann was an important part of the development of modern embryology – indeed, he is often called the 'father' of

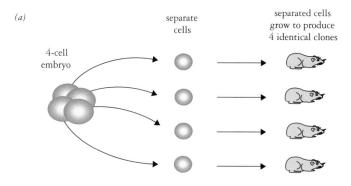

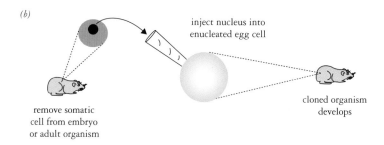

Fig. 14.2 Two methods for animal cloning. (*a*) Embryo splitting. Cells from an early embryo (shown here as a 4-cell embryo) are separated and allowed to continue development. Each cell directs the process of development to produce a new individual. The four organisms are genetically identical clones. (*b*) The technique of nuclear transfer. A nucleus from a somatic cell (from an embryo or an adult) is transplanted into an enucleated egg cell. If development can be sustained, a clone develops. Note that in this case the clone is not *absolutely* identical to its 'parent'. Mitochondrial DNA is inherited from the cytoplasm and will, therefore, have been derived from the egg cell. This is known as a maternal inheritance pattern.

The developmental status of cells is important, with the concept of nuclear totipotency an important part of the thinking that led to experiments in SCNT-based cloning.

this discipline. He had proposed nuclear transfer and had demonstrated cloning by **embryo splitting**. The basis of these two techniques is shown in Fig. 14.2. The experiments with embryo splitting that had refuted Weismann's ideas showed that embryo cells retain the capacity to form all cell types. This became known as the concept of **nuclear totipotency**, which is now a fundamental part of developmental genetics. A cell is said to be totipotent if it can direct the formation of all cells in the organism. If it can direct a more limited number of cell types, it is said to be **pluripotent** or **multipotent**. Extending this along the developmental timeline, a cell that is not capable of directing development under appropriate conditions is said to be **irreversibly differentiated**.

Nuclear totipotency is in many ways self-evident, as an adult organism has many different types of cell. The original zygote genome, passed on by successive mitotic divisions, must have the capacity to generate these different cells. However, the key in developing cloning

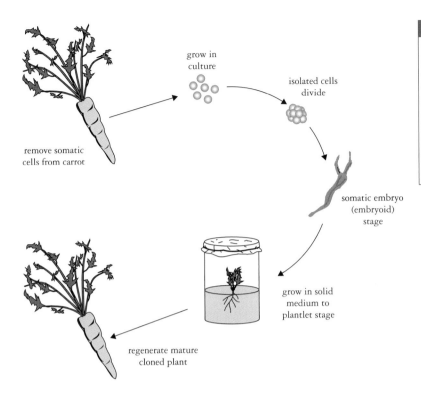

Fig. 14.3 Cloning of carrots. Somatic cells can be removed and grown in culture. Under appropriate conditions the cells begin to divide, then develop into somatic embryos known as embryoids. These can be transferred to a solid growth medium for plantlet development. The final stage is regeneration of the complete organism.

techniques was not that this idea was disputed, but rather that it was centred around attempts to see when embryo cells became irreversibly differentiated, and perhaps lost the *capacity* (but not the *genes*) to be totipotent. The next experiments to shed light on this area were carried out in the 1950s.

14.2 | Frogs and toads and carrots

Plant development is somewhat simpler than animal development, largely because there are fewer types of cell to arrange in the developing structure. However, the concept of nuclear totipotency is just as valid in plants as it is in animals. In fact, one of the early unequivocal experimental demonstrations of nuclear totipotency was provided by the humble carrot in the late 1950s. Work by F. C. Steward and his colleagues at Cornell University showed that carrot plants could be regenerated from somatic (body) tissue, as shown in Fig. 14.3. This technique is now often used in the propagation of valuable plants in agriculture. The ability of plants to regenerate has of course been exploited for many years by taking cuttings and grafting – these are essentially asexual 'cloning' procedures.

Amphibians provide useful systems for embryological research in that the eggs are relatively large and plentiful, and development

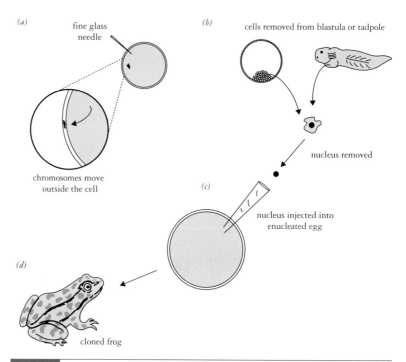

(a) fine glass needle

chromosomes move outside the cell

(b) cells removed from blastula or tadpole

nucleus removed

(c) nucleus injected into enucleated egg

(d) cloned frog

Fig. 14.4 Amphibian cloning. This is the type of procedure used by Briggs and King in 1952. (*a*) Egg cells are pricked with a fine glass needle, which activates the egg and causes the chromosomes to move outside the cytoplasm. Donor cell nuclei are removed from early embryos or tadpoles, as shown in (*b*). The donor nucleus is injected into the enucleated egg (*c*), and development can proceed. Early embryo donor nuclei gave better results than nuclei taken from later stages of development.

The frog *Rana pipiens* was used in animal nuclear transfer experiments in the early 1950s.

proceeds under less stringent environmental conditions than would be required for mammalian embryos. The frog *Rana pipiens* was used by Robert Briggs and Thomas King around 1952 to carry out Spemann's 'fantastical experiment'. Using nuclei isolated from blastula cells, they were able to generate cloned embryos, some of which developed into tadpoles (Fig. 14.4). As more work was done, it became apparent that early embryo cells could direct development, but that cell nuclei isolated from cells of older embryos were much less likely to generate clones. It was becoming clear that there was a point at which the cell DNA could not easily be 'deprogrammed' and used to direct the development of a new organism, and this remained the limiting factor in cloning research for many years. The work was later extended by John Gurdon at Oxford, using the toad *Xenopus laevis*. Some success was achieved, with fertile adult toads being generated from intestinal cell nuclei. In some of the experiments serial transfers were used, with the implanted nuclei allowed to develop; then these cells used for the isolation of nuclei for further transfer. There was, however, a good deal of debate as to whether Gurdon had in fact used fully differentiated cells or if contamination with primordial reproductive cells had produced the results.

14.3 | A famous sheep – the breakthrough achieved

Despite the uncertainty surrounding Gurdon's experiments, the work with frogs and toads seemed to demonstrate that animal cloning by nuclear transfer was indeed a feasible proposition, if the right conditions could be established. Although the work seemed promising, cloning from a fully differentiated adult cell remained elusive. It was suspected that the manipulations used in the experiments could be damaging the donor nuclei, and by the 1970s it was clear that further development of the techniques was required. Reports of mice cloned from early embryo nuclei appeared in 1977, but the work was again somewhat inconclusive and difficult to repeat.

The work with frogs and toads established the feasibility of SCNT-based cloning in animals and opened the way for the work that would result in the cloning of mammals.

Development of the techniques for nuclear transfer cloning continued, with sheep and cattle as the main targets because of their potential for biotechnological applications in agriculture and in the production of therapeutic proteins. By the mid 1980s several groups had success with nuclear transfer from early embryos. A key figure at this time was Steen Willadsen, a Danish scientist who in 1984 achieved the first cloning of a sheep, using nuclear transfer from early embryo cells. In 1985, Willadsen cloned a cow from embryo cells, and in 1986 he achieved the same feat using cells from older embryos. The work with the older embryos (at the 64- to 128-cell stage) was not published, but Willadsen had demonstrated that cloning from older cells might not be impossible, as most people thought. By this stage several companies had become involved with cloning technology, particularly in cattle, and the future looked promising for the technology, the scientists who could do it, and the industry. However, as with the Flavr Savr transgenic tomato, there were problems in establishing cloning on a commercial footing, and by the early 1990s the promise had all but evaporated.

At around the same time that commercial interests were developing around cloning, a scientist called Ian Wilmut, working near Edinburgh in what would become the Roslin Institute, was busy generating transgenic sheep. He was keen to try to improve the efficiency of this somewhat hit-or-miss procedure, and cloning seemed an attractive way of doing this. If cells could be grown in culture, and the target transgenes added to these cells rather than being injected into fertilised eggs, the transgenic cells could be selected and used to clone the organism. This approach had been successful in mice, using embryonic stem cells, although it proved impossible to isolate the equivalent cells from sheep, cattle, or pigs. However, older cells, derived from a foetus or adult organism, could be grown easily in culture. Thus, a frustrating impasse existed – if only the older cells could be coaxed into directing development when used in a nuclear transfer experiment, then the process would work. This was the step that informed opinion said was impossible.

The use of older mature cells as donors in SCNT experiments was proving to be the problem area for reproductive cloning in the 1980s and early 1990s.

In 1986 Wilmut attended a scientific meeting in Dublin, where he heard about the 64- to128-cell cattle cloning experiments from a

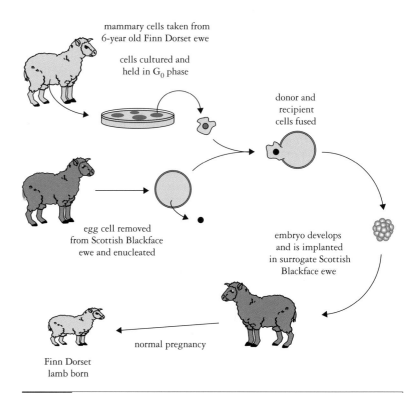

Fig. 14.5 The cloning method by which Dolly was produced. Mammary cells from an adult ewe were isolated, cultured, and held in the G_0 phase of the cell cycle. These were the donor cells. Egg cells taken from a different breed were enucleated to act as recipients. The donor and recipient cells were fused and cultured. The embryos were then implanted into surrogate mothers and pregnancies established. Dolly was born some 5 months later.

vet who had worked with Willadsen. This encouraged him to continue with the cloning work. The key developments came when Keith Campbell joined Wilmut's group in 1990. Campbell was an expert on the **cell cycle** and was able to develop techniques for growing cells in culture and then causing the cells to enter a quiescent phase of the cycle known as **G_0 phase**. Wilmut and Campbell thought that this might be a critical factor, perhaps the key to success. They were proved correct when, in 1995, Megan and Morag were born. These were lambs that had been produced by nuclear transfer using cultured cells derived from early embryos. Extensions to the work were planned, supported by PPL Therapeutics, the biotechnology firm set up in 1987 to commercialise transgenic sheep technology. Wilmut and Campbell devised a complex experiment in which they would use embryo cells, foetal cells, and adult cells in a cloning procedure. A strange quirk of history appears at this point – the adult cells came from a vial that had been stored frozen at PPL for three years, and their source animal was long forgotten. However, it was known that the cells were from the udder of a 6-year-old Finn Dorset ewe. The cloning process is summarised in Fig. 14.5.

Fig. 14.6 Dolly – the first mammal to be cloned from a fully differentiated somatic adult cell. Photograph courtesy of the Roslin Institute. Reproduced with permission.

From the adult cell work, 29 embryos were produced from 277 udder cells. These were implanted into Scottish Blackface surrogate mothers, and some 148 days later, on 5 July 1996, one lamb was born. She was named Dolly, after the singer Dolly Parton (make the connection yourself!). Dolly is shown in Fig. 14.6. What had once seemed impossible had been achieved.

14.4 | Beyond Dolly

The birth of Dolly demonstrated that adult differentiated cells could, under the appropriate conditions, give rise to clones. Whilst the magnitude of this scientific achievement was appreciated by Wilmut and his colleagues at the time, the extent of the public reaction caught them a little by surprise. Suddenly Wilmut, Campbell, and Ron James of PPL were in the limelight, unfamiliar ground for scientists. Dolly herself became something of a celebrity in the media, with both the serious science press and the popular press maintaining an interest. She was mated with a Welsh mountain ram in 1997 and gave birth to a female lamb called Bonnie in April 1998, thus demonstrating that generation of an organism by reproductive cloning appeared to have no effect on normal reproductive processes in the clone.

Being a celebrity sheep, Dolly quite naturally was subjected to detailed scrutiny, and thus when signs of possible premature ageing

The issue of telomere length and ageing is an interesting area of debate that has been highlighted by reproductive cloning.

Cloning technology is becoming more established but still presents enormous technical challenges.

were noted when she was around 5 years old, there was some cause for concern. One possible reason for this could be the progressive shortening of chromosomal **telomeres** that occurs during successive rounds of cell division, as this is thought to be associated with the ageing process. As the donor cell that created Dolly was 6 years old, she could be considered as effectively six when she was born (at least from a cellular genetics point of view); thus, perhaps this had an impact on her normal lifespan. There were also signs of arthritis, but there is doubt as to whether this was in any way associated with the fact that she was a clone.

Dolly died on 14 February 2003, from a progressive retroviral lung disease that sheep are susceptible to. Her death did not appear to be a consequence of her clone status, and her remains were preserved and are displayed in the Royal Museum in Edinburgh (part of the National Museum of Scotland). Some commentators have made much of the fact that she attained approximately half a normal lifespan, and that this was offset by having 'lived the other half already' as the cell from the donor organism. This is probably much too simplistic, and more research is needed to establish the physiological and genetic features of cloned organisms that may arise as a consequence of the cloning process. As work has progressed it has become apparent that there is much complexity associated with this type of procedure, and that donor cell DNA may not in fact be completely reprogrammed during an SCNT process. This can lead to difficulties in that gene expression may not fully reflect the requirement of embryological development in the clone. It also seems that, in around one-third of cases, cloned organisms are larger than normal at birth, which is termed **large offspring syndrome**.

Having achieved success with Dolly, Wilmut and his colleagues went on to produce the first transgenic cloned sheep, a Poll Dorset clone (named Polly) carrying the gene for Factor IX. Thus, the goal of producing transgenics using nuclear transfer technology had been achieved by the late 1990s and offers great potential for the future.

Cloning of a range of organisms, including cats, mice, goats, cattle, rabbits, horses, donkeys, and monkeys has now been achieved, using either the embryo-splitting technique or nuclear transfer. Cloning of human embryos by embryo splitting was reported as far back as 1993, although most scientists consider that nuclear transfer cloning of humans should not be attempted. However, proposals to clone humans have been put forward, notably by Richard Seed in 1997, and by an Italian reproductive expert who stated that he was setting out to establish cloning of humans. As with so many aspects of modern genetics, the future will no doubt hold many contentious and interesting developments in the field of organismal cloning. Regardless of how the application of cloning technology develops, when the history is written the central character will be a sheep named after a Country and Western singer – bizarre but strangely appropriate in a field that stirs the imagination like few others in science.

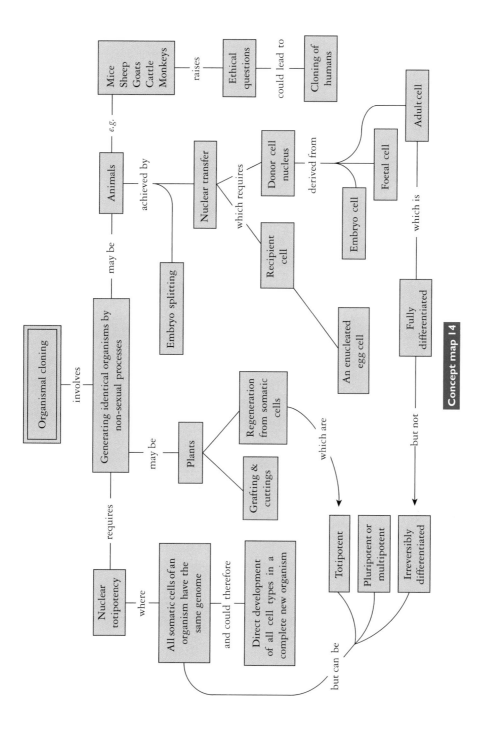

Concept map 14

Chapter 15

Brave new world or genetic nightmare?

This final chapter is short. It does not answer any questions but simply raises them for consideration. There are no 'correct' answers to these questions, as each must be addressed from the perspective of the individual, family, society, race, or nation that is facing up to the situation. There are no diagrams or photographs, and very little factual information. However, the topics discussed are probably the most important that a student of genetic engineering can consider. In practical terms, relatively few people will ever go on to work in science and technology, but we will all have to cope with the consequences of gene-based research and its applications. Informed and vigorous debate is the only way that the developments of gene manipulation technology can become accepted and established.

15.1 Is science ethically and morally neutral?

It is often said that science *per se* is neither 'good' nor 'bad', and that it is therefore ethically and morally neutral. Whilst this may be true of science as a *process*, it is the developments and applications that arise from the scientific process that pose the ethical questions. The example that is often quoted is the development of the atomic bomb – the science was interesting and novel, and of itself ethically neutral, but the application (*i.e.* use of the devices in conflict) posed a completely different set of moral and ethical questions. Also, science is of course carried out by *scientists*, who are most definitely not ethically and morally neutral, as they demonstrate the same breadth and range of opinion as the rest of the human race. An assumption often made by the layman is that scientists and the scientific process are the same thing, which is unfortunate.

Despite the purist argument that science is in some way immune from ethical considerations, I believe that to separate the process from its applications is an artificial distinction. In the developed world, we live in societies shaped by technology, which is derived from the application of scientific discoveries. We must all share the responsibility of policing the new genetic technology.

The question of the moral and ethical neutrality of science is not easily answered, with many different aspects to be considered.

15.2 | Elements of the ethics debate

Where benefits are clear, ethical concerns tend to be dealt with more easily than when there is some doubt as to the need for, or the effect of, a particular process or product.

Advances in the basic science of genetics usually pose few problems from an ethical standpoint. The major concerns are usually separate from the actual experiments – perhaps the use of animals in research, or the potential for transgenic crops to contaminate non-transgenic or wild populations. We will consider some of the ethical problems in medicine, biotechnology, transgenic organisms, and organismal cloning – thus following the arrangement of this book. However, there is considerable overlap in many of these areas, and it is again some-what artificial to separate these topics.

In medicine, few would argue against the development of new drugs and therapies, where clear benefit is obtained. Perhaps the one area in the medically related applications of genetic research that is difficult is the human genome information. Genetic screen-ing, and thus the possibility of genetic discrimination, is an area of active debate at the moment. The molecular diagnosis of genetically based disease is now well established, and the major ethical dilem-mas tend to centre around whether or not a foetus should be aborted if a disease-causing trait is detected. If and when it becomes possible to screen routinely for polygenic and multifactorial traits, perhaps involving personality and predisposition to behavioural problems, the ethical picture will become even more complex. This whole area of predisposition, as opposed to a confirmed causal link between genes and disease, is a difficult area in which to establish any ethical rules, as many of the potential problems are as yet hypothetical.

The possibility of genetic discrimination based on genome sequence knowledge has opened up a new area of concern, where predisposition becomes an active issue in addition to actual conditions that a person might suffer from.

The **biotechnology industry** is a difficult area to define, as the applications of gene technology in biotech applications are very diverse. The one ethical thread linking disparate applications is the influence of commercial interests. Patenting gene sequences raises questions, as does the production of products such as bovine soma-totropin. Many people see biotechnology applications as driven by commercial pressures, and some are uneasy with this. Similar ques-tions can be asked of any manufacturing process, but the use of bio-logical material seems to set up a different attitude in many people. In some cases the apparent arrogance of biotechnology companies upsets people who might actually agree with the overall aims of the company, and several biotech companies have found that this public opinion can be a potent force in determining the success or failure of a product.

Public opinion, particularly with regard to consumer acceptance of a product, can have a very powerful effect on commercial success.

The development and use of transgenic organisms poses several ethical questions. The one thing that has been a little surprising is the reversal of the usual plant/animal debate as far as transgenesis is concerned. Traditionally, animal welfare has been the major source of difficulty among pressure groups, concerned individuals, scientists, and regulators. Plants were largely ignored in the ethical debate until the late 1990s, when the public backlash against genetically modified (GM) foods began to influence what biotech companies were doing.

The issue of GM foods is an emotive matter for many people and continues to be an area of controversy.

Concerns were in two areas – the effect of GMO-derived foods on health, and the effect on the environment. The environmental debate in particular has been driven by many different groups, who claim that an ecological disaster might be waiting to emerge from GM plant technology because of cross-pollination. It is impossible to predict what might happen in such cases, although the protagonists of GM crops claim to have evaluated the risks. The simple answer is that we do not know what the long-term ecological effects might be.

Transgenic animals have not posed as big a problem as plants. This is a little surprising but can be explained by the fact that animals are much easier to identify and contain (they need to copulate rather than pollinate!) and, therefore, any risk of transgenic traits getting into wild populations is much lower than that for crop plants. Also, transgenic animals are often used in a medical context, where the benefits are often more tangible and generally appreciated. Welfare issues are still a concern, but the standard of care for animals in transgenic research is very high, and tightly monitored in most countries. Overall the acceptance of transgenic mouse models for disease, and transgenic animals as bioreactors, seems to pose less of a problem generally than aspects of plant biotechnology. The possibility of xenotransplantation offers hope but also raises questions, particularly for groups with specific moral or religious objections to this type of application.

Organismal cloning is perhaps the most difficult area from an ethical viewpoint, as the possibility of human cloning exists and is being taken seriously by some. There is debate about the unique nature of personality, character, soul, nature, or whatever it might be called; what would a cloned individual actually *be*? Views range from those who think that he or she would be in some sort of limbo, to those who see essentially no difference between a clone and a normal individual, with the biochemical and physiological characteristics of the individual being the determinants of what that individual is and will become.

Human cloning – fact, fiction, fantasy, or fabrication? This is the ethical minefield that is challenging many people at present, and the topic of organismal cloning is likely to remain at the forefront of the ethics debate for many years to come.

Cloning pets is perhaps less serious an issue than human cloning, but there are strongly held views about this aspect of cloning that is currently a reality. Cloning a much-loved but deceased pet may well become a financial/commercial issue rather than an ethical one, so this presents a rather different moral dilemma than some of the other aspects of gene manipulation and organismal cloning.

In a similar way to the progression of the Dolly research, and its extension into transgenic cloning (Polly), many people fear that the possibility of generating transgenic 'designer babies' may become reality at some point in the future. Normal conception, genetic screening, transgenesis to replace defective genes, implantation, and development to full term? Or perhaps selection of a set of characteristics from a list, and production of the desired phenotype by manipulating the genome? This all sounds very fanciful at present, even absurd. In the early 1990s informed opinion said the same thing about cloning from adult cells. Dolly arrived in 1996.

15.2.1 The role of the scientist

In considering all the areas just outlined, we can consider that there are two major constituencies involved – the scientists who carry out the research and development work, and the society within which this is done. Each of these two groups has an important role to play in the debate.

The scientist, and also others who work in the applied industries that arise from basic research and development, is often seen as the expert in the debate. However, two points should be noted here. First, as already noted, scientists are of course part of society, with all the attendant opportunities and responsibilities that this brings. Second, scientists may be expert in their particular fields but may not be expert in formal ethics or moral philosophy. Thus, the position of the scientist is not necessarily a simple one. All that we can ask of our scientific community is that its members carry out their work carefully, honestly, and with a sense of moral responsibility. Most scientists are also more than willing to consider the impact of their work in a wider context, often engaging with other experts in complementary fields to ensure full and informed debate.

Unfortunately, as in all walks of life, sometimes things in science do not go according to plan. Often progress is slow and can follow unproductive lines of investigation, but this is of course an inherent part of the scientific method and is accepted by all who are involved (although the general public sometimes set expectations rather too high and expects scientists to be magicians). However, much more serious issues arise when scientists are found to have been falsifying data or otherwise compromising the integrity of the discipline. Thankfully these instances are rare, but they do happen and cause significant damage to the field when they come to light. The one positive aspect of this area of science is that the scientific community is very vigilant, and when cases do appear they are dealt with appropriately.

15.2.2 The role of society

If the scientists have a duty to act responsibly, the wider community also carries some of this responsibility, as well as the right to accurate and balanced information. Society is a complex mixture of many different types of people, groups, and businesses, and presents many different points of view. Some of these can be extreme and often antagonistic, and care must be taken to ensure that a forceful stance taken by a minority of people does not compromise the level of serious debate that is needed to ensure that appropriate decisions are taken by those who regulate and govern.

Balancing the need to ensure open dialogue, there is also a need to guard against alienating various groups whose members may have genuine concerns. There is too often a tendency for a debate to become polarised, with the attendant risk of oversimplification of the issues. Given that we are now functioning in the 'global village',

> The science community has a major role to play in presenting the arguments of the ethical and moral aspects of modern genetics.

> Modern communication mechanisms mean that information placed in the public domain is immediately available worldwide, which has changed the dynamics of interactive debate.

where information is available worldwide through various media, issues are immediately placed in this global dimension and, thus, attract a much wider range of responses than would have previously been the case.

In summing up the elements of the ethics debate, it is important to accept that all members of societies have a role to play in determining what policies and procedures are put in place to monitor, regulate, and advise on the use of gene manipulation and its many applications. It is to the credit of those involved that, in most cases, full and frank discussion can enable consensus to be reached. Not everyone will necessarily be in agreement with the decisions made in this way, but most people accept that this is the most suitable way to proceed.

15.3 Does Frankenstein's monster live inside Pandora's box?

The rather awful question that forms the title for this final section illustrates how two phrases that would not usually exist in one sentence can be joined together. This of course is exactly the essence of gene manipulation – splicing pieces of DNA together to generate recombinant molecules that would not exist in nature. In this book we have looked at the range of gene manipulation technology, from the basic techniques up to advanced applications. In trying to answer the question 'is it good or bad', we have seen that there can be no answer to this. As with any branch of human activity, the responsibility for using genetic technology lies with those who discover, adapt, implement, and regulate it. However, the pressures that exist when commercial development of genetic engineering is undertaken can sometimes change the balance of responsibility. Most scientists ply their trade with honesty and integrity and would not dream of falsifying results or inventing data. They take a special pride in what they do, and in a curious paradox remain detached from it whilst being totally involved with it. Once the science becomes a technology, things are not quite so clear-cut, and corporate responsibility is sometimes not quite so easy to define as individual responsibility.

When writing the second edition of this book, I stated that developments in genetics and gene manipulation continue to progress at a staggering pace. This is even more evident as I write the conclusion to this edition, with more sophisticated applications of gene manipulation, genetics, and cell biology emerging almost daily. In fact, as I was working on the final pre-publication files for the text (November 2007), the first report of cloning primate embryonic stem cells by nuclear transfer had just been published. In addition, 'reprogramming' human skin cells to become pluripotent was also reported. These two achievements bring closer the possibility of regenerative medicine in humans, which many see as a major and welcome

You should be prepared to play your part in the debates that have yet to take place, so that appropriate and due consideration is given to any new developments that have the potential to have an impact on all our lives.

development. Others see this as fraught with difficulties, both technical and ethical.

Even those working at the forefront of the various disciplines associated with genetics and gene manipulation have often been surprised at how fast things have progressed, particularly with respect to genome sequencing projects. By the time this book is published, many more developments will have taken place. I hope that by reading this book you will be a little better prepared to assess these new developments, and take your part in the debate that we all must engage in to ensure that gene technology is used as a force for good.

Suggestions for further reading

In this section I have outlined some of the sources that may be useful in gaining access to additional material on the topics covered in the text. A departure from previous editions is that I have not listed any specific texts. This is not because there are none available; indeed there are many excellent texts that will take your study of the topic further. However, the World Wide Web now provides easy access to so many sources of information that it has almost become redundant to list texts on paper. In addition, a browse through a good academic bookshop or library is often the best way to spot texts that may be of interest. I have not referred to the primary literature, as this is accessible from some of the texts and review journals that you will come across.

Books

General search engines on the Web will identify academic publishers and other sources of texts. Specific sites such as booksellers Amazon (http://www.amazon.co.uk or http://www.amazon.com) provide internal search facilities so that texts on various topics can be identified. If you select carefully, additional texts can provide much useful information, and often extend the ideas presented in this book or treat the topics with a different emphasis. All should be accessible to readers of this book. More speculative (or even fictional) texts are often of interest if taken in context, and some of the historical treatments of various fields can provide an insight into how science works and outline the central personalities involved.

Review journals

Review journals are a good source of detailed information about specific topics and often provide extensive lists of relevant primary literature references. Journals may appear annually, monthly, or weekly. Some examples are *Scientific American* and *New Scientist*, which publish readable articles on a wide range of topics. The *Annual Reviews* series offers volumes in *Biochemistry and Genetics*, and the *Trends* series publishes *Trends in Genetics*, *Biotechnology*, and *Biochemical Sciences*. All should be available from good academic libraries and some large public libraries.

In addition to review journals, many front-line research journals also publish useful review articles and news about research fields. The best known are *Nature* (with versions such as *Nature Genetics* and *Nature Biotechnology*) and *Science*. Most journals are now available electronically through academic libraries.

Using the World Wide Web

Over a very short timeframe, the Internet and World Wide Web (WWW) have had an enormous impact on how we access and use information. The term *Internet* is generally used to describe the network of computers that together provide the means to publish and share information, whilst the term *World Wide Web* is a more general description of the information that is published using the Internet. However, the two terms are often used interchangeably, and the phrases 'surfing the net' and 'surfing the web' have become part of modern-day language.

If you are already a competent Web user, you won't need this section. If you are a newcomer, I have provided some information to get you started. The Web provides an exciting and easy-to-use medium that is both a goldmine and a minefield, so get to grips with the basics and have a go. It's fun!

Getting started

You need a computer and a connection to the Internet. Computer access to the Web may be provided by your place of work or study, or you can set up a personal account from home. Universities, colleges, research institutes, and companies will usually have sophisticated IT (information technology) resources, and often their own intranet system and website. Access to and from such sites is usually easy, fast, and free from personal expense if you are a student or employee. Most institutions have rules about using the Web from their computers, often called acceptable use policies. This is to prevent misuse of the system, such as downloading inappropriate material, accessing non-work-related chatrooms, playing games on the Web, and so on.

To set up a personal account you need to have a computer with a modem, and you need to register with an Internet service provider (ISP). Usually there are costs involved, and these may be for access and/or telephone charges. Most ISP companies offer a range of packages to suit your likely usage pattern, so it pays to examine the packages on offer to make sure that you get the best value. It is very easy to run up substantial time online, and if you are anything other than a very occasional user, a package with unmetered access is probably the best option to choose. A broadband connection is better than a dial-up option, and these days most geographical areas can access broadband facilities. Costs vary, and many ISP companies provide their own extensive website and news services, plus e-mail facilities. To actually look at Web pages, you need to have a Web browser on your computer. The two most common versions are *Netscape* and *Internet Explorer*. One or the other is usually supplied with your computer, or they can be obtained on CD-ROM or downloaded from the Web if you have access to another computer that already has an online connection.

Finding websites is generally straightforward if you know where you are going. Each website has an 'address' known as a uniform resource locator or URL. Some of these have been listed in the text already. A URL generally begins with http://www. followed by the specific address. Many recent textbooks have associated websites, as do research groups, university departments, companies, and so forth.

If you don't know the URL, or want to search for a particular type of website, your ISP will normally have its own search engine and often enables access to several of these. A search engine enables you to look for information using a range of terms, and it is astonishing what you can find. However, caution is recommended.

Caution!

There is an awful lot of information on the Web, and there is a lot of awful information there as well! It is very easy to get sidetracked and end up wasting a lot of time searching through sites that are of no value (but may be interesting nonetheless!). I have just typed in some search terms using the search engine supplied by a typical ISP. The number of 'hits' for the various terms were as follows:

•	genetics	5 563 801
•	cloning	2 244 013
•	cystic fibrosis	966 406
•	DNA cloning	499 824
•	sheep cloning	127 707
•	plasmid vector	71 228
•	homopolymer tailing	123

Two points are clear – first, the number of retrieved items is often far too large to be useful, and second, the more specific, defined, or restricted the search, the better. However, even with something like 'sheep cloning' or 'plasmid vector' there is still too much information to look through, so learning to use the advanced search facilities provided by various search engines is well worth the effort.

Types of website

The main point to be aware of when using the Web is that there is very little restriction on what gets published, and by whom. The peer-review system that is used in science is not often applied to Web-based information, so not all information may be accurate. The type of website can often give some indication as to how reliable the information may be. The last part of the URL usually indicates the type of organisation, although this does vary from country to country. Examples are '.gov' for government site, '.ac.uk' for UK academic sites ('.edu' is used in some other countries), and '.com' or '.co' for commercial or company sites. The characteristics of these types of site can be described as follows:

- Government sites – usually accurate information, but this can have a particular emphasis. Political influences and 'spin' can sometimes make certain types of information highly suspect for some government sites – a knowledge of the local situation is often useful.
- Research institutes and universities – very often the most useful sites, as these tend to be run by people who are used to publishing information by the peer-review mechanism, and therefore are aware of the need to be accurate and unbiased.
- Company sites – again can provide a wealth of information, but this is obviously presented to support company policy in most cases. Many responsible companies take care to present a balanced view of any contentious issues, often including links to other sources of information.
- Publishers – range from book publishers to online journals. Often provide excellent resource material, but for some journal sites access may require registration and payment of a fee for each article downloaded.
- Information 'gateway' sites – many specialist research areas have associated sites that collate and distribute information about the latest developments. Often it is possible to register for e-mail updates to be sent directly when they are published. As with journals, some of these sites require registration and subscription to access the full range of facilities.
- Pressure group sites – can provide a lot of useful information, very often gathered to counteract some particular claim made by another party. Despite the obvious dangers of accepting the views presented, many of these sites take great care to present accurate information. Can range from moderate to extreme in terms of the views presented.
- Personal sites – prepared by individuals, therefore likely to be the most variable in terms of content and viewpoint. Range from highly respected sites by well-known experts to the sort of site that is best avoided as far as serious study is concerned.

The difficulty in weeding out the less useful (or downright weird) sites is that anyone with a flair for design can produce a website that looks impressive and authoritative. Conversely, some extremely useful and respected sites can look a little dull if a good Web designer has not been involved. However, as long as you remember that critical evaluation is essential, you should have few problems.

One fairly recent interesting addition to the genre of Web-based publishing is the interactive encyclopaedia Wikipedia (http:// en.wikipedia.org), which was set up in 2001. This is an active encyclopaedia that is 'editable' by its contributors, and thus represents a novel publishing phenomenon. The theory is that it is effectively a self-regulating project that grows as more articles are added and refined. In general the articles are written by authors who are expert in their fields and who take care to ensure accuracy and objectivity. Wikipedia is therefore a useful resource, but it should be backed up by further validation of the content if serious study is undertaken on a particular topic. However, it remains an exciting and innovative source of information that has set a new benchmark for Web-based publishing.

Dive in and enjoy!

Part of the fun of using the Web is that it *is* such a diverse and extensive medium, and it is constantly changing and expanding. Although I had contemplated providing a list of 'useful URLs', I decided against this for a number of reasons. First, some URLs are given in particular parts of the text and provide an entry point for particular topics. Second, if you are using this book as a course text, your tutor should be able to direct you to any specific websites that he or she uses to support the course. Third, using a search engine is the most effective way of getting up-to-date information about what is available, and by finding a few relevant sites, you can quickly build up a set of links that take you into other related sites. Good surfing!

Glossary

Abundance class Refers to the relative abundance of different mRNA molecules in a cell at any given time.

Acquired immune deficiency syndrome (AIDS) Disease of the immune system, causative agent the HIV virus (q.v.), characterised by deficiency in T-lymphocytes rendering sufferers susceptible to infection and other clinical conditions.

Activase™ Recombinant-derived tissue plasminogen activator, the first rDNA therepeutic to be produced from mammalian cells in the early 1980s.

ADA Adenosine deaminase, deficiency results in SCIDS (q.v.).

Adaptor A synthetic single-stranded non-self-complementary oligonucleotide used in conjunction with a linker to add cohesive ends to DNA molecules.

Adenine (A) Nitrogenous base found in DNA and RNA.

Adenosine Nucleoside composed of ribose and adenine.

Adenosine deaminase Enzyme that converts adenosine to inosine. Deficiency causes toxic nucleosides to build up, which causes immune system deficiency and results in SCIDS (q.v.).

Adeno-associated virus Virus used in gene therapy delivery methods.

Adenovirus Virus that can infect through nasal passages, used in gene therapy delivery methods.

Adsorption (1) Adhesion of molecules or macromolecules to a substrate or matrix, (2) also used to describe the attachment of a bacteriophage to the bacterial cell surface.

Aetiology Of disease; relating to the causes of the disease.

Affinity chromatography Technique used to separate cellular components by selective binding and release using an appropriate medium to which the component of interest will bind. An example is the selection of mRNA from total RNA using oligo(dT)-cellulose.

Age-related macular degeneration Disease condition of the retina where over-promotion of blood vessel growth leads to impaired visual function. Potential target for RNAi (q.v.) therapy.

Agrobacterium tumefaciens Bacterium that infects plants and causes crown gall disease (q.v.). Carries a plasmid (the Ti plasmid) used for gene manipulation in plants.

Agarose Jelly-like matrix, extracted from seaweed, used as a support in the separation of nucleic acids by gel electrophoresis.

AIDS See *acquired immune deficiency syndrome.*

Alcohol oxidase (AOX) Enzyme of alcohol metabolism. The gene promoter can be used for the construction of expression vectors in yeast.

Alkaline phosphatase An enzyme that removes 5′ phosphate groups from the ends of DNA molecules, leaving 5′ hydroxyl groups.

Allele One of two or more variants of a particular gene.

Allele-specific oligonucleotide (ASO) Oligonucleotide with a sequence that can be matched precisely to a particular allele by using stringent hybridisation conditions.

Alpha-1-antitrypsin Protease inhibitor used in the treatment of cystic fibrosis.

Alpha peptide Part of the ß-galactosidase protein, encoded by the *lacZ'* gene fragment.

Alzheimer disease Neurodegenerative disease, characterised by progressive deterioration in cognitive function.

Ampicillin (Ap) A semisynthetic ß-lactam antibiotic.

Amplicon Refers to the product of a polymerase chain reaction.

Aneuploidy Variation in chromosome number where single chromosomes are affected; thus, the chromosome complement is not an exact multiple of the haploid chromosome number.

Animal model Usually a transgenic mouse in which a disease state has been engineered. See also *knockout mouse, knockin mouse, oncomouse.*

Antibody An immunoglobulin that specifically recognises and binds to an antigenic determinant on an antigen.

Anticodon The three bases on a tRNA molecule that are complementary to the codon on the mRNA.

Antigen A molecule that is bound by an antibody. Also used to describe molecules that can induce an immune response, although these are more properly described as immunogens.

Antiparallel The arrangement of complementary DNA strands, which run in different directions with respect to their $5' \rightarrow 3'$ polarity.

Antisense RNA Produced from a gene sequence inserted in the opposite orientation, so that the transcript is complementary to the normal mRNA and can therefore bind to it and prevent translation.

AP-PCR See *arbitrarily primed PCR.*

Arabidopsis thaliana Small plant favoured as a research organism for plant molecular biologists.

Arbitrarily primed PCR PCR using low-stringency random primers, useful in the technique of RAPD analysis (q.v.) for genomic fingerprinting.

ASO See *allele-specific oligonucleotide.*

Aspergillus niger Ascomycete fungus useful for genetics research and biotechnology.

Autoradiograph Image produced on X-ray film in response to the emission of radioactive particles.

Autosome A chromosome that is not a sex chromosome.

Auxotroph A cell that requires nutritional supplements for growth.

BAC Bacterial artificial chromosome, used for cloning large pieces of genomic DNA.

Bacillus thuringiensis Bacterium used in crop protection and in the generation of Bt plants that are resistant to insect attack. The bacterium produces a toxin that affects the insect.

Bacteriophage A bacterial virus.

Baculovirus A particular type of virus that infects insect cells, producing large inclusions in the infected cells.

Bacterial alkaline phosphatase (BAP) See *alkaline phosphatase*.

Bal 31 nuclease An exonuclease that degrades both strands of a DNA molecule at the same time.

Beta galactosidase An enzyme encoded by the *lacZ* gene. Splits lactose into glucose and galactose.

Beta lactoglobulin (BLG) The major whey protein in cow's milk. The gene promoter can be used in the construction of expression vectors.

Binary vector A vector system in which two plasmids are used. Each provides specific functions that complement each other to provide full functionality.

Bioinformatics The emerging discipline of collating and analysing biological information, especially genome sequence information.

Biolistic Refers to a method of introducing DNA into cells by bombarding them with microprojectiles, which carry the DNA.

Biological containment Refers to the biological properties of a cell or vector system that makes it unlikely that the strain will survive outside the laboratory environment. Cf. *physical containment*.

Biological data set Refers to any data set of biological origin, but in context is taken to mean the large databases of sequence and interaction data that have arisen as a consequence of modern molecular biology investigations.

Biotechnology Generic word to describe the application of bioscience for the benefit of humankind. Encompasses a wide range of disciplines and procedures but is often mistakenly thought to refer exclusively to the industrial-scale use of genetically modified microorganisms.

BLG See *beta lactoglobulin*.

Blunt ends DNA termini without overhanging 3′ or 5′ ends. Also known as *flush ends*.

Bovine somatotropin (BST) Bovine growth hormone, produced as rBST for use in dairy cattle to increase milk production.

BST See *bovine somatotropin*.

Bt plants Transgenic plants carrying the toxin-producing gene from *Bacillus thuringiensis* as a means to protect the plant from insect attack.

CAAT box A sequence located approximately 75 base pairs upstream from eukaryotic transcription start sites. This sequence is one of those that enhance binding of RNA polymerase.

Caenorhabditis elegans A nematode worm used as a model organism in developmental and molecular studies.

Calf intestinal phosphatase (CIP) See *alkaline phosphatase*.

Cap A chemical modification that is added to the 5′ end of a eukaryotic mRNA molecule during post-transcriptional processing of the primary transcript.

Capsid The protein coat of a virus.

Carbohydrate Molecule containing carbon, hydrogen, and oxygen, empirical formula CH_2O. Important in energy storage and conversion reactions in the cell; examples include glucose, fructose, and lactose. Polymers of carbohydrates are known as polysaccharides and are used as energy storage compounds.

cDNA DNA that is made by copying mRNA using the enzyme reverse transcriptase.

cDNA library A collection of clones prepared from the mRNA of a given cell or tissue type, representing the genetic information expressed by such cells.

Cell membrane Lipid bilayer–based structure containing proteins embedded in or on the membrane. Acts as a selectively permeable barrier that separates the cell from its environment.

Cell wall Found in bacteria, fungi, algae, and plants, the cell wall is a rigid structure that encloses the cell membrane and contents. Composed of a variety of polysaccharide-based components such as peptidoglycan (bacteria), chitin (fungi), and cellulose (plants, algae, and fungi).

Central Dogma Statement regarding the unidirectional transfer of information from DNA to RNA to protein.

Centrifugation Method of separating components by spinning at high speed. The g-forces cause materials to pellet or move through the centrifugation medium. Uses include spinning down whole cells, cell debris, precipitated nucleic acids, or other components. Also used in ultracentrifuges for separating macromolecules under gradient centrifugation.

CFTR gene (protein) Cystic fibrosis transmembrane conductance regulator, the gene and protein involved in defective ion transport that causes cystic fibrosis.

Chimaera An organism (usually transgenic) composed of cells with different genotypes.

ChIP assay See *chromatin immunoprecipitation assay*.

Chromatin immunoprecipitation assay Method of identifying regions of DNA that are associated with particular proteins by immunoprecipitation of the protein/DNA region.

Chromatography Method of separating various types of molecules based on their affinity or physical behaviour when passed through a matrix and eluted with a suitable solvent.

Chromosomal abnormalities (aberrations) Term used to describe genetic conditions that involve gain or loss of whole chromosomes or parts of chromosomes.

Chromosome A DNA molecule carrying a set of genes. There may be a single chromosome, as in bacteria, or multiple chromosomes, as in eukaryotic organisms.

Chromosome jumping Technique used to isolate non-contiguous regions of DNA by 'jumping' across gaps that may appear as a consequence of uncloned regions of DNA in a gene library.

Chromosome walking Technique used to isolate contiguous cloned DNA fragments by using each fragment as a probe to isolate adjacent cloned regions.

Chymosin (chymase) Enzyme used in cheese production, available as a recombinant product.

***Cis*-acting element** A DNA sequence that exerts its effect only when on the same DNA molecule as the sequence it acts on. For example, the CAAT box (q.v.) is a *cis*-acting element for transcription in eukaryotes.

Cistron A sequence of bases in DNA that specifies one polypeptide.

Clone (1) A colony of identical organisms; often used to describe a cell carrying a recombinant DNA fragment. (2) Used as a verb to describe the generation of recombinants. (3) A complex organism (*e.g.* sheep) generated from a totipotent cell nucleus by nuclear transfer into an enucleated ovum.

Clone bank See *cDNA library, genomic library*.

Clone contig Refers to contiguous (q.v.) cloned sequences. A series of contiguous clones represents the arrangement of the DNA sequence as it would be found *in vivo*.

Co-dominance Alleles are said to be co-dominant if they are expressed equally in terms of their contribution to the phenotype.

Codon The three bases in mRNA that specify a particular amino acid during translation.

Cohesive ends Those ends (termini) of DNA molecules that have short complementary sequences that can stick together to join two DNA molecules. Often generated by restriction enzymes.

Cointegration Formation of a single DNA molecule from two components; usually refers to the formation of a larger plasmid vector from two smaller component plasmids.

Comparative genome analysis Use of genome analysis techniques to establish similarities and differences between genomes of organisms at different taxonomic levels.

Competent Refers to bacterial cells that are able to take up exogenous DNA.

Competitor RT-PCR Technique used to quantify the amount of PCR product by spiking samples with known amounts of a competitor sequence.

Complementary DNA See *cDNA*.

Complementation Process by which genes on different DNA molecules interact. Usually a protein product is involved, as this is a diffusible molecule that can exert its effect away from the DNA itself. For example, a $lacZ^+$ gene on a plasmid can complement a mutant ($lacZ^-$) gene on the chromosome by enabling the synthesis of ß-galactosidase.

Computer hardware The computer machinery that provides the infrastructure for running the software programmes.

Computer software The programmes that run on computer hardware. Software packages enable manipulation and processing of various sorts of data, text, and graphics.

Concatemer A DNA molecule composed of a number of individual pieces joined together *via* cohesive ends (q.v.).

Congenital Present at birth, usually used to describe genetically derived abnormalities.

Conjugation Plasmid-mediated transfer of genetic material from a 'male' donor bacterium to a 'female' recipient.

Consensus sequence A sequence that is found in most examples of a particular genetic element, and which shows a high degree of conservation. An example is the CAAT box (q.v.).

Containment See *Biological containment, physical containment*.

Contiguous In molecular biology, refers to DNA sequences that follow or join on to each other.

Copy number (1) The number of plasmid molecules in a bacterial cell. (2) The number of copies of a gene in the genome of an organism.

cos **site** The region generated when the cohesive ends of lambda DNA join together.

Cosmid A hybrid vector made up of plasmid sequences and the cohesive ends (*cos* sites) of bacteriophage lambda.

Covalent bond Relatively strong molecular bond in which the electron configurations of the constituent atoms is satisfied by sharing electrons.

Crown gall disease Plant disease caused by the Ti plasmid of *Agrobacterium tumefaciens*, in which a 'crown gall' of tissue is produced after infection.

Crystallisation Method of concentrating a solute by forming crystalline structures, usually following from the evaporation of solvent.

C terminus Carboxyl terminus, defined by the – COOH group of an amino acid or protein.

Cyanogen bromide Chemical used to cleave a fusion protein product from the N-terminal vector-encoded sequence after synthesis.

Cystic fibrosis Disease affecting lungs and other tissues, caused by ion transport defects in the CFTR gene (q.v.).

Cytidine Nucleoside composed of ribose and cytosine.

Cytosine (C) Nitrogenous base found in DNA and RNA.

Data mining Refers to the technique of searching databases (of various types) for information.

Data warehouse Term to describe a data storage 'facility', usually storage space on a computer where the data are held in electronic formats of various types.

Database A collection of related data entries held in electronic format; examples include DNA and protein sequence databases.

DDBJ DNA Data Bank of Japan, the central Japanese DNA sequence database.

Dehydration synthesis A form of condensation reaction where two molecules are joined by the removal of water.

Deletion Change to the genetic material caused by removal of part of the sequence of bases in DNA.

Deoxynucleoside triphosphate (dNTP) Triphosphorylated ('high-energy') precursor required for synthesis of DNA, where N refers to one of the four bases (A, G, T, or C).

Deoxyribonucleic acid (DNA) A condensation heteropolymer composed of nucleotides. DNA is the primary genetic material in all organisms apart from some RNA viruses. Usually double-stranded.

Deoxyribose The sugar found in DNA.

Deoxyribonuclease (DNase) A nuclease enzyme that hydrolyses (degrades) single- and double-stranded DNA.

Deproteinisation Removal of protein contaminants from a preparation of nucleic acid.

Diabetes mellitus (DM) Condition where high levels of blood glucose exist because of problems in regulation of glucose levels. May be caused by insulin deficiency or other defects in the glucose regulation system.

Dicer Enzyme involved in RNA interference (RNAi, q.v.)

Dideoxynucleoside triphosphate (ddNTP) A modified form of dNTP used as a chain terminator in DNA sequencing.

Diploid Having two sets of chromosomes. *Cf. haploid.*

Directed evolution Method of engineering changes in protein structure by producing random changes and selecting the most useful, analogous to an evolutionary process. Cf. *rational design.*

Disarmed vector A vector in which some characteristic (*e.g.* conjugation) has been disabled.

DMD See *Duchenne muscular dystrophy.*

DNA See *Deoxyribonucleic acid.*

DNA chip A DNA microarray used in the analysis of gene structure and expression. Consists of oligonucleotide sequences immobilised on a 'chip' array.

DNA fingerprinting See *genetic fingerprinting.*

DNA footprinting Method of identifying regions of DNA to which regulatory proteins will bind.

DNA ligase Enzyme used for joining DNA molecules by the formation of a phosphodiester bond between a 5′ phosphate and a 3′ OH group.

DNA microarray See *DNA chip.*

DNA polymerase An enzyme that synthesises a copy of a DNA template.

DNA profiling Term used to describe the various methods for analysing DNA to establish identity of an individual.

DNA shuffling Refers to the generation of different combinations of gene sequence components by using the exon sequences in different combinations. Can be used in protein engineering, although also occurs *in vivo*, where it is usually referred to as exon shuffling.

DNase protection Method used to determine protein binding regions on DNA sequences, where the protein protects the DNA from nuclease digestion.

Dominant An allele that is expressed and appears in the phenotype in heterozygous individuals. *Cf. Recessive.*

Dot blot Technique in which small spots, or 'dots', of nucleic acid are immobilised on a nitrocellulose or nylon membrane for hybridisation. *Cf. Slot blot.*

Downstream processing Refers to the procedures used to purify products (usually proteins) after they have been expressed in bacterial, fungal, or mammalian cells.

Down syndrome Clinical condition resulting from trisomy-21 (three copies of chromosome 21), a consequence of non-disjunction during meiosis.

Draft sequence A DNA (or protein) sequence in unfinished form, often incomplete or with some unresolved anomalies.

Drosophila melanogaster Fruit fly used as a model organism in genetic, developmental, and molecular studies.

Duchenne muscular dystrophy X-linked (q.v.) muscle-wasting disease caused by defects in the gene for the protein dystrophin (q.v.).

Dystrophin Large protein linking the cytoskeleton to the muscle cell membrane, defects in which cause muscular dystrophy.

EBI European Bioinformatics Institute.

Edman degradation Method used for determining the amino acid sequence of a polypeptide.

Electroporation Technique for introducing DNA into cells by giving a transient electric pulse.

ELISA See *Enzyme-linked immunosorbent assay*.

ELSI Sometimes used as shorthand to describe the ethical, legal, and social implications of genetic engineering.

EMBL European Molecular Biology Laboratory.

Embryo splitting Technique used to clone organisms by separating cells in the early embryo, which then go on to direct development and produce identical copies of the organism.

Emergent properties Refers to the apperance of new characteristics as complexity increases, often not predictable from knowledge of the component parts that make up the next level of complexity. Sometimes described as 'the whole is greater than the sum of the parts'.

End labelling Adding a radioactive molecule onto the end(s) of a polynucleotide.

Endonuclease An enzyme that cuts within a nucleic acid molecule, as opposed to an exonuclease (q.v.), which digests DNA from one or both ends.

Endothia parasitica Fungus used in biotechnology applications.

Enhancer A sequence that enhances transcription from the promoter of a eukaryotic gene. May be several thousand base pairs away from the promoter.

Ensembl Browser for looking at DNA sequence data from genomic sequencing projects. Ensembl was developed by EMBL and the Wellcome Trust Sanger Institute.

Enzyme A protein that catalyses a specific reaction.

Enzyme-linked immunosorbent assay (ELISA) Technique for detection of specific antigens by using an antibody linked to an enzyme that generates a coloured product. The antigens are fixed onto a surface (usually a 96-well plastic plate) and thus large numbers of samples can be screened at the same time.

Enzyme replacement therapy Therapeutic procedure in which a defective enzyme function is restored by replacing the enzyme itself. *Cf. Gene therapy*.

Epigenesis Theory of development that regards the process as an iterative series of steps, in which the various signals and control events interact to regulate development.

EPSP synthase An enzyme (full name 5-enolpyruvylshikimate-3-phosphate synthase) of amino acid biosynthesis that is inhibited by the herbicide glyphosate (the active ingredient of Roundup™ and Tumbleweed™).

Error-prone PCR PCR carried out under low-stringency conditions, thus generating variants of PCR products. Can be used in directed mutagenesis techniques.

Escherichia coli The most commonly used bacterium in molecular biology.

Ethidium bromide A molecule that binds to DNA and fluoresces when viewed under ultraviolet light. Used as a stain for DNA.

Eukaryotic The property of having a membrane-bound nucleus.

Exon Region of a eukaryotic gene that is expressed *via* mRNA.

Exon shuffling See *DNA shuffling*.

Exonuclease An enzyme that digests a nucleic acid molecule from one or both ends.

Expressivity The degree to which a particular genotype generates its effect in the phenotype. *Cf. Penetrance.*

Expressome Refers to the entire complement of expressed components in any given cell –includes primary transcripts, mature mRNAs, and proteins.

Extrachromosomal element A DNA molecule that is not part of the host cell chromosome.

Ex vivo Outside the body. Usually used to describe gene therapy procedure in which the manipulations are performed outside the body, and the altered cells returned after processing. *Cf. In vivo, in vitro.*

Factor IX One of a family of blood clotting factors; produced in transgenic sheep.

False negative Results when a test does not identify the true situation, as can be the case in AIDS testing where antibodies may be undetectable in the early stages. Thus, a test at this stage can give a negative even if the person is infected.

False positive A spurious result where a positive is obtained that is not correct. Can affect a number of different assay and experimental techniques.

Fibrin Insoluble protein involved in blood clot formation.

Fibrinogen Precursor of fibrin, converted to fibrin by the action of thrombin.

Filtration Method of separating soild and liquid components of a suspension by passing through a filter.

Finished sequence Refers to a completed DNA or protein sequence in which anomalies and missing regions have been resolved.

Flavr Savr (*sic*) Transgenic tomato in which polygalacturonase (q.v.) synthesis is restricted using antisense technology. Despite the novel science, the Flavr Savr was not a commercial success.

Flush ends See *Blunt ends.*

Foldback DNA Class of DNA that has palindromic or inverted repeat regions that re-anneal rapidly when duplex DNA is denatured.

Food and Drug Administration (FDA) Regulatory body in the USA responsible for approval of medicines and foodstuffs.

Formulation Used to describe the 'recipe' used for the production of a pharmaceutical.

Freeze-drying Technique for concentrating solutes by removal of water under vacuum at low temperature.

FTP File Transfer Protocol, specification of which is important in transfer of data between computer systems.

Fusion protein A hybrid recombinant protein that contains vector-encoded amino acid residues at the N terminus.

Gamete Refers to the haploid male (sperm) and female (egg) cells that fuse to produce the diploid zygote (q.v.) during sexual reproduction.

Gel electrophoresis Technique for separating nucleic acid molecules on the basis of their movement through a gel matrix under the influence of an electric field. See *Agarose* and *Polyacrylamide*.

Gel retardation Method of determining protein-binding sites on DNA fragments on the basis of their reduced mobility, relative to unbound DNA, in gel electrophoresis experiments.

GenBank The original DNA sequence database, now one of the main repositories for genome sequence data.

Gene The unit of inheritance, located on a chromosome. In molecular terms, usually taken to mean a region of DNA that encodes one function. Broadly, therefore, one gene encodes one protein.

Gene addition therapy Form of gene therapy where a gene function is added to the complement in the genome, with no absolute requirement for genetic exchange to replace a defective gene.

Gene bank See *Genomic library*.

Gene cloning The isolation of individual genes by generating recombinant DNA molecules, which are then propagated in a host cell that produces a clone that contains a single fragment of the target DNA.

Gene protection technology Range of techniques used to ensure that particular commercially derived recombinant constructs cannot be used without some sort of control or process, usually supplied by the company marketing the recombinant. Also known as genetic use restriction technology and genetic trait control technology.

Gene therapy The use of cloned genes in the treatment of genetically derived malfunctions. May be delivered *in vivo* or *ex vivo*. May be offered as gene addition or gene replacement versions.

Genetic code The triplet codons that determine the types of amino acid that are inserted into a polypeptide during translation. There are 61 codons for 20 amino acids (plus 3 stop codons), and the code is therefore referred to as *degenerate*.

Genetic fingerprinting A method that uses radioactive probes to identify bands derived from hypervariable regions of DNA (q.v.). The band pattern is unique for an individual and can be used to establish identity or family relationships.

Genetic mapping Low-resolution method to assign gene locations (loci) to their position on the chromosome. *Cf. recombination frequency mapping, physical mapping.*

Genetic marker A phenotypic characteristic that can be ascribed to a particular gene.

Genetic trait control technology Version of gene protection technology (q.v.), sometimes called 'traitor technology'.

Genetic use restriction technology (GURT) See *Gene protection technology*.

Genetically modified food (GMF) A GMO (q.v.) specifically for use as a foodstuff.

Genetically modified organism (GMO) An organism in which a genetic change has been engineered. Usually used to describe transgenic plants and animals.

Genome Used to describe the complete genetic complement of a virus, cell, or organism.

Genomics The study of genomes, particularly genome sequencing.

Genomic library A collection of clones that together represent the entire genome of an organism.

Genotype The genetic constitution of an organism. *Cf. Phenotype.*

Germ line Gamete producing (reproductive) cells that give rise to eggs and sperm.

GFP See *Green fluorescent protein.*

Gigabase (Gb) 10^9 bases or base pairs.

GIGO effect From computer science, meaning 'garbage in, garbage out'. Refers to the need to ensure that input data are robust and accurate if valid results are to be obtained when processed by various computer applications.

Glyphosate Herbicide that inhibits EPSP synthase (q.v.).

GMF See *Genetically modified food.*

GMO See *Genetically modified organism.*

Green fluorescent protein Protein useful as a marker for various processes in cell biology and rDNA techniques. As the name suggests, GFP-expressing tissues fluoresce with a green colour.

Guanine (G) Nitrogenous base found in DNA and RNA.

Guanosine Nucleoside composed of ribose and guanine.

GURT See *Genetic use restriction technology, Gene protection technology.*

Hansela polymorpha Species of yeast used in biotechnology applications.

Haploid Having one set of chromosomes. *Cf. Diploid.*

Heterologous Refers to gene sequences that are not identical, but show variable degrees of similarity.

Heteropolymer A polymer composed of different types of monomer. Most protein and nucleic acid molecules are heteropolymers.

Heterozygous Refers to a diploid organism (cell or nucleus) that has two different alleles at a particular locus.

HGP Human Genome Project (q.v.).

HIV See *Human immunodeficiency virus.*

Hogness box See *TATA box.*

Homologous (1) Refers to paired chromosomes in diploid organisms. (2) Used to strictly describe DNA sequences that are identical; however, the percentage homology between related sequences is sometimes quoted.

Homologue One of a pair of homologous chromosomes.

Homopolymer A polymer composed of only one type of monomer, such as polyphenylalanine (protein) or polyadenine nucleic acid.

Homozygous Refers to a diploid organism (cell or nucleus) that has identical alleles at a particular locus.

Host A cell used to propagate recombinant DNA molecules.

HPV See *Human papilloma virus.*

HUGO Human Genome Organisation.

Human Genome Project The multinational collaborative effort to determine the DNA sequence of the human genome.

Human immunodeficiency virus (HIV) Retrovirus, causative agent of acquired immune deficiency syndrome (q.v.).

Human papilloma virus (HPV) Causative agent of warts, genital warts, and linked with cervical cancer. Over 100 variants have been described.

Humulin™ Recombinant-derived human insulin.

Huntington disease Middle-age onset autosomal dominant degenerative condition. Sometimes known as Huntington's chorea, so called because of the characteristic physical movements of affected individuals.

HUPO The Human Proteome Organisation, an international consortium of proteomics researchers analogous in concept to the Human Genome Organisation (HUGO).

Hybrid-arrest translation Techniques used to identify the protein product of a cloned gene, in which translation of its mRNA is prevented by the formation of a DNA–mRNA hybrid.

Hybrid-release translation Technique in which a particular mRNA is selected by hybridisation with its homologous, cloned DNA sequence, and is then translated to give a protein product that can be identified.

Hybridisation The joining together of artificially separated nucleic acid molecules *via* hydrogen bonding between complementary bases.

Hydrolysis Reaction where two covalently joined molecules are split apart by the addition of the elements of water. In effect the reversal of a dehydration synthesis reaction.

Hyperchromic effect Change in absorbance of nucleic acids, depending on the relative amounts of single-stranded and double-stranded forms. Used as a measurement in denaturation/renaturation studies.

Hypervariable region (HVR) A region in a genome that is composed of a variable number of repeated sequences and is diagnostic for the individual. See *Genetic fingerprinting*.

Ice-minus bacteria Bacteria engineered to disrupt the normal ice-forming process, used to protect plants from frost damage.

IFA See *Indirect immunofluorescent assay*.

IGF-1 See *Insulin-like growth factor*.

In silico Used to describe virtual experiments carried out on a computer, often by manipulation of sequence data. One example is to predict the effect on a protein's structure of altering specific parts of the gene sequence.

In vitro Literally 'in glass', meaning in the test tube, rather than in the cell or organism.

In vivo Literally 'in life', meaning the natural situation, within a cell or organism.

Indirect immunofluorescence assay (IFA) Assay using secondary antibody detection methods to identify antigens.

Information technology (IT) Commonly (and somewhat loosely) used to describe computer-based manipulation of data sets of various types.

Inosine Nucleoside found in tRNA, sometimes used in synthetic oligonucleotides at degenerate positions as it can pair with all the other DNA bases.

INSDC International nucleotide sequence database collaboration, a network set up to help manage sequence databases.

Insertion vector A bacteriophage vector that has a single cloning site into which DNA is inserted.

Insulin Protein hormone involved in the regulation of blood glucose levels. Has been available in recombinant form since the 1980s.

Insulin-like growth factor (IGF-1) Polypeptide hormone, synthesis of which is stimulated by growth hormone. Implicated in some concerns about the safety of using recombinant bovine growth hormone in cattle to increase milk yields.

Interactome Term used to describe the set of interactions between cellular components such as proteins and other metabolites. *Cf. Genome, Proteome, Transcriptome, Metabolome.*

Intervening sequence Region in a eukaryotic gene that is not expressed *via* the processed mRNA.

Intron See *Intervening sequence.*

Inverse PCR Method for using PCR to amplify DNA for which there are no sequence data available for primer design. Works by circularising and inverting the target sequence so that primers within a known sequence area can be used.

Inverted repeat A short sequence of DNA that is repeated, usually at the ends of a longer sequence, in a reverse orientation.

IPTG *iso*-propyl-thiogalactoside, a gratuitous inducer that de-represses transcription of the *lac* operon.

IT See *Information technology.*

Kilobase (kb) 10^3 bases or base pairs, used as a unit for measuring or specifying the length of DNA or RNA molecules.

Klenow fragment A fragment of DNA polymerase I that lacks the $5' \rightarrow 3'$ exonuclease activity.

Kluyveromyces lactis Species of yeast useful in biotechnology applications.

Knockdown Refers to a reduction in gene expression, often as the result of modification of the target gene sequence (or control sequences) using rDNA techniques.

Knockin mouse A transgenic mouse in which a gene function has been added or 'knocked in'. Used primarily to generate animal models for the study of human disease. *Cf. Knockout mouse.*

Knockout mouse A transgenic mouse in which a gene function has been disrupted or 'knocked out'. Used primarily to generate animal models for the study of human disease (*e.g.* cystic fibrosis). *Cf. Knockin mouse.*

Lambda Bacteriophage used in vector construction, commonly for the generation of insertion and replacement vectors. Replacement (substitution) vectors are often used for genomic library construction and can accept DNA fragments up to some 23 kb in length.

Linkage mapping Genetic mapping (q.v.) technique used to establish the degree of linkage between genes. See also *Recombination frequency mapping.*

Linker A synthetic self-complementary oligonucleotide that contains a restriction enzyme recognition site. Used to add cohesive ends (q.v.) to DNA molecules that have blunt ends (q.v.).

Lipase Enzyme that hydrolyses fats (lipids).

Liposome (lipoplex) Lipid-based method for delivering gene therapy.

Locus The site at which a gene is located on a chromosome.

Lysogenic Refers to bacteriophage infection that does not cause lysis of the host cell.

Lytic Refers to bacteriophage infection that causes lysis of the host cell.

Macromolecule Large polymeric molecule made up of monomeric units, commonly used to describe proteins (monomers are amino acids) and nucleic acids (monomers are nucleotides).

Maternal inheritance Pattern of inheritance from female cytoplasm. Mitochondrial genes are inherited in this way, as the mitochondria are inherited with the ovum.

Mega (M) SI prefix, 10^6.

Megabase (Mb) 10^6 bases or base pairs.

Megahectare (MHa) 10^6 hectares.

Messenger RNA (mRNA) The ribonucleic acid molecule transcribed from DNA that carries the codons specifying the sequence of amino acids in a protein.

Metabolome Refers to the population of metabolites in a cell, which, together with proteomics and transcriptomics (q.v.), can give a snapshot of the cell's activity at any given point in time.

Micro (μ) SI prefix, 10^{-6}.

Microinjection Introduction of DNA into the nucleus or cytoplasm of a cell by insertion of a microcapillary and direct injection.

Micro RNAs (miRNAs) Short RNA molecules synthesised as part of the RNA interference mechanism (q.v.).

Microsatellite DNA Type of sequence repeated many times in the genome. Based on dinucleotide repeats, microsatellites are highly variable and can be used in mapping and profiling studies.

Milli (m) SI prefix, 10^{-3}.

Minisatellite DNA Type of sequence based on variable number tandem repeats (VNTRs; q.v.). Used in genetic mapping and profiling studies.

Mini-Ti vector Vector for cloning in plant cells, based on part of the Ti plasmid of *Agrobacterium tumefaciens* (q.v.).

miRNA See *Micro RNAs*.

Molecular cloning Alternative term for gene cloning.

Molecular ecology Use of molecular biology and recombinant DNA techniques in studying ecological topics.

Molecular paleontology Use of molecular techniques to investigate the past, as in DNA profiling from mummified or fossilised samples.

Monocistronic Refers to an RNA molecule encoding one function.

Monogenic Trait caused by a single gene. *Cf. Polygenic.*

Monomer The unit that makes up a polymer. Nucleotides and amino acids are the monomers for nucleic acids and proteins, respectively.

Monosomy Diploid cells in which one of a homologous pair of chromosomes has been lost. *Cf. Trisomy.*

Monozygotic Refers to identical twins, generated from the splitting of a single embryo at an early stage.

mRNA See *Messenger RNA.*

Mosaic An embryo or organism in which not all the cells carry identical genomes.

Multifactorial Caused by many factors (*e.g.* genetic trait in which many genes and environmental influences may be involved).

Multi-locus probe DNA probe used to identify several bands in a DNA fingerprint or profile. Generates the 'bar code' pattern in a genetic fingerprint.

Multiple cloning site (MCS) A short region of DNA in a vector that has recognition sites for several restriction enzymes.

Multipotent Cell that can give rise to a range of differentiated cells. *Cf. Totipotent, Pluripotent.*

Mus musculus The mouse, a major model organism for molecular genetic studies. Many different types of transgenic mice are available.

Mutagenesis The process of inducing mutations in DNA.

Mutagenesis *in vitro* Introduction of defined mutations in a cloned sequence by manipulation in the test tube. See also *Oligonucleotide-directed mutagenesis.*

Mutant An organism (or gene) carrying a genetic mutation.

Mutation An alteration to the sequence of bases in DNA. May be caused by insertion, deletion, or modification of bases.

Muteins Refers to proteins that have been engineered by the incorporation of mutational changes.

Nano (n) SI prefix, 10^{-9}.

Native protein A recombinant protein that is synthesised from its own N terminus, rather than from an N terminus supplied by the cloning vector.

NCBI National Centre for Biotechnology Information.

Nested fragments A series of nucleic acid fragments that differ from each other (in terms of length) by one or only a few nucleotides.

Nested PCR Form of PCR where two sets of primers are used, one pair being internal (nested) with respect to the other pair.

Neutral molecular polymorphism A molecular polymorphism that has no adverse effect and can be used to tag a gene sequence that it co-segregates with.

Nick translation Method for labelling DNA with radioactive dNTPs.

Non-disjunction Failure of chromsomes to separate during meiosis, resulting in aneuploidy. An example is Down syndrome, caused by non-disjunction resulting in trisomy-21.

Northern blotting Transfer of RNA molecules onto membranes for the detection of specific sequences by hybridisation.

N terminus Amino terminus, defined by the $-NH_2$ group of an amino acid or proFein.

Nuclear transfer Method for cloning organisms in which a donor nucleus is taken from a somatic cell and transferred to the recipient ovum.

Nuclease An enzyme that hydrolyses phosphodiester bonds.

Nucleoside A nitrogenous base bound to a sugar.

Nucleotide A nucleoside bound to a phosphate group.

Nucleoid Region of a bacterial cell in which the genetic material is located.

Nucleus Membrane-bound region in a eukaryotic cell that contains the genetic material.

Oligo Prefix meaning few, as in oligonucleotide or oligopeptide.

Oligo(dT)-cellulose Short sequence of deoxythymidine residues linked to a cellulose matrix, used in the purification of eukaryotic mRNA.

Oligolabelling See *Primer extension*.

Oligomer General term for a short sequence of monomers.

Oligonucleotide A short sequence of nucleotides.

Oligonucleotide-directed mutagenesis Process by which a defined alteration is made to DNA using a synthetic oligonucleotide.

OMIM Online Mendelian Inheritance in Man, a database of genetically based conditions.

Oncomouse Transgenic mouse engineered to be susceptible to cancer.

Oocyte Stage in development of the female gamete or ovum (egg). Often the terms oocyte and ovum are used interchangeably.

Operator Region of an operon, close to the promoter, to which a repressor protein binds.

Operon A cluster of bacterial genes under the control of a single regulatory region.

Organismal cloning The production of an identical copy of an individual organism by techniques such as embryo splitting or nuclear transfer. Used to distinguish the process from molecular cloning (q.v.).

Ornithine transcarbamylase deficiency (OTCD) Enzyme deficiency condition targetted for gene therapy.

Ovum The mature female gamete or egg cell, derived from the oocyte. Often the terms ovum and oocyte are used interchangeably.

Palindrome A DNA sequence that reads the same on both strands when read in the same (*e.g.* $5' \rightarrow 3'$) direction. Examples include many restriction enzyme recognition sites.

Pedigree analysis Determination of the transmission characteristics of a particular gene by examination of family histories.

Penetrance The proportion of individuals with a particular genotype that show the genotypic characteristic in the phenotype. *Cf. Expressivity.*

PCR See *Polymerase chain reaction.*

Phage See *Bacteriophage.*

Phagemid A vector containing plasmid and phage sequences.

Pharm animal Transgenic animal used for the production of pharmaceuticals.

Phenotype The observable characteristics of an organism, determined both by its genotype (q.v.) and its environment.

Phosphodiester bond A bond formed between the 5′ phosphate and the 3′ hydroxyl groups of two nucleotides.

Physical containment Refers to engineering proctective measures to contain GM experiments and ensure that no host/vector systems can escape into the environment. Includes rooms under negative pressure, filtration, and secure access and egress in laboratory design. *Cf. Biological containment.*

Physical mapping Mapping genes with reference to their physical location on the chromosome. Generates the next level of detail compared to genetic mapping (q.v.).

Physical marker A sequence-based tag that labels a region of the genome. There are several such tags that can be used in mapping studies. *Cf. RFLP, STS.*

Pichia pastoris Species of yeast used in biotechnology applications.

Pico (p) SI prefix, 10^{-12}.

Plaque A cleared area on a bacterial lawn caused by infection by a lytic bacteriophage.

Plasmid A circular extrachromosomal element found naturally in bacteria and some other organisms. Engineered plasmids are used extensively as vectors for cloning.

Plasmin Active form of the protease derived from plasminogen that acts to hydrolyse fibrin, the constituent of blood clots.

Plasminogen Precursor of plasmin, converted to the active form by tissue plasminogen activator (q.v.).

Ploidy number Refers to the number of sets of chromosomes (*e.g.* haploid, diploid, triploid, *etc.*).

Pluripotent Cell that can give rise to a range of differentiated cells. *Cf. Multipotent, Totipotent.*

Polyacrylamide A cross-linked matrix for gel electrophoresis (q.v.) of small fragments of nucleic acids, primarily used for electrophoresis of DNA. Also used for electrophoresis of proteins.

Polyadenylic acid A string of adenine residues. Poly(A) tails are found at the 3′ ends of most eukaryotic mRNA molecules.

Polycistronic Refers to an RNA molecule encoding more that one function. Many bacterial operons are expressed *via* polycistronic mRNAs.

Polygalacturonase Enzyme involved in pectin degradation. Target for antisense control in the Flavr Savr tomato (q.v.).

Polygenic trait A trait determined by the interaction of more than one gene (*e.g.* eye colour in humans).

Polyhedra Capsid structures in baculoviruses, composed of the protein polyhedrin.

Polylinker See *Multiple cloning site.*

Polymer A long sequence of monomers.

Polymerase An enzyme that synthesises a copy of a nucleic acid.

Polymerase chain reaction (PCR) A method for the selective amplification of DNA sequences. Several variants exist for different applications.

Polymorphism Refers to the occurrence of many allelic variants of a particular gene or DNA sequence motif. Can be used to identify individuals by genetic mapping and DNA profiling techniques.

Polynucleotide A polymer made up of nucleotide monomers.

Polynucleotide kinase (PNK) An enzyme that catalyses the transfer of a phosphate group onto a 5′ hydroxyl group.

Polypeptide A chain of amino acid residues. *Cf. Protein.*

Polystuffer An expendable stuffer fragment in a vector that is composed of many repeated sequences.

Posilac™ Commercially available recombinant bovine somatotropin (rBST).

Positional cloning Cloning genes for which little information is available apart from their location on the chromosome.

Post-transcriptional gene silencing (PTGS) Alternative name for RNA interference (RNAi, q.v.).

Post-translational modification Modification of a protein after it has been synthesised. An example would be the addition of sugar residues to form a glycoprotein.

Preformationism Refers to the idea that all development is pre-coded in the zygote, and that development is simply the unfolding of this information. Now considered too simplistic. *Cf. Epigenesis.*

Pribnow box Sequence found in prokaryotic promoters that is required for transcription initiation. The consensus sequence (q.v.) is TATAAT.

Primary database Database in which data are deposited without, or with minimal, manipulation. May have sophisticated search and process tools.

Primary transcript The initial, and often very large, product of transcription of a eukaryotic gene. Subjected to processing to produce the mature mRNA molecule.

Primer extension Synthesis of a copy of a nucleic acid from a primer. Used in labelling DNA and in determining the start site of transcription.

Probe A labelled molecule used in hybridisation procedures.

Processivity Refers to the sequential addition of components by an enzyme; for example, DNA and RNA polymerases are processive enzymes.

Proinsulin Precursor of insulin that includes an extra polypeptide sequence that is cleaved to generate the active insulin molecule.

Prokaryotic The property of lacking a membrane-bound nucleus (*e.g.* in bacteria such as *E. coli*).

Promoter DNA sequence(s) lying upstream from a gene, to which RNA polymerase binds.

Pronucleus One of the nuclei in a fertilised egg prior to fusion of the gametes.

Prophage A bacteriophage maintained in the lysogenic state in a cell.

Protease Enzyme that hydrolyses polypeptides.

Protein A condensation (dehydration) heteropolymer composed of amino acid residues linked together by peptide bonds to give a polypeptide.

Proteome Refers to the population of proteins produced by a cell. *Cf. Genome, Transcriptome.*

Protoplast A cell from which the cell wall has been removed.

Prototroph A cell that can grow in an unsupplemented growth medium.

PTGS See *Post-transcriptional gene silencing.*

Purine A double-ring nitrogenous base such as adenine and guanine.

Pyrimidine A single-ring nitrogenous base such as cytosine, thymine, and uracil.

Radiolabelling Short for radioactive labelling; method used to incorporate radioactive isotopes into biological molecules. An example is labelling nucleic acids with ^{32}P-dNTPs to prepare high-specific-activity probes for use in hybridisation experiments.

Random amplified polymorphic DNA (RAPD) PCR-based method of DNA profiling that involved amplification of sequences using random primers. Generates a type of genetic fingerprint that can be used to identify individuals.

RAPD See *Random amplified polymorphic DNA.*

Rational design Process of engineering changes in protein structure by using existing knowledge to design such changes. *Cf. Directed evolution.*

Reading frame The pattern of triplet codon sequences in a gene. There are three reading frames, depending on which nucleotide is the start point. Insertion and deletion mutations can disrupt the reading frame and have serious consequences, as often the entire coding sequence becomes non-sense after the point of mutation.

Recessive An allele where the expression is masked in the phenotype in heterozygous individuals. *Cf. Dominant.*

Recombinant DNA A DNA molecule made up of sequences that are not normally joined together.

Recombination frequency mapping Method of genetic mapping that uses the number of crossover events that occur during meiosis to estimate the distance between genes. *Cf. Physical mapping.*

Redundancy In molecular biology, refers to the fact that some amino acids can be specified by multiple codons; thus, the third base (the 'wobble' position in the codon) is effectively redundant from an informatics point of view.

Regulatory gene A gene that exerts its effect by controlling the expression of another gene.

Renaturation kinetics Method of analysing the complexity of genomes by studying the patterns obtained when DNA is denatured and allowed to renature.

Repetitive sequence A sequence that is repeated a number of times in the genome.

Replacement vector A bacteriophage vector in which the cloning sites are arranged in pairs, so that the section of the genome between these sites can be replaced with insert DNA.

Replication Copying the genetic material during the cell cycle. Also refers to the synthesis of new phage DNA during phage multiplication.

Replicon A piece of DNA carrying an origin of replication.

Restriction enzyme An endonuclease that cuts DNA at sites defined by its recognition sequence.

Restriction fragment A piece of DNA produced by digestion with a restriction enzyme.

Restriction fragment length polymorphism (RFLP) A variation in the locations of restriction sites bounding a particular region of DNA, such that the fragment defined by the restriction sites may be of different lengths in different individuals.

Restriction mapping Technique used to determine the location of restriction sites in a DNA molecule.

Retrovirus A virus that has an RNA genome that is copied into DNA during the infection.

Reverse transcriptase An RNA-dependent DNA polymerase found in retroviruses, used *in vitro* for the synthesis of cDNA.

RFLP See *Restriction fragment length polymorphism*.

Rhizomucor miehei Fungus used in biotechnology applications.

Rhizomucor pusillus Fungus used in biotechnology applications.

Ribonuclease (RNase) An enzyme that hydrolyses RNA.

Ribonucleic acid (RNA) A condensation heteropolymer composed of ribonucleotides.

Ribosomal RNA (rRNA) RNA that is part of the structure of ribosomes.

Ribosome The 'jig' that is the site of protein synthesis. Composed of rRNA and proteins.

Ribosome-binding site A region on an mRNA molecule that is involved in the binding of ribosomes during translation.

RISC See *RNA-induced silencing complex*.

RNA See *Ribonucleic acid*.

RNA-induced silencing complex (RISC) Complex of protein and the nuclease slicer, involved in RNA interference (q.v.).

RNA interference (RNAi) Complex *in vivo* process involved in regulating gene expression post-transcriptionally by 'interfering' with transcript availability.

RNA processing The formation of functional RNA from a primary transcript (q.v.). In mRNA production this involves removal of introns, addition of a 5' cap, and polyadenylation.

RT-PCR Reverse transcriptase (transcription) PCR, where a cDNA copy of mRNA is made and then amplified using PCR.

S_1 mapping Technique for determining the start point of transcription.

S_1 nuclease An enzyme that hydrolyses (degrades) single-stranded DNA.

Saccharomyces cerevisiae Unicellular yeast (baker's yeast, also known as budding yeast) that is extensively used as a model microbial eukaryote in molecular studies. Also used in the biotechnology industry for a range of applications, as well as in brewing and bread-making.

Schizosaccharomyces pombe Fission yeast, used in a variety of applications in basic research and biotechnology.

SCIDS Severe combined immunodeficiency syndrome, a condition that results from a defective enzyme (adenosine deaminase, q.v.).

Scintillation counter (spectrometer) A machine for determining the amount of radioactivity in a sample. Detects the emission of light from a fluor that is excited by radioactive disintegrations. Varies in efficiency according to the energy of the isotope, thus producing an estimate as counts per minute rather than actual disintegrations per minute.

Screening Identification of a clone in a genomic or cDNA library (q.v.) by using a method that discriminates between different clones.

Secondary database A database in which the information has been derived by manipulation of an existing data set.

Seedcorn funding Early stage capital funding to enable a project or new company to become established. Often provided by public or private bodies who are willing to take a risk on the capital, or by charity foundations.

Selection Exploitation of the genetics of a recombinant organism to enable desirable, recombinant genomes to be selected over non-recombinants during growth.

Sequence tagged site Refers to a DNA sequence that is unique in the genome and that can be used in mapping studies. Usually identified by PCR amplification.

Sex chromosome The non-autosomal X and Y chromosomes in humans that determine the sex of the individual. Males are XY, females XX.

Sex-linked Refers to pattern of inheritance where the allele is located on a sex chromosome. *Cf. X-linked*.

Shine–Dalgarno sequence See *Ribosome-binding site*.

Short interfering RNAs (siRNAs) RNAs involved in gene regulation via the process of RNA interference (RNAi, q.v.).

Short tandem repeat (STR) Short (2–5 base pairs) repeat elements in the genome, also known as microsatellites. Useful for mapping and tagging genomes. *Cf. VNTRs*.

Shotgun cloning Refers to a method of cloning where random fragments are cloned, with no attempt to order or arrange contiguous fragments as part of the cloning stage.

Single-locus probe Probe used in DNA fingerprinting that identifies a single sequence in the genome. Diploid organisms therefore usually show two bands in a fingerprint, one allelic variant from each parent.

Single nucleotide polymorphism Polymorphic pattern at a single base, essentially the smallest polymorphic unit that can be identified.

Single-pass sequence data Sequence data that have been produced by a single read; often contains anomalies or errors. *Cf. Draft sequence, Finished sequence*.

siRNAs See *Short interfering RNAs*.

Site-directed mutagenesis See *Oligonucleotide-directed mutagenesis*.

Slicer Nuclease involved in RNA interference (RNAi, q.v.) as part of the RISC complex (q.v.).

Slot-blot Similar to dot blotting (q.v.) but uses slots in a template apparatus rather than circular dots.

SME Small to medium enterprise; term used to describe a company with a few too many employees, but which has not yet reached 'large' size. Many successful SMEs grow into larger companies as they expand their range of products or services.

SNP See *Single nucleotide polymorphism*.

Somatic cell Body cell, as opposed to germ line cell.

Somatotropin Growth hormone, see also *Bovine somatotropin*.

Southern blotting Method for transferring DNA fragments onto a membrane for detection of specific sequences by hybridisation.

Specific activity The amount of radioactivity per unit material; for example, a labelled probe might have a specific activity of 10^6 counts/minute per microgram. Also used to quantify the activity of an enzyme.

Sperm The mature male gamete. *Cf. Ovum, oocyte*.

Sticky ends See *Cohesive ends*.

Structural gene A gene that encodes a protein product.

STR See *Short tandem repeat*.

STS See *Sequence tagged site*.

Stuffer fragment The section in a replacement vector (q.v.) that is removed and replaced with insert DNA. See *Polystuffer*.

Substitution vector See *Replacement vector*.

SwissProt One of the major protein sequence databases.

T-DNA Region of Ti plasmid of *Agrobacterium tumefaciens* that can be used to deliver recombinant DNA into the plant cell genome.

Tandem repeat A repeat composed of an array of sequences repeated contiguously in the same orientation.

Taq **polymerase** Thermostable DNA polymerase from the themophilic bacterium *Thermus aquaticus*. Used in the polymerase chain reaction (q.v.).

TATA box Sequence found in eukaryotic promoters. Also known as the Hogness box, it is similar to the Pribnow box (q.v.) found in prokaryotes, and has the consensus sequence TATAAAT.

Technology transfer In biotechnology, refers to the process of moving from laboratory scale science to commercial production. Often a complex and risky stage in product development.

Temperate Refers to bacteriophages that can undergo lysogenic infection of the host cell.

Terminal transferase An enzyme that adds nucleotide residues to the 3′ terminus of an oligo- or polynucleotide.

Tetracycline (Tc) A commonly used antibiotic.

Text mining Using bibliographic databases to search for information, analogous to searching sequence databases. *Cf. Data mining*.

Thermal cycler Heating/cooling system for PCR applications. Enables denaturation, primer binding, and extension cycles to be programmed and automated.

Thermus aquaticus Thermophilic bacterium from which *Taq* polymerase (q.v.) is purified. Other bacteria from this genus include *Thermus flavus* and *Thermus thermophilus*.

Thymine (T) Nitrogenous base found in DNA only.

Thymidine Nucleoside composed of deoxyribose and thymine.

Ti-plasmid Plasmid of *Agrobacterium tumefaciens* that causes crown gall disease (q.v.).

Tissue plasminogen activator (TPA) A protease that occurs naturally, and functions in breaking down blood clots. Acts on an inactive precursor (plasminogen), which is converted to the active form (plasmin). This attacks the clot by breaking up fibrin, the protein involved in clot formation.

Totipotent A cell that can give rise to all cell types in an organism. Totipotency has been demonstrated by cloning carrots from somatic cells, and by nuclear transfer experiments in animals.

TPA See *Tissue plasminogen activator*.

Traitor technology See *Genetic trait control technology*.

***Trans*-acting element** A genetic element that can exert its effect without having to be on the same molecule as a target sequence. Usually such an element encodes a protein product (perhaps an enzyme or a regulatory protein) that can diffuse to the site of action.

Transcription (T_C) The synthesis of RNA from a DNA template.

Transcriptional unit The DNA sequence that encodes the RNA molecule (*i.e.* from the transcription start site to the stop site).

Transcriptome The population of RNA molecules (usually mRNAs) that is expressed by a particular cell type. *Cf. Genome, Proteome*.

Transfection Introduction of purified phage or virus DNA into cells.

Transfer RNA (tRNA) A small RNA ($\sim$75–85 bases) that carries the anticodon and the amino acid residue required for protein synthesis.

Transformant A cell that has been transformed by exogenous DNA.

Transformation The process of introducing DNA (usually plasmid DNA) into cells. Also used to describe the change in growth characteristics when a cell becomes cancerous.

Transgene The target gene involved in the generation of a transgenic (q.v.) organism.

Transgenic An organism that carries DNA sequences that it would not normally have in its genome.

Translation (T_L) The synthesis of protein from an mRNA template.

Transposable element A genetic element that carries the information that allows it to integrate at various sites in the genome. Transposable elements are sometimes called 'jumping genes'.

TREMBL Stands for 'Translated EMBL' and is a tool for deriving protein sequence data from nucleotide sequence databases.

Trisomy Aneuploid (q.v.) condition where an extra chromosome is present. Common example is the trisomy-21 condition that causes Down syndrome.

Uracil (U) Nitrogenous base found in RNA only.

Uniform resource locator (URL) A consequence of the 'dot-com' era, a URL is the 'address' of a website, usually prefixed by *http://* and/or *www*. URLs often end in a generic term (such as *.co, .com, .ac, .gov*) and may have a country identifier also (thus, *.ac.uk* is an academic institution in the United Kingdom).

UniProt Protein sequence database resource.

URL See *Uniform resource locator*.

Variable number tandem repeat (VNTR) Repetitive DNA composed of a number of copies of a short sequence, involved in the generation of polymorphic loci that are useful in genetic fingerprinting. Also known as hypervariable regions. See also *Minisatellite* and *Microsatellite DNA*.

Vector A DNA molecule that is capable of replication in a host organism and can act as a carrier molecule for the construction of recombinant DNA.

Venture capital Funding used to capitalise young companies to enable development to a more secure stage. Often following seedcorn funding, venture capital is often medium to high risk for the provider.

Virulent Refers to bacteriophages that cause lysis of the host cell.

Virus An infectious agent that cannot replicate without a host cell.

VNTR See *Variable number tandem repeat*.

WAP See *Whey acid protein*.

Western blotting Transfer of electrophoretically separated proteins onto a membrane for probing with antibody.

WGS See *Whole genome shotgun*.

Whey acid protein Milk protein, the gene promoter of which can be used for expression vectors to enable production of rDNA products in the milk of lactating animals.

Whole genome shotgun Method for large-scale sequencing of genomes. *Cf. Shotgun cloning*.

Wobble Refers to the degenerate (q.v.) nature of the third base in many codons.

World Wide Web (WWW) The information resource hosted on the Internet that enables almost instantaneous access to a vast amount of different types of information. Has become an essential part of everyday life and is extensively used in all areas of science and technology.

Xenotransplantation The use of tissues or organs from a non-human source for transplantation.

X-gal 5-Bromo-4-chloro-3-indolyl-ß-D-galactopyranoside: a chromogenic substrate for ß-galactosidase; on cleavage it yields a blue-coloured product.

X-linked Pattern of inheritance where the allele is located on the X chromosome. In humans, this can result in males expressing recessive characteristics that would normally be masked in an autosomal heterozygote.

YAC Yeast artificial chromosome, a vector for cloning very large pieces of DNA in yeast.

Yarrowia lipolytica Species of yeast useful in biotechnology applications.

YCp Yeast centromere plasmid.

YEp Yeast episomal plasmid.

YIp Yeast integrative plasmid.

YRp Yeast replicative plasmid.

Zygote Single-celled product of the fusion of a male and a female gamete (q.v.). Develops into an embryo by successive mitotic divisions.

Index

Note: "f" and "t" after page numbers refer to figures and tables, respectively.